Graduate Texts in Physics

Graduate Texts in Physics

Graduate Texts in Physics publishes core learning/teaching material for graduate- and advanced-level undergraduate courses on topics of current and emerging fields within physics, both pure and applied. These textbooks serve students at the MS- or PhD-level and their instructors as comprehensive sources of principles, definitions, derivations, experiments and applications (as relevant) for their mastery and teaching, respectively. International in scope and relevance, the textbooks correspond to course syllabi sufficiently to serve as required reading. Their didactic style, comprehensiveness and coverage of fundamental material also make them suitable as introductions or references for scientists entering, or requiring timely knowledge of, a research field.

More information about this series at http://www.springer.com/series/8431

Massimiliano Bonamente

Statistics and Analysis of Scientific Data

Second Edition

 Springer

Massimiliano Bonamente
University of Alabama
Huntsville
Alabama, USA

ISSN 1868-4513   ISSN 1868-4521   (electronic)
Graduate Texts in Physics
ISBN 978-1-4939-8239-4   ISBN 978-1-4939-6572-4   (eBook)
DOI 10.1007/978-1-4939-6572-4

This Springer imprint is published by Springer Nature
The registered company is Springer Science+Business Media LLC
The registered company address is: 233 Spring Street, New York, NY 10013, U.S.A

To Giorgio and Alida, who taught me the value of a book.

To Carlo and Gaia, to whom I teach the same.

And to Kerry, with whom I share the love of books, and everything else.

Preface to the First Edition

Across all sciences, a quantitative analysis of data is necessary to assess the significance of experiments, observations, and calculations. This book was written over a period of 10 years, as I developed an introductory graduate course on statistics and data analysis at the University of Alabama in Huntsville. My goal was to put together the material that a student needs for the analysis and statistical interpretation of data, including an extensive set of applications and problems that illustrate the practice of statistical data analysis.

The literature offers a variety of books on statistical methods and probability theory. Some are primarily on the mathematical foundations of statistics, some are purely on the theory of probability, and others focus on advanced statistical methods for specific sciences. This textbook contains the foundations of probability, statistics, and data analysis methods that are applicable to a variety of fields—from astronomy to biology, business sciences, chemistry, engineering, physics, and more—with equal emphasis on mathematics and applications. The book is therefore not specific to a given discipline, nor does it attempt to describe every possible statistical method. Instead, it focuses on the fundamental methods that are used across the sciences and that are at the basis of more specific techniques that can be found in more specialized textbooks or research articles.

This textbook covers probability theory and random variables, maximum-likelihood methods for single variables and two-variable datasets, and more complex topics of data fitting, estimation of parameters, and confidence intervals. Among the topics that have recently become mainstream, Monte Carlo Markov chains occupy a special role. The last chapter of the book provides a comprehensive overview of Markov chains and Monte Carlo Markov chains, from theory to implementation.

I believe that a description of the mathematical properties of statistical tests is necessary to understand their applicability. This book therefore contains mathematical derivations that I considered particularly useful for a thorough understanding of the subject; the book refers the reader to other sources in case of mathematics that goes beyond that of basic calculus. The reader who is not familiar with calculus may skip those derivations and continue with the applications.

Nonetheless, statistics is necessarily slanted toward applications. To highlight the relevance of the statistical methods described, I have reported original data from four fundamental scientific experiments from the past two centuries: J.J. Thomson's experiment that led to the discovery of the electron, G. Mendel's data on plant characteristics that led to the law of independent assortment of species, E. Hubble's observation of nebulae that uncovered the expansion of the universe, and K. Pearson's collection of biometric characteristics in the UK in the early twentieth century. These experiments are used throughout the book to illustrate how statistical methods are applied to actual data and are used in several end-of-chapter problems. The reader will therefore have an opportunity to see *statistics in action* on these classic experiments and several additional examples.

The material presented in this book is aimed at upper-level undergraduate students or beginning graduate students. The reader is expected to be familiar with basic calculus, and no prior knowledge of statistics or probability is assumed. Professional scientists and researchers will find it a useful reference for fundamental methods such as maximum-likelihood fit, error propagation formulas, goodness of fit and model comparison, Monte Carlo methods such as the jackknife and bootstrap, Monte Carlo Markov chains, Kolmogorov-Smirnov tests, and more. All subjects are complemented by an extensive set of numerical tables that make the book completely self-contained.

The material presented in this book can be comfortably covered in a one-semester course and has several problems at the end of each chapter that are suitable as homework assignments or exam questions. Problems are both of theoretical and numerical nature, so that emphasis is equally placed on conceptual and practical understanding of the subject. Several datasets, including those in the four "classic experiments," are used across several chapters, and the students can therefore use them in applications of increasing difficulty.

Huntsville, AL, USA   Massimiliano Bonamente

Preface to the Second Edition

The second edition of *Statistics and Analysis of Scientific Data* was motivated by the overall goal to provide a textbook that is mathematically rigorous *and* easy to read and use as a reference at the same time. Basically, it is a book for both the student who wants to learn in detail the mathematical underpinnings of statistics and the reader who wants to just find the practical description on how to apply a given statistical method or use the book as a reference.

To this end, first I decided that a more clear demarcation between theoretical and practical topics would improve the readability of the book. As a result, several pages (i.e., mathematical derivations) are now clearly marked throughout the book with a vertical line, to indicate material that is primarily aimed to those readers who seek a more thorough mathematical understanding. Those parts are not required to learn how to apply the statistical methods presented in the book. For the reader who uses this book as a reference, this makes it easy to skip such sections and go directly to the main results. At the end of each chapter, I also provide a *summary of key concepts*, intended for a quick look-up of the results of each chapter.

Secondly, certain existing material needed substantial re-organization and expansion. The second edition is now comprised of 16 chapters, versus ten of the first edition. A few chapters (Chap. 6 on mean, median, and averages, Chap. 9 on multi-variable regression, and Chap. 11 on systematic errors and intrinsic scatter) contain material that is substantially new. In particular, the topic of multi-variable regression was introduced because of its use in many fields such as business and economics, where it is common to apply the regression method to many independent variables. Other chapters originate from re-arranging existing material more effectively. Some of the numerical tables in both the main body and the appendix have been expanded and re-arranged, so that the reader will find it even easier to use them for a variety of applications and as a reference.

The second edition also contains a new *classic experiment*, that of the measurement of *iris* characteristics by R.A. Fisher and E. Anderson. These new data are used to illustrate primarily the method of regression with many independent variables. The textbook now features a total of five classic experiments (including G. Mendel's data on the independent assortment of species, J.J. Thomson's data on the discovery of the electron, K. Pearson's collection of data of biometric characteristics, and E. Hubble's measurements of the expansion of the universe). These data and their analysis provide a unique way to learn the statistical methods presented in the book and a resource for the student and the teacher alike. Many of the end-of-chapter problems are based on these experimental data.

Finally, the new edition contains corrections to a number of typos that had inadvertently entered the manuscript. I am very much in debt to many of my students at the *University of Alabama in Huntsville* for pointing out these typos to me over the past few years, in particular, to Zachary Robinson, who has patiently gone through much of the text to find typographical errors.

Huntsville, AL, USA   Massimiliano Bonamente

Acknowledgments

In my early postdoc years, I was struggling to solve a complex data analysis problem. My longtime colleague and good friend Dr. Marshall Joy of NASA's Marshall Space Flight Center one day walked down to my office and said something like "Max, I have a friend in Chicago who told me that there is a method that maybe can help us with our problem. I don't understand any of it, but here's a paper that talks about Monte Carlo Markov chains. See if it can help us." That conversation led to the appreciation of one of statistics and data analysis, most powerful tools and opened the door for virtually all the research papers that I wrote ever since. For over a decade, Marshall taught me how to be careful in the analysis of data and interpretation of results—and always used a red felt-tip marker to write comments on my papers.

The journey leading to this book started about 10 years ago, when Prof. A. Gordon Emslie, currently provost at Western Kentucky University, and I decided to offer a new course in data analysis and statistics for graduate students in our department. Gordon's uncanny ability to solve virtually any problem presented to him—and likewise make even the experienced scientist stumble with his questions—has been a great source of inspiration for this book.

Some of the material presented in this book is derived from Prof. Kyle Siegrist's lectures on probability and stochastic processes at the University of Alabama in Huntsville. Kyle reinforced my love for mathematics and motivated my desire to emphasize both mathematics and applications for the material presented in this book.

Contents

Chapter 1
Theory of Probability

Abstract The theory of probability is the mathematical framework for the study of the probability of occurrence of events. The first step is to establish a method to assign the probability of an event, for example, the probability that a coin lands heads up after a toss. The *frequentist*—or empirical—approach and the *subjective*— or Bayesian— approach are two methods that can be used to calculate probabilities. The fact that there is more than one method available for this purpose should not be viewed as a limitation of the theory, but rather as the fact that for certain parts of the theory of probability, and even more so for statistics, there is an element of subjectivity that enters the analysis and the interpretation of the results. It is therefore the task of the statistician to keep track of any assumptions made in the analysis, and to account for them in the interpretation of the results. Once a method for assigning probabilities is established, the Kolmogorov axioms are introduced as the "rules" required to manipulate probabilities. Fundamental results known as Bayes' theorem and the theorem of total probability are used to define and interpret the concepts of statistical independence and of conditional probability, which play a central role in much of the material presented in this book.

1.1 Experiments, Events, and the Sample Space

Every experiment has a number of possible outcomes. For example, the experiment consisting of the roll of a die can have six possible outcomes, according to the number that shows after the die lands. The *sample space* Ω is defined as the set of all possible outcomes of the experiment, in this case $\Omega = \{1, 2, 3, 4, 5, 6\}$. An *event* A is a subset of Ω, $A \subset \Omega$, and it represents a number of possible outcomes for the experiment. For example, the event "even number" is represented by $A = \{2, 4, 6\}$, and the event "odd number" as $B = \{1, 3, 5\}$. For each experiment, two events always exist: the sample space itself, Ω, comprising all possible outcomes, and $A = \emptyset$, called the *impossible event*, or the event that contains no possible outcome.

© Springer Science+Busines Media New York 2017
M. Bonamente, *Statistics and Analysis of Scientific Data*, Graduate Texts in Physics, DOI 10.1007/978-1-4939-6572-4_1

Events are conveniently studied using set theory, and the following definitions are very common in theory of probability:

- The complementary $\overline{A}$ of an event A is the set of all possible outcomes except those in A. For example, the complementary of the event "odd number" is the event "even number."
- Given two events A and B, the union $C = A \cup B$ is the event comprising all outcomes of A and those of B. In the roll of a die, the union of odd and even numbers is the sample space itself, consisting of all possible outcomes.
- The intersection of two events $C = A \cap B$ is the event comprising all outcomes of A that are also outcomes of B. When $A \cap B = \emptyset$, the events are said to be *mutually exclusive*. The union and intersection can be naturally extended to more than two events.
- A number of events A_i are said to be a *partition* of the sample space if they are mutually exclusive, and if their union is the sample space itself, $\cup A_i = \Omega$.
- When all outcomes in A are comprised in B, we will say that $A \subset B$ or $B \supset A$.

1.2   Probability of Events

The probability P of an event describes the odds of occurrence of an event in a single trial of the experiment. The probability is a number between 0 and 1, where $P = 0$ corresponds to an impossible event, and $P = 1$ to a certain event. Therefore the operation of "probability" can be thought of as a function that transforms each possible event into a real number between 0 and 1.

1.2.1   The Kolmogorov Axioms

The first step to determine the probability of the events associated with a given experiment is to establish a number of basic rules that capture the meaning of probability. The probability of an event is required to satisfy the three axioms defined by Kolmogorov [26]:

1. The probability of an event A is a non-negative number, $P(A) \geq 0$;
2. The probability of all possible outcomes, or sample space, is normalized to the value of unity, $P(\Omega) = 1$;
3. If $A \subset \Omega$ and $B \subset \Omega$ are *mutually exclusive* events, then

$$P(A \cup B) = P(A) + P(B) \tag{1.1}$$

Figure 1.1 illustrates this property using set diagrams. For events that are not mutually exclusive, this property does not apply. The probability of the union is

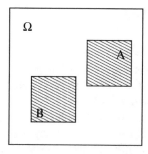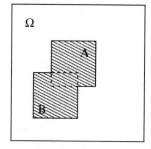

Fig. 1.1   The probability of the event $P(A \cup B)$ is the sum of the two individual probabilities, only if the two events are mutually exclusive. This property enables the interpretation of probability as the "area" of a given event within the sample space

represented by the area of $A \cup B$, and the outcomes that overlap both events are not double-counted.

These axioms should be regarded as the basic "ground rules" of probability, but they provide no unique specification on how event probabilities should be assigned. Two major avenues are available for the assignment of probabilities. One is based on the repetition of the experiments a large number of times under the same conditions, and goes under the name of the frequentist or classical method. The other is based on a more theoretical knowledge of the experiment, but without the experimental requirement, and is referred to as the Bayesian approach.

1.2.2   Frequentist or Classical Method

Consider performing an experiment for a number $N \gg 1$ of times, under the same experimental conditions, and measuring the occurrence of the event A as the number $N(A)$. The probability of event A is given by

$$P(A) = \lim_{N \to \infty} \frac{N(A)}{N};$$ (1.2)

that is, the probability is the relative frequency of occurrence of a given event from many repetitions of the same experiment. The obvious limitation of this definition is the need to perform the experiment an infinite number of times, which is not only time consuming, but also requires the experiment to be repeatable in the first place, which may or may not be possible.

The limitation of this method is evident by considering a coin toss: no matter the number of tosses, the occurrence of heads up will never be exactly 50 %, which is what one would expect based on a knowledge of the experiment at hand.

1.2.3   Bayesian or Empirical Method

Another method to assign probabilities is to use the knowledge of the experiment and the event, and the probability one assigns represents the degree of belief that the event will occur in a given try of the experiment. This method implies an element of subjectivity, which will become more evident in Bayes' theorem (see Sect. 1.7). The Bayesian probability is assigned based on a quantitative understanding of the nature of the experiment, and in accord with the Kolmogorov axioms. It is sometimes referred to as *empirical* probability, in recognition of the fact that sometimes the probability of an event is assigned based upon a practical knowledge of the experiment, although without the classical requirement of repeating the experiment for a large number of times. This method is named after the Rev. Thomas Bayes, who pioneered the development of the theory of probability [3].

Example 1.1 (Coin Toss Experiment)   In the coin toss experiment, the determination of the empirical probability for events "heads up" or "tails up" relies on the knowledge that the coin is unbiased, and that therefore it must be true that $P(tails) = P(heads)$. This empirical statement signifies the use of the Bayesian method to determine probabilities. With this information, we can then simply use the Kolmogorov axioms to state that $P(tails) + P(heads) = 1$, and therefore obtain the intuitive result that $P(tails) = P(heads) = 1/2$.   ◇

1.3   Fundamental Properties of Probability

The following properties are useful to improve our ability to assign and manipulate event probabilities. They are somewhat intuitive, but it is instructive to derive them formally from the Kolmogorov axioms.

1. The probability of the null event is zero, $P(\emptyset) = 0$.

   Proof Start with the mutually exclusive events $\emptyset$ and Ω. Since their union is Ω, it follows from the Third Axiom that $P(\Omega) = P(\Omega) + P(\emptyset)$. From the Second Axiom we know that $P(\Omega) = 1$, from this it follows that $P(\emptyset) = 0$.   □

   The following property is a generalization of the one described above:

2. The probability of the complementary event $\overline{A}$ satisfies the property

$$P(\overline{A}) = 1 - P(A). \tag{1.3}$$

   Proof By definition, it is true that $A \cup \overline{A} = \Omega$, and that $A, \overline{A}$ are mutually exclusive. Using the Second and Third axiom, $P(A \cup \overline{A}) = P(A) + P(\overline{A}) = 1$, from which it follows that $P(\overline{A}) = 1 - P(A)$.   □

3. The probability of the union of two events satisfies the general property that

$$P(A \cup B) = P(A) + P(B) - P(A \cap B). \tag{1.4}$$

This property generalizes the Third Kolmogorov axiom, and can be interpreted as the fact that outcomes in the overlap region of the two events should be counted only once, as illustrated in Fig. 1.1.

Proof First, realize that the event $A \cup B$ can be written as the union of three mutually exclusive sets, $A \cup B = (A \cap \bar{B}) \cup (B \cap \bar{A}) \cup (A \cap B)$, see Fig. 1.1. Therefore, using the Third axiom, $P(A \cup B) = P(A \cap \bar{B}) + P(B \cap \bar{A}) + P(A \cap B)$.
   Then, notice that for any event A and B, it is true that $A = (A \cap \bar{B}) \cup (A \cap B)$, since $\{B, \bar{B}\}$ is a partition of Ω. This implies that $P(A) = P(A \cap B) + P(A \cap \bar{B})$ due to the fact that the two sets are again mutually exclusive, and likewise for event B. It thus follows that $P(A \cup B) = P(A) - P(A \cap B) + P(B) - P(B \cap A) + P(A \cap B) = P(A) + P(B) - P(A \cap B)$. $\qquad\square$

Example 1.2 An experiment consists of drawing a number between 1 and 100 at random. Calculate the probability of the event: "drawing either a number greater than 50, or an odd number, at each try."
   The sample space for this experiment is the set of numbers $i = 1, \ldots, 100$, and the probability of drawing number i is $P(A_i) = 1/100$, since we expect that each number will have the same probability of being drawn at each try. A_i is the event that consists of drawing number i. If we call B the event consisting of all numbers greater than 50, and C the event with all odd numbers, it is clear that $P(B) = 0.5$, and likewise $P(C) = 0.5$. The event $A \cap B$ contains all odd numbers greater than 50, and therefore $P(A \cap B) = 0.25$. Using (1.4), we find that the probability of drawing either a number greater than 50, or an odd number, is 0.75. This can be confirmed by a direct count of the possible outcomes. $\qquad\diamond$

1.4   Statistical Independence

Statistical independence among events means that the occurrence of one event has no influence on the occurrence of other events. Consider, for example, rolling two dice, one after the other: the outcome of one die is independent of the other, and the two tosses are said to be statistically independent. On the other hand, consider the following pair of events: the first is the roll of die 1, and the second is the roll of die 1 *and* die 2, so that for the second event we are interested in the sum of the two tosses. It is clear that the outcome of the second event—e.g., the sum of both dice—depends on the first toss, and the two events are not independent.
   Two events A and B are said to be statistically independent if and only if

$$P(A \cap B) = P(A) \cdot P(B). \tag{1.5}$$

At this point, it is not obvious that the concept of statistical independence is embodied by (1.5). A few examples will illustrate the meaning of this definition, which will be explored further in the following section on conditional probability.

Example 1.3 Determine the probability of obtaining two 3 when rolling two dice. This event can be decomposed in two events: $A = \{$die 1 shows 3 and die 2 shows any number$\}$ and $B = \{$die 2 shows 3 and die 1 shows any number$\}$.

It is natural to assume that $P(A) = 1/6$, $P(B) = 1/6$ and state that the two events A and B are independent by nature, since each event involves a different die, which has no knowledge of the other one. The event we are interested in is $C = A \cap B$ and the definition of probability of two statistically independent events leads to $P(C) = P(A \cap B) = P(A) \cdot P(B) = 1/36$. This result can be confirmed by the fact that there is only one combination out of 36 that gives rise to two consecutive 3. ◇

The example above highlights the importance of a proper, and sometimes extended, definition of an event. The more careful the description of the event and of the experiment that it is drawn from, the easier it is to make probabilistic calculation and the assessment of statistical independence.

Example 1.4 Consider the events $A = \{$die 1 shows 3 and die 2 shows any number$\}$ and $B = \{$the sum of the two dice is 9$\}$. Determine whether they are statistically independent.

In this case, we will calculate the probability of the two events, and then check whether they obey (1.5) or not. This calculation will illustrate that the two events are *not* statistically independent.

Event A has a probability $P(A) = 1/6$; in order to calculate the probability of event B, we realize that a sum of 9 is given by the following combinations of outcomes of the two rolls: (3,6), (4,5), (5,4) and (6,3). Therefore, $P(B) = 1/9$. The event $A \cap B$ is the situation in which *both* event A and B occur, which corresponds to the single combination (3,6); therefore, $P(A \cap B) = 1/36$. Since $P(A) \cdot P(B) = 1/6 \cdot 1/9 = 1/54 \neq P(A \cap B) = 1/36$, we conclude that the two events are not statistically independent. This conclusion means that one event influences the other, since a 3 in the first toss has certainly an influence on the possibility of both tosses having a total of 9. ◇

There are two important necessary (but not sufficient) conditions for statistical independence between two events. These properties can help identify whether two events are independent.

1. If $A \cap B = \emptyset$, A and B *cannot* be independent, unless one is the empty set. This property states that there must be some overlap between the two events, or else it is not possible for the events to be independent.

   Proof For A and B to be independent, it must be true that $P(A \cap B) = P(A) \cdot P(B)$, which is zero by hypothesis. This can be true only if $P(A) = 0$ or $P(B) = 0$, which in turn means $A = \emptyset$ or $B = \emptyset$ as a consequence of the Kolmogorov axioms.   □

2. If $A \subset B$, then A and B *cannot* be independent, unless B is the entire sample space. This property states that the overlap between two events cannot be such that one event is included in the other, in order for statistical independence to be possible.

Proof In order for A and B to be independent, it must be that $P(A \cap B) = P(A) \cdot P(B) = P(A)$, given that $A \subset B$. This can only be true if $B = \Omega$, since $P(\Omega) = 1$.
$\square$

Example 1.5 Consider the above Example 1.3 of the roll of two dice; each event was formulated in terms of the outcome of both rolls, to show that there was in fact overlap between two events that are independent of one another. ◇

Example 1.6 Consider the following two events: $A = \{$die 1 shows 3 and die 2 shows any number$\}$ and $B = \{$die 1 shows 3 or 2 and die 2 shows any number$\}$. It is clear that $A \subset B$, $P(A) = 1/6$ and $P(B) = 1/3$. The event $A \cap B$ is identical to A and $P(A \cap B) = 1/6$. Therefore $P(A \cap B) \neq P(A) \cdot P(B)$ and the two events are not statistically independent. This result can be easily explained by the fact that the occurrence of A implies the occurrence of B, which is a strong statement of dependence between the two events. The dependence between the two events can also be expressed with the fact that the non-occurrence of B implies the non-occurrence of A. ◇

1.5 Conditional Probability

The conditional probability describes the probability of occurrence of an event A *given* that another event B has occurred and it is indicated as $P(A/B)$. The symbol "/" indicates the statement *given that* or *knowing that*. It states that the event after the symbol is known to have occurred. When two or more events are not independent, the probability of a given event will in general depend on the occurrence of another event. For example, if one is interested in obtaining a 12 in two consecutive rolls of a die, the probability of such event does rely on the fact that the first roll was (or was not) a 6.

The following relationship defines the conditional probability:

$$P(A \cap B) = P(A/B) \cdot P(B) = P(B/A) \cdot P(A); \qquad (1.6)$$

Equation (1.6) can be equivalently expressed as

$$P(A/B) = \begin{cases} \dfrac{P(A \cap B)}{P(B)} & \text{if } P(B) \neq 0 \\ 0 & \text{if } P(B) = 0. \end{cases} \qquad (1.7)$$

A justification for this definition is that the occurrence of B means that the probability of occurrence of A is that of $A \cap B$. The denominator of the conditional probability is $P(B)$ because B is the set of all possible outcomes that are known to have happened. The situation is also depicted in the right-hand side panel of Fig. 1.1: knowing that B has occurred, leaves the probability of occurrence of A to the occurrence of the intersection $A \cap B$, out of all outcomes in B. It follows directly from (1.6) that if A and B are statistically independent, then the conditional probability is $P(A/B) = P(A)$, i.e., the occurrence of B has no influence on the occurrence of A. This observation further justifies the definition of statistical independence according to (1.5).

Example 1.7 Calculate the probability of obtaining 8 as the sum of two rolls of a die, given that the first roll was a 3.

Call event A the sum of 8 in two separate rolls of a die and event B the event that the first roll is a 3. Event A is given by the probability of having tosses (2,6), (3,5), (4,4), (5,3), (6,2). Since each such combination has a probability of 1/36, $P(A) = 5/36$. The probability of event B is $P(B) = 1/6$. Also, the probability of $A \cap B$ is the probability that the first roll is a 3 and the sum is 8, which can clearly occur only if a sequence of (3,5) takes place, with probability $P(A \cap B) = 1/36$.

According to the definition of conditional probability, $P(A/B) = P(A \cap B)/P(B) = 6/36 = 1/6$, and in fact only combination (5,3)—of the six available with 3 as the outcome of the second toss—gives rise to a sum of 8. The occurrence of 3 in the first roll has therefore increased the probability of A from $P(A) = 5/36$ to $P(A/B) = 1/6$, since not any outcome of the first roll would be equally conducive to a sum of 8 in two rolls.   ◇

1.6   A Classic Experiment: Mendel's Law of Heredity and the Independent Assortment of Species

The experiments performed in the nineteenth century by Gregor Mendel in the monastery of Brno led to the discovery that certain properties of plants, such as seed shape and color, are determined by a pair of genes. This pair of genes, or *genotype*, is formed by the inheritance of one gene from each of the parent plants.

Mendel began by crossing two pure lines of pea plants which differed in one single characteristic. The first generation of hybrids displayed only one of the two characteristics, called the *dominant* character. For example, the first-generation plants all had round seed, although they were bred from a population of pure round seed plants and one with wrinkled seed. When the first-generation was allowed to self-fertilize itself, Mendel observed the data shown in Table 1.1 [31].

(continued)

Table 1.1 Data from G. Mendel's experiment

Character	No. of dominant	No. of recessive	Fract. of dominant
Round vs. wrinkled seed	5474	1850	0.747
Yellow vs. green seed	6022	2001	0.751
Violet-red vs. white flower	705	224	0.759
Inflated vs. constricted pod	882	299	0.747
Green vs. yellow unripe pod	428	152	0.738
Axial vs. terminal flower	651	207	0.759
Long vs. short stem	787	277	0.740

Table 1.2 Data from G. Mendel's experiment for plants with two different characters

	Yellow seed	Green seed
Round seed	315	108
Wrinkled seed	101	32

In addition, Mendel performed experiments in which two pure lines that differed by two characteristics were crossed. In particular, a line with yellow and round seed was crossed with one that had green and wrinkled seeds. As in the previous case, the first-generation plants had a 100 % occurrence of the dominant characteristics, while the second-generation was distributed according to the data in Table 1.2.

One of the key results of these experiments goes under the name of *Law of independent assortment*, stating that a daughter plant inherits one gene from each parent plant independently of the other parent. If we denote the genotype of the dominant parent as DD (a pair of dominant genes) and that of the recessive parent as RR, then the data accumulated by Mendel support the hypothesis that the first-generation plants will have the genotype DR (the order of genes in the genome is irrelevant) and the second generation plants will have the following four genotypes: DD, DR, RD and RR, in equal proportions. Since the first three genomes will display the dominant characteristic, the ratio of appearance of the dominant characteristic is expected to be 0.75. The data appear to support in full this hypothesis.

In probabilistic terms, one expects that each second-generation plant has $P(D) = 0.5$ of drawing a dominant first gene from each parent and $P(R) = 0.5$ of drawing a recessive gene from each parent. Therefore, according to the

(continued)

hypothesis of independence in the inheritance of genes, we have

$$\begin{cases} P(DD) = P(D) \cdot P(D) = 0.25 \\ P(DR) = P(D) \cdot P(R) = 0.25 \\ P(RD) = P(R) \cdot P(D) = 0.25 \\ P(RR) = P(R) \cdot P(R) = 0.25. \end{cases} \qquad (1.8)$$

When plants differing by two characteristics are crossed, as in the case of the data in Table 1.2, then each of the four events in (1.8) is independently mixed between the two characters. Therefore, there is a total of 16 possibilities, which give rise to 4 possible combinations of the two characters. For example, a display of both recessive characters will have a probability of $1/16 = 0.0625$. The data seemingly support this hypothesis with a measurement of a fraction of 0.0576.

1.7   The Total Probability Theorem and Bayes' Theorem

In this section we describe two theorems that are of great importance in a number of practical situations. They make use of a partition of the sample space Ω, consisting of n events A_i that satisfy the following two properties:

$$A_i \cap A_j = \emptyset, \ \forall i \neq j$$

$$\bigcup_{i=1}^{n} A_i = \Omega. \qquad (1.9)$$

For example, the outcomes 1, 2, 3, 4, 5 and 6 for the roll of a die partition the sample space into a number of events that cover all possible outcomes, without any overlap among each other.

Theorem 1.1 (Total Probability Theorem) *Given an event B and a set of events A_i with the properties (1.9),*

$$P(B) = \sum_{i=1}^{n} P(B \cap A_i) = \sum_{i=1}^{n} P(B/A_i) \cdot P(A_i). \qquad (1.10)$$

Proof The first equation is immediately verified given that the $B \cap A_i$ are mutually exclusive events such that $B = \cup_i (B \cap A_i)$. The second equation derives from the application of the definition of conditional probability.   □

The total probability theorem is useful when the probability of an event B cannot be easily calculated and it is easier to calculate the conditional probability B/A_i given a suitable set of conditions A_i. The example at the end of Sect. 1.7 illustrates one such situation.

Theorem 1.2 (Bayes' Theorem) *Given an event B and a set of events A_i with properties (1.9),*

$$P(A_i/B) = \frac{P(B/A_i)P(A_i)}{P(B)} = \frac{P(B/A_i)P(A_i)}{\sum_{i=1}^{n} P(B \cap A_i)} \qquad (1.11)$$

Proof The proof is an immediate consequence of the definition of conditional probability, (1.6), and of the Total Probability theorem, (1.10). □

Bayes' theorem is often written in a simpler form by taking into account two events only, $A_i = A$ and B:

$$P(A/B) = \frac{P(B/A)P(A)}{P(B)} \qquad (1.12)$$

In this form, Bayes' theorem is just a statement of how the order of conditioning between two events can be inverted.

Equation (1.12) plays a central role in probability and statistics. What is especially important is the interpretation that each term assumes within the context of a specific experiment. Consider B as the *data* collected in a given experiment— these data can be considered as an event, containing the outcome of the experiment. The event A is a *model* that is used to describe the data. The model can be considered as an ideal outcome of the experiment, therefore both A and B are events associated with the same experiment. Following this interpretation, the quantities involved in Bayes' theorem can be interpreted as in the following:

- $P(B/A)$ is the probability, or *likelihood*, of the data given the specified model, and indicated as $\mathscr{L}$. The likelihood represents the probability of making the measurement B given that the model A is a correct description of the experiment.
- $P(A)$ is the probability of the model A, without any knowledge of the data. This term is interpreted as a *prior probability*, or the degree belief that the model is true before the measurements are made. Prior probabilities should be based upon quantitative knowledge of the experiment, but can also reflect the subjective belief of the analyst. This step in the interpretation of Bayes' theorem explicitly introduces an element of subjectivity that is characteristic of Bayesian statistics.
- $P(B)$ is the probability of collecting the dataset B. In practice, this probability acts as a normalization constant and its numerical value is typically of no practical consequence.
- Finally, $P(A/B)$ is the probability of the model after the data have been collected. This is referred to as the *posterior probability* of the model. The posterior

probability is the ultimate goal of a statistical analysis, since it describes the probability of the model based on the collection of data. According to the value of the posterior probability, a model can be accepted or discarded.

This interpretation of Bayes' theorem is the foundation of Bayesian statistics. Models of an experiment are usually described in terms of a number of parameters. One of the most common problems of statistical data analysis is to estimate what values for the parameters are permitted by the data collected from the experiment. Bayes' theorem provides a way to update the prior knowledge on the model parameters given the measurements, leading to posterior estimates of parameters. One key feature of Bayesian statistics is that the calculation of probabilities are based on a prior probability, which may rely on a subjective interpretation of what is known about the experiment before any measurements are made. Therefore, great attention must be paid to the assignment of prior probabilities and the effect of priors on the final results of the analysis.

Example 1.8 Consider a box in which there are red and blue balls, for a total of $N = 10$ balls. What is known a priori is just the total number of balls in the box. Of the first 3 balls drawn from the box, 2 are red and 1 is blue (drawing is done with re-placement of balls after drawing). We want to use Bayes' theorem to make inferences on the number of red balls (i) present in the box, i.e., we seek $P(A_i/B)$, the probability of having i red balls in the box, given that we performed the measurement $B = \{$Two red balls were drawn in the first three trials$\}$.

Initially, we may assume that $P(A_i) = 1/11$, meaning that there is an equal probability of having $0, 1, \ldots$ or 10 red balls in the box (for a total of 11 possibilities) before we make any measurements. Although this is a subjective statement, a uniform distribution is normally the logical assumption in the absence of other information. We can use basic combinatorial mathematics to determine that the likelihood of drawing $D = 2$ red balls out of $T = 3$ trials, given that there are i red balls (also called event A_i):

$$P(B/A_i) = \binom{T}{D} p^D q^{T-D}. \tag{1.13}$$

In this equation p is the probability of drawing one of the red balls in a given drawing assuming that there are i red balls, $p = i/N$, and q is the probability of drawing one of the blue balls, $q = 1 - p = (N - i)/N$. The distribution in (1.13) is known as the binomial distribution and it will be derived and explained in more detail in Sect. 3.1. The likelihood $P(B/A_i)$ can therefore be rewritten as

$$P(B/A_i) = \binom{3}{2} \left(\frac{i}{N}\right)^2 \left(\frac{N-i}{N}\right) \tag{1.14}$$

The probability $P(B)$ is the probability of drawing $D = 2$ red balls out of $T = 3$ trial, for all possible values of the true number of red balls, $i = 0, \ldots, 10$. This probability can be calculated from the Total Probability theorem,

$$P(B) = \sum_{i=0}^{N} P(B/A_i) \cdot P(A_i) \tag{1.15}$$

We can now put all the pieces together and determine the *posterior probability* of having i red balls, $P(A_i/B)$, using Bayes' theorem, $P(A_i/B) = P(B/A_i)P(A_i)/P(B)$.

The equation above is clearly a function of i, the true number of red balls. Consider the case of $i = 0$, i.e., what is the *posterior probability* of having no red balls in the box. Since

$$P(B/A_0) = \binom{3}{2} \left(\frac{0}{N}\right)^2 \left(\frac{N-0}{N}\right) = 0,$$

it follows that $P(A_o/B) = 0$, i.e., it is impossible that there are no red balls. This is obvious, since two times a red ball was in fact drawn, meaning that there is at least one red ball in the box. Other posterior probabilities can be calculated in a similar way.   ◇

Summary of Key Concepts for this Chapter

☐ *Event:* A set of possible outcomes of an experiment.
☐ *Sample space*: All possible outcomes of an experiment.
☐ *Probability of an Event:* A number between 0 and 1 that follows the Kolmogorov axioms.
☐ *Frequentist or Classical approach:* A method to determine the probability of an event based on many repetitions of the experiment.
☐ *Bayesian or Empirical approach:* A method to determine probabilities that uses prior knowledge of the experiment.
☐ *Statistical independence*: Two events are statistically independent when the occurrence of one has no influence on the occurrence of the other, $P(A \cap B) = P(A)P(B)$.
☐ *Conditional probability:* Probability of occurrence of an event given that another event is known to have occurred, $P(A/B) = P(A \cap B)/P(B)$.
☐ *Total Probability theorem:* A relationship among probabilities of events that form a partition of the sample space, $P(B) = \sum P(B/A_i)P(A_i)$.
☐ *Bayes' theorem*: A relationship among conditional probabilities that enables the change in the order of conditioning of the events, $P(A/B) = P(B/A)P(A)/P(B)$.

Problems

1.1 Describe the sample space of the experiment consisting of flipping four coins simultaneously. Assign the probability to the event consisting of "two heads up and two tails up." In this experiment it is irrelevant to know which specific coin shows heads up or tails up.

1.2 An experiment consists of rolling two dice simultaneously and independently of one another. Find the probability of the event consisting of having either an odd number in the first roll or a total of 9 in both rolls.

1.3 In the roll of a die, find the probability of the event consisting of having either an even number or a number greater than 4.

1.4 An experiment consists of rolling two dice simultaneously and independently of one another. Show that the two events, "the sum of the two rolls is 8" and "the first roll shows 5" are not statistically independent.

1.5 An experiment consists of rolling two dice simultaneously and independently of one another. Show that the two events, "first roll is even" and "second roll is even" are statistically independent.

1.6 A box contains 5 balls, of which 3 are red and 2 are blue. Calculate (a) the probability of drawing two consecutive red balls and (b) the probability of drawing two consecutive red balls, given that the first draw is known to be a red ball. Assume that after each draw the ball is replaced in the box.

1.7 A box contains 10 balls that can be either red or blue. Of the first three draws, done with replacement, two result in the draw of a red ball. Calculate the ratio of the probability that there are 2 or just 1 red ball in the box and the ratio of probability that there are 5 or 1 red balls.

1.8 In the game of baseball a player at bat either reaches base or is retired. Consider three baseball players: player A was at bat 200 times and reached base 0.310 of times; player B was at bat 250 times, with an on-base percentage of 0.296; player C was at bat 300 times, with an on-base percentage 0.260. Find (a) the probability that when either player A, B, or C were at bat, he reached base, (b) the probability that, given that a player reached base, it was A, B, or C.

1.9 An experiment consists of rolling two dice simultaneously and independently of one another. Calculate (a) the probability of the first roll being a 1, given that the sum of both rolls was 5, (b) the probability of the sum being 5, given that the first roll was a 1 and (c) the probability of the first roll being a 1 and the sum being 5. Finally, (d) verify your results with Bayes' theorem.

1.10 Four coins labeled 1 through 4 are tossed simultaneously and independently of one another. Calculate (a) the probability of having an ordered combination heads-tails-heads-tails in the four coins, (b) the probability of having the same ordered combination given that any two coins are known to have landed heads-up and (c) the probability of having two coins land heads up given that the sequence heads-tails-heads-tails has occurred.

Chapter 2
Random Variables and Their Distributions

Abstract The purpose of performing experiments and collecting data is to gain information on certain quantities of interest called random variables. The exact value of these quantities cannot be known with absolute precision, but rather we can constrain the variable to a given range of values, narrower or wider according to the nature of the variable itself and the type of experiment performed. Random variables are described by a distribution function, which is the theoretical expectation for the outcome of experiments aimed to measure it. Other measures of the random variable are the mean, variance, and higher-order moments.

2.1 Random Variables

A random variable is a quantity of interest whose true value is unknown. To gain information on a random variable we design and conduct experiments. It is inherent to any experiment that the random variable of interest will never be known exactly. Instead, the variable will be characterized by a *probability distribution function*, which determines what is the probability that a given value of the random variable occurs. Repeating the measurement typically increases the knowledge we gain of the distribution of the variable. This is the reason for wanting to measure the quantity as many times as possible.

As an example of random variable, consider the gravitational constant G. Despite the label of "constant", we only know it to have a range of possible values in the approximate interval $G = 6.67428 \pm 0.00067$ (in the standard S.I. units). This means that we don't know the true value of G, but we estimate the range of possible values by means of experiments. The random nature of virtually all quantities lies primarily in the fact that no quantity is known exactly to us without performing an experiment and that any experiment is never perfect because of practical or even theoretical limitations. Among the practical reasons are, for example, limitations in the precision of the measuring apparatus. Theoretical reasons depend on the nature of the variable. For example, the measurement of the position and velocity of a subatomic particle is limited by the Heisenberg uncertainty principle, which forbids an exact knowledge even in the presence of a perfect measuring apparatus.

The general method for gaining information on a random variable X starts with set of measurements x_i, ensuring that measurements are performed under the same

© Springer Science+Busines Media New York 2017

M. Bonamente, *Statistics and Analysis of Scientific Data*, Graduate Texts in Physics, DOI 10.1007/978-1-4939-6572-4_2

Fig. 2.1 Example of data collected to measure a random variable X. The 500 measurements were binned according to their value to construct the sample distribution. The shape of the distribution depends on the nature of the experiment and of the number of measurements

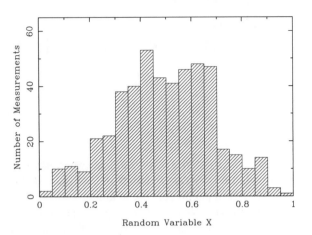

experimental conditions. Throughout the book we will reserve uppercase letters for the name of the variable itself and lowercase letters for the actual measurements. From these measurements, one obtains a histogram corresponding to the frequency of occurrence of all values of X (Fig. 2.1). The measurements x_i form the *sample distribution* of the quantity, which describes the empirical distribution of values collected in the experiment. On the other hand, random variables are typically expected to have a theoretical distribution, e.g., Gaussian, Poisson, etc., known as the *parent distribution*. The parent distribution represents the belief that there is an ideal description of a random variable and its form depends on the nature of the variable itself and the method of measurement. The sample distribution is expected to become the parent distribution if an infinite number of measurements are performed, in such a way that the randomness associated with a small number of measurements is eliminated.

Example 2.1 In Sect. 3.3 we will show that a discrete variable (e.g., one that can only take integer values) that describes a counting experiment follows a Poisson function,

$$P(n) = \frac{\mu^n}{n!} e^{-\mu}$$

in which μ is the mean value of the random variable (for short, its true-yet-unknown value) and n is the actual value measured for the variable. $P(n)$ indicates the probability of measuring the value n, given that the true value is μ. Consider the experiment of counting the number of photons reaching Earth from a given star; due to a number of factors, the count may not always be the same every time the experiment is performed, and if only one experiment is performed, one would obtain a sample distribution that has a single "bar" at the location of the measured value and this sample distribution would not match well a Poisson function. After a small number of measurements, the distribution may appear similar to that in Fig. 2.1

and the distribution will then become smoother and closer to the parent distribution as the number of measurements increases. Repeating the experiment therefore will help in the effort to estimate as precisely as possible the parameter μ that determines the Poisson distribution.

$\diamond$

2.2   Probability Distribution Functions

It is convenient to describe random variables with an analytic function that determines the probability of the random variable to have a given value. Discrete random variables are described by a *probability mass function* $f(x_i)$, where $f(x_i)$ represents the probability of the variable to have an exact value of x_i. Continuous variables are described by a *probability distribution function* $f(x)$, such that $f(x)dx$ is the probability of the variable to have values in the interval $[x, x + dx]$. For simplicity we will refer to both types of distributions as probability distribution functions throughout the book.

Probability distribution functions have the following properties:

1. They are normalized to 1. For continuous variables this means

$$\int_{-\infty}^{+\infty} f(x)dx = 1. \tag{2.1}$$

For variables that are defined in a subset of the real numbers, e.g., only values $x \geq 0$ or in a finite interval, $f(x)$ is set to zero outside the domain of definition of the function. For discrete variables, hereafter the integrals are replaced by a sum over all values that the function of integration can have.

2. The probability distribution can never be negative, $f(x) \geq 0$. This is a consequence of the Kolmogorov axiom that requires a probability to be non-negative.

3. The function $F(x)$, called the *(cumulative) distribution function*,

$$F(x) = \int_{-\infty}^{x} f(\tau)d\tau, \tag{2.2}$$

represents the probability that the variable has any value less or equal than x. $F(x)$ is a non-decreasing function of x that starts at zero and has its highest value of one.

Example 2.2 The *exponential random variable* follows the probability distribution function defined by

$$f(x) = \lambda e^{-\lambda x}, \ x \geq 0, \tag{2.3}$$

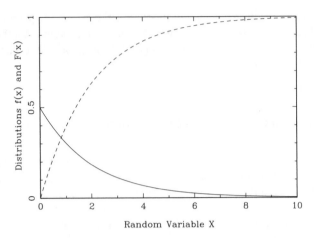

Fig. 2.2 The distribution function $f(x)$ (*solid line*) and the cumulative distribution function $F(x)$ (*dashed line*) for an exponential variable with $\lambda = 0.5$

where λ is an adjustable parameter that must be positive. The probability distribution function is therefore $f(x) = 0$ for negative values of the variable. The cumulative distribution function is given by

$$F(x) = 1 - e^{-\lambda x}. \tag{2.4}$$

In Fig. 2.2 are drawn the probability distribution function $f(x)$ and the cumulative distribution function $F(x)$ for an exponential variable with $\lambda = 0.5$. $\diamond$

2.3   Moments of a Distribution Function

The probability distribution function $f(x)$ provides a complete description of the random variable. It is convenient to find a few quantities that describe the salient features of the distribution. The *moment* of order n, μ_n, is defined as

$$\mu_n = E[X^n] \equiv \int f(x)x^n dx. \tag{2.5}$$

The moment μ_n is also represented as $E[X^n]$, the *expectation* of the function X^n. It is possible to demonstrate, although mathematically beyond the scope of this book, that the knowledge of moments of all orders is sufficient to determine uniquely the distribution function [42]. This is an important fact, since it shifts the problem of determining the distribution function to that of determining at least some of its moments. Moreover, a number of distribution functions only have a few non-zero moments, and this renders the task even more manageable.

The moments or expectations of a distribution are theoretical quantities that can be calculated from the probability distribution $f(x)$. They are *parent* quantities that we wish to estimate via measurements. In the following we describe the two main expectations, the mean and the variance, and the *sample* quantities that approximate them, the sample mean and the sample variance. Chapter 5 describes a method to justify the estimates of parent quantities via sample quantities.

2.3.1 The Mean and the Sample Mean

The moment of the first order is also known as the *mean* or *expectation* of the random variable,

$$\mu = E[X] = \int_{-\infty}^{+\infty} xf(x)dx. \tag{2.6}$$

The expectation is a linear operation and therefore satisfies the property that, e.g.,

$$E[aX + bY] = aE[X] + bE[Y], \tag{2.7}$$

where a and b are constants. This is a convenient property to keep in mind when evaluating expectations of complex functions of a random variable X.

To estimate the mean of a random variable, consider N measurements x_i and define the *sample mean* as

$$\bar{x} = \frac{1}{N} \sum_{i=1}^{N} x_i. \tag{2.8}$$

To illustrate that the sample mean $\bar{x}$ defined by (2.8) is equivalent to the mean μ, consider a discrete variable, for which

$$E[X] = \sum_{j=1}^{M} f(x_j)x_j, \tag{2.9}$$

where $f(x_j)$ is the probability distribution function and we have assumed that the variable can only have M possible values. According to the classical interpretation of the probability, the distribution function is given by

$$f(x_j) = \lim_{N\to\infty} \frac{N(x_j)}{N},$$

in which $N(x_j)$ is the number of occurrence of the value x_j. Since $\Sigma N(x_j)x_j$ is the value obtained in N measurements, it is equivalent to Σx_i. Therefore the sample mean will be identical to the parent mean in the limit of an infinite number of measurements,

$$\lim_{N \to \infty} \bar{x} = \lim_{N \to \infty} \frac{1}{N} \sum_{i=1}^{N} x_i = \lim_{N \to \infty} \frac{1}{N} \sum_{j=1}^{M} N(x_j)x_j = \sum_{j=1}^{M} f(x_j)x_j = E[X].$$

A proof that the sample mean provides an unbiased estimate of the mean will be given in Chap. 5 for Gaussian and Poisson variables.

The sample mean is therefore a representative value of the random variable that estimates the parent mean using a finite number of measurements. Other measures of a random variable include the *mode*, defined as the value of maximum probability, and the *median*, defined as the value that separates the lower 50 % and the upper 50 % of the distribution function. For distributions that are symmetric with respect to the peak value, as is the case for the Gaussian distribution defined below in Sect. 3.2, the peak value coincides with the mean, median, and mode. A more detailed analysis of the various measures of the "average" value of a variable is described in Chap. 6.

2.3.2   The Variance and the Sample Variance

The *variance* is the expectation of the square of the deviation of X from its mean:

$$Var(X) = E[(X - \mu)^2] = \int_{-\infty}^{+\infty} (x - \mu)^2 f(x)dx = \sigma^2. \tag{2.10}$$

The square root of the variance is referred to as the *standard deviation* or *standard error* σ and it is a common measure of the average difference of a given measurement x_i from the mean of the random variable. Notice that from the point of view of physical dimensions of the moments defined by (2.5), moments of the n-th order have the dimensions of the random variable to the n-th power. For example, if X is measured in meters, the variance is measured in meters square (m^2), thus the need to use the square root of the variance as a measure of the standard deviation of the variable from its mean.

The main reason for defining the average difference of a measurement from its mean in terms of a moment of the second order is that the expectation of the *deviation* $X - \mu$ is always zero, as can be immediately seen using the linearity property of the expectation. The deviation of a random variable is therefore not of common use in statistics, since its expectation is null.

The *sample variance* is defined as

$$s^2 = \frac{1}{N-1} \sum_{i=1}^{N} (x_i - \bar{x})^2 \tag{2.11}$$

and a proof that this quantity is an unbiased estimate of the parent variance will be provided in Chap. 5. The presence of a factor of $N-1$, and not just N, in the denominator of the sample variance, is caused by the fact that the sample variance requires also an estimate of the sample mean, since the exact value of the parent mean is unknown. This result will be explained further in Sect. 5.1.2.

Using the linear property of the expectation, it is straightforward to show that the following property applies:

$$Var(X) = E[X^2] - \mu^2. \tag{2.12}$$

This relationship is very convenient to calculate the variance from the moments of the first and second order. The deviation and the variance are moments calculated with respect to the mean, also referred to as *central moments*.

Another useful property of the variance, which follows from the fact that the variance is a moment of the second order, is

$$Var(aX) = a^2 Var(X) \tag{2.13}$$

where a is a constant.

2.4 A Classic Experiment: J.J. Thomson's Discovery of the Electron

A set of experiments by J.J. Thomson in the late nineteenth century were aimed at the measurement of the ratio between the mass and charge of a new lightweight particle, which was later named *electron*. The experiment was truly groundbreaking not just for the method used, but also because it revolutionized our understanding of physics and natural sciences by proving that the new particle was considerably lighter than the previously known charge carrier, the proton.

The experiment described in this book was reported by Thomson in [39]. It consists of measuring the deflection of negatively charged cathode rays by a magnetic field H in a tube. Thomson wanted to measure the mass m of the charged particles that constituted these cathode rays. The experiment is based on the measurement of the following quantities: W is the kinetic energy of the

(continued)

particles, $Q = Ne$ is the amount of electricity carried by the particles (N is the number of particles and e the charge of each particle) and $I = HR$, where R is the radius of curvature of the path of these rays in a magnetic field H. The measurements performed by Thomson were used to infer the ratio m/e and the speed v of the new lightweight particle according to

$$v = \frac{2W}{QI};$$

$$\frac{m}{e} = \frac{I^2Q}{2W}.$$

(2.14)

For the purpose of the data analysis of this experiment, it is only necessary to know that W/Q and I are the primary quantities being measured, and inferences on the secondary quantities of interest are based on (2.14). For the proton, the mass-to-charge ratio was known to be approximately 1×10^{-4} g per electromagnetic (EMU) charge unit, where the EMU charge unit is equivalent to 10^{-10} electrostatic charge units, or ESU (a more common unit of measure for charge). In Thomson's units, the accepted value of the mass to charge ratio of the electron is now 5.7×10^{-8}. Some of the experimental data collected by Thomson are reported in Tables 2.1 and 2.2, in which "gas" refers to the gas used in the tubes he used for the experiment.

Some of Thomson's conclusions are reported here:

(a) *"It will be seen from these tables that the value of m/e is independent of the nature of the gas"*;
(b) *"the values of m/e were, however, the same in the two tubes."*;
(c) *"for the first tube, the mean for air is 0.40×10^{-7}, for hydrogen 0.42×10^{-7} and for carbonic acid 0.4×10^{-7}"*;
(d) *"for the second tube, the mean for air is 0.52×10^{-7}, for hydrogen 0.50×10^{-7} and for carbonic acid 0.54×10^{-7}"*.

Using the equations for sample mean and variance explained in Sect. 2.3, we are already in a position to measure the sample means and variances in air as $\overline{m/e}_1 = 0.42$ and $s_1^2 = 0.005$ for Tube 1, $\bar{x}_2 = 0.52$ and $s_2^2 = 0.003$ for Tube 2. These statistics can be reported as a measurement of 0.42 ± 0.07 for Tube 1 and 0.52 ± 0.06 for Tube 2. To make more quantitative statements on the statistical agreement between the two measurements, we need to know what is the probability distribution function of the sample mean. The test to determine whether the two measurements are consistent with each other will be explained in Sect. 7.5. For now, we simply point out that the fact that the range of the two measurements overlap, is an indication of the statistical agreement of the two measurements.

(continued)

Note: The three measurements marked with a star appear to have value of v or m/e that are inconsistent with the formulas to calculate them from W/Q and I. They may be typographical errors in the original publication. The first appears to be a typo in W/Q (6×10^{12} should be 6×10^{11}), the corrected value is assumed throughout this book. The second has an inconsistent value for v (should be 6.5×10^9, not 7.5×10^9), the third has inconsistent values for both v and m/e, but no correction was applied in these cases to the data in the tables.

Table 2.1 Data from Thomson's measurements of Tube 1

Gas	W/Q	I	m/e	v
Tube 1				
Air	4.6×10^{11}	230	0.57×10^{-7}	4×10^9
Air	1.8×10^{12}	350	0.34×10^{-7}	1×10^{10}
Air	6.1×10^{11}	230	0.43×10^{-7}	5.4×10^9
Air	2.5×10^{12}	400	0.32×10^{-7}	1.2×10^{10}
Air	5.5×10^{11}	230	0.48×10^{-7}	4.8×10^9
Air	1×10^{12}	285	0.4×10^{-7}	7×10^9
Air	1×10^{12}	285	0.4×10^{-7}	7×10^9
Hydrogen* .	6×10^{12}	205	0.35×10^{-7}	6×10^9
Hydrogen ..	2.1×10^{12}	460	0.5×10^{-7}	9.2×10^9
Carbonic acid*	8.4×10^{11}	260	0.4×10^{-7}	7.5×10^9
Carbonic acid	1.47×10^{12}	340	0.4×10^{-7}	8.5×10^9
Carbonic acid	3.0×10^{12}	480	0.39×10^{-7}	1.3×10^{10}

See Note for meaning of ⋆

Table 2.2 Data from Thomson's measurements of Tube 2

Gas	W/Q	I	m/e	v
Tube 2				
Air	2.8×10^{11}	175	0.53×10^{-7}	3.3×10^9
Air*	2.8×10^{11}	175	0.47×10^{-7}	4.1×10^9
Air	3.5×10^{11}	181	0.47×10^{-7}	3.8×10^9
Hydrogen .	2.8×10^{11}	175	0.53×10^{-7}	3.3×10^9
Air	2.5×10^{11}	160	0.51×10^{-7}	3.1×10^9
Carbonic acid	2.0×10^{11}	148	0.54×10^{-7}	2.5×10^9
Air	1.8×10^{11}	151	0.63×10^{-7}	2.3×10^9
Hydrogen .	2.8×10^{11}	175	0.53×10^{-7}	3.3×10^9
Hydrogen .	4.4×10^{11}	201	0.46×10^{-7}	4.4×10^9
Air	2.5×10^{11}	176	0.61×10^{-7}	2.8×10^9
Air	4.2×10^{11}	200	0.48×10^{-7}	4.1×10^9

See Note for meaning of ⋆

2.5   Covariance and Correlation Between Random Variables

It is common to measure more than one random variable in a given experiment. The variables are often related to one another and it is therefore necessary to define a measure of how one variable affects the measurement of the others. Consider the case in which we wish to measure both the length of one side of a square and the area; it is clear that the two quantities are related in a way that the change of one quantity affects the other in the same manner, i.e., a positive change of the length of the side results in a positive change of the area. In this case, the length and the area will be said to have a positive correlation. In this section we introduce the mathematical definition of the degree of correlation between variables.

2.5.1   Joint Distribution and Moments of Two Random Variables

When two (or more) variables are measured at the same time via a given experiment, we are interested in knowing what is the probability of a given pair of measurements for the two variables. This information is provided by the *joint probability distribution function*, indicated as $h(x, y)$, with the meaning that $h(x, y)dxdy$ is the probability that the two variables X and Y are in a two-dimensional interval of size $dxdy$ around the value (x, y). This two-dimensional function can be determined experimentally via its sample distribution, in the same way as one-dimensional distributions.

It is usually convenient to describe one variable at a time, even if the experiment features more than just one variable. In this case, the expectation of each variable (for example, X) is defined as

$$E[X] = \int_{-\infty}^{+\infty} \int_{-\infty}^{+\infty} xh(x, y)dxdy = \mu_x \qquad (2.15)$$

and the variance is similarly defined as

$$E[(X - \mu_x)^2] = \int_{-\infty}^{+\infty} \int_{-\infty}^{+\infty} (x - \mu_x)^2 h(x, y)dxdy = \sigma_x^2. \qquad (2.16)$$

These equations recognize the fact that the other variable, in this case Y, is indeed part of the experiment, but is considered *uninteresting* for the calculation at hand. Therefore the uninteresting variable is integrated over, weighted by its probability distribution function.

The *covariance* of two random variables is defined as

$$Cov(X, Y) \equiv E[(X - \mu_x)(Y - \mu_y)] =$$
$$\int_{-\infty}^{+\infty} \int_{-\infty}^{+\infty} (x - \mu_x)(y - \mu_y)h(x, y)dxdy = \sigma_{xy}^2. \tag{2.17}$$

The covariance is the expectation of the product of the deviations of the two variables. Unlike the deviation of a single variable, whose expectation is always zero, this quantity will be positive if, on average, a positive deviation of X is accompanied by a positive deviation of Y, or if two negative deviations are likely to occur simultaneously, so that the integrand is a positive quantity. If, on the other hand, the two variables tend to have deviations of opposite sign, the covariance will be negative. The covariance, like the mean and variance, is a parent quantity that can be calculated from the theoretical distribution of the random variables.

The *sample covariance* for a collection of N pairs of measurements is calculated as

$$s_{xy}^2 = \frac{1}{N-1} \sum_{i=1}^{N} (x_i - \bar{x})(y_i - \bar{y}), \tag{2.18}$$

using a similar equation to the sample variance.

The *correlation coefficient* ρ is simply a normalized version of the covariance,

$$\rho(X, Y) = \frac{Cov(X, Y)}{\sigma_x \sigma_y}. \tag{2.19}$$

The correlation coefficient is a number between -1 and $+1$. When the correlation is zero, the two variables are said to be *uncorrelated*. The fact that the correlation coefficient is normalized to within the values ± 1 derives from (2.10) and the properties of the joint distribution function.

The *sample correlation coefficient* is naturally defined as

$$r = \frac{s_{xy}^2}{s_x s_y} \tag{2.20}$$

in which s_x^2 and s_y^2 are the sample variances of the two variables.

The covariance between two random variables is very important in evaluating the variance in the sum (or any other function) of two random variables, as explained in detail in Chap. 4. The following examples illustrate the calculation of the covariance and the sample covariance.

Example 2.3 (Variance of Sum of Variables) Consider the random variables X, Y and the sum $Z = X + Y$: the variance is given by

$$Var(Z) = \int \int (x + y - (\mu_x + \mu_y))^2 h(x, y) dx dy =$$

$$Var(X) + Var(Y) + 2Cov(X, Y)$$

which can also be written in the compact form $\sigma_z^2 = \sigma_x^2 + \sigma_y^2 + 2\sigma_{xy}^2$. This shows that variances add linearly only if the two random variables are uncorrelated. Failure to check for correlation will result in errors in the calculation of the variance of the sum of two random variables.   ◇

Example 2.4 Consider the measurement of the following pairs of variables: (0, 2), (2, 5), (1, 4), (−3, −1). We can calculate the *sample* covariance by means of the following equation:

$$s_{xy}^2 = \frac{1}{3} \sum_{i=1}^{4} (x_i - \bar{x})(y_i - \bar{y}) = \frac{17}{3}$$

where $\bar{x} = 0$ and $\bar{y} = 2.5$. Also, the individual variances are calculated as

$$s_x^2 = \frac{1}{3} \sum_{i=1}^{4} (x_i - \bar{x})^2 = \frac{14}{3}$$

$$s_y^2 = \frac{1}{3} \sum (y_i - \bar{y})^2 = \frac{21}{3}$$

which results in the sample correlation coefficient between the two random variables of

$$r = \frac{17}{\sqrt{14 \times 21}} = 0.99.$$

This is in fact an example of nearly perfect correlation between the two variables. In fact, positive deviations of one variable from the sample mean are accompanied by positive deviations of the other by nearly the same amount.   ◇

2.5.2 Statistical Independence of Random Variables

The independence between events was described and quantified in Chap. 1, where it was shown that two events are independent only when the probability of their intersection is the product of the individual probabilities. The concept is extended here to random variables by defining two random variables as *independent* if and only if the joint probability distribution function can be factored in the following form:

$$h(x, y) = f(x) \cdot g(y), \tag{2.21}$$

where $f(x)$ and $g(y)$ are the probability distribution functions of the two random variables. When two variables are independent, the individual probability distribution function of each variable is obtained via *marginalization* of the joint distribution with respect to the other variable, e.g.,

$$f(x) = \int_{-\infty}^{+\infty} h(x, y) dy. \tag{2.22}$$

It is important to remark that independence between random variables and uncorrelation are not equivalent properties. Independence, which is a property of the distribution functions, is a much stronger property than uncorrelation, which is based on a statement that involves only moments. It can be proven that independence implies uncorrelation, but not vice versa.

Proof The fact that independence implies uncorrelation is shown by calculating the covariance of two independent random variables of joint distribution function $h(x, y)$. The covariance is

$$\sigma_{xy}^2 = \int_{-\infty}^{+\infty} \int_{-\infty}^{+\infty} (x - \mu_x)(y - \mu_y) h(x, y) dx dy =$$

$$\int_{-\infty}^{+\infty} (x - \mu_x) f(x) dx \int_{-\infty}^{+\infty} (y - \mu_y) g(y) dy = 0,$$

since each integral vanishes as the expectation of the deviation of a random variable. ☐

As a counter-example of the fact that dependent variables can have non-zero correlation factor, consider the case of a random variable X with a distribution $f(x)$ that is symmetric around the origin, and another variable $Y = X^2$. They cannot be independent since they are functionally related, but it will be shown that their covariance is zero. Symmetry about zero implies $\mu_x = 0$. The mean of Y is $E[Y] = E[X^2] = \sigma_x^2$ since the mean of X is null. From this, the covariance is given by

$$Cov(X, Y) = E[X(Y - \sigma_X^2)] = E[X^3 - X\sigma_x^2] = E[X^3] = 0$$

due to the symmetry of $f(x)$. Therefore the two variables X and X^2 are uncorrelated, yet they are not independent.

Example 2.5 (Photon Counting Experiment) A photon-counting experiment consists of measuring the total number of photons in a given time interval and the number of background events detected by the receiver in the same time interval. The experiment is repeated six times, by measuring simultaneously the total number of counts T as $(10, 13, 11, 8, 10, 14)$ and the number of background counts B as $(2, 3, 2, 1, 1, 3)$. We want to estimate the mean number of source photons and its standard error.

The random variable we seek to measure is $S = T - B$ and the mean and variance of this random variable can be easily shown to be

$$\mu_S = \mu_T - \mu_B$$
$$\sigma_S^2 = \sigma_T^2 + \sigma_B^2 - 2\sigma_{TB}^2$$

(the derivation is similar to that of Example 2.3). From the data, we measure the sample means and variances as $\overline{T} = 11.0$, $\overline{B} = 2.0$, $s_T^2 = 4.8$, $s_B^2 = 0.8$ and the sample covariance as $s_{TB}^2 = +1.6$.

Notice that the correlation coefficient between T and S, as estimated via the measurements, is then given by $corr(T, B) = 1.6/\sqrt{4.8 \times 0.8} = 0.92$, indicating a strong degree of correlation between the two measurements. The measurements can be summarized as

$$\mu_S = 11.0 - 2.0 = 9.0$$
$$\sigma_S^2 = 4.8 + 0.8 - 2 \times 1.6 = 2.4$$

and be reported as $S = 9.00 \pm 1.55$ counts (per time interval). Notice that if the correlation between the two measurements had been neglected, then one would (erroneously) report $S = 9.00 \pm 2.37$, e.g., the standard deviation would be largely overestimated. The correlation between total counts and background counts in this example has a significant impact in the calculation of the variance of S and needs to be taken into account.   ◇

2.6   A Classic Experiment: Pearson's Collection of Data on Biometric Characteristics

In 1903 K. Pearson published the analysis of a collection of biometric data on more than 1000 families in the United Kingdom, with the goal of establishing how certain characters, such as height, are correlated and inherited [33]. Prof. Pearson is also the inventor of the χ^2 test and a central figure in the development of the modern science of statistics.

Pearson asked a number of families, composed of at least the father, mother, and one son or daughter, to perform measurements of height, span of arms and length of left forearm. This collection of data resulted in a number

(continued)

of tables, including some for which Pearson provides the distribution of two measurements at a time. One such table is that reporting the mother's height versus the father's height, Table 2.3.

The data reported in Table 2.3 represent the joint probability distribution of the two physical characters, binned in one-inch intervals. When a non-integer count is reported (e.g., a value of 0.25, 0.5 or 0.75), we interpret it as meaning that the original measurement fell exactly at the boundary between two cells, although Pearson does not provide an explanation for non-integer values.

For every column and row it is also reported the sum of all counts. The bottom row in the table is therefore the distribution of the father's height, irrespective of the mother's height, likewise the rightmost column is the distribution of the mother's height, regardless of the father's height. The process of obtaining a one-dimensional distribution from a multi-dimensional illustrates the *marginalization* over certain variables that are not of interest. In the case of the bottom column, the marginalization of the distribution was done over the mother's height, to obtain the distribution of father's height.

From Table 2.3 it is not possible to determine whether there is a correlation between father's and mother's heights. In fact, according to (2.18), we would need all 1079 pairs of height measurements originally collected by Pearson to calculate the covariance. Since Pearson did not report these *raw* (i.e, unprocessed) data, we cannot calculate either the covariance or the correlation coefficient. The measurements reported by Pearson are in a format that goes under the name of *contingency table*, consisting of a table with measurements that are binned into suitable two-dimensional intervals.

Summary of Key Concepts for this Chapter

☐ *Random variable:* A quantity that is not known exactly and is described by a probability distribution function $f(x)$.

☐ *Moments of a distribution:* Expectations for the random variable or functions of the random variable, such as the mean $\mu = E[X]$ and the variance $\sigma^2 = E[(X - \mu)^2]$.

☐ *Sample mean and sample variance*: Quantities calculated from the measurements that are intended to approximate the corresponding parent quantities (mean and variance).

☐ *Joint distribution function*: The distribution of probabilities for a pair of variables.

☐ *Covariance:* A measure of the tendency of two variables to follow one another, $Cov(X, Y) = E[(X - \mu_X)(Y - \mu_Y)]$.

☐ *Correlation coefficient:* A normalized version of the covariance that takes values between -1 (perfect anti-correlation) and +1 (perfect correlation).

☐ *Statistically independent variables:* Two variables whose joint probability distribution function can be factored as $h(x, y) = f(x)g(y)$.

Table 2.3 Joint distribution of father's height (columns) and mother's height (rows) from Pearson's experiment, in inches

Father's height

Mother's height	58–	59–	60–	61–	62–	63–	64–	65–	66–	67–	68–	69–	70–	71–	72–	73–	74–	75–	
52–53	0	0	0	0	0	0	0	1	0.5	0	0	0	0	0	0	0	0	0	1.5
53–54	0	0	0	0	0	0	0	0	0.5	0	0	0	0	0	0	0	0	0	0.5
54–55	0	0	0	0	0	0	0	0.25	0.25	0	0.5	0	0	0	0	0	0	0	1
55–56	0	0	0	0.5	1	0	0	0.25	0.25	0	0.5	0	0	0	0	0	0	0	2.5
56–57	0	0	0	0	0.75	1.25	0	1	1.75	1.75	0	0	0	0	0	0	0	0	6.5
57–58	0	0	0	0.25	1	1.25	1.5	4	3.25	2.5	3	1.25	0.5	0	0	0	0	0	18.5
58–59	0	0.25	0.75	1.25	1.25	2.75	4	7	5.75	4.5	3.75	1.25	2	0	0	0	0	0	34.5
59–60	0	1.25	1.25	1	4	4.5	7.75	10	15	16.75	9	5.5	3.25	1.25	1	0.5	0	0	82
60–61	0.25	0.25	0.5	2	4.25	4.5	18	16	24	14.75	23.25	12.75	7.25	5.75	4.25	0.75	0	0	138.5
61–62	0.25	0.25	0	0	8	8.25	15	17.25	25	20.75	24	14.25	14.25	10	4	0.75	0.5	0	162.5
62–63	0	0.5	0.5	1.25	4.75	7.75	10	26	21.25	28	28	23	14.25	10.75	4.5	2	1	0.5	184
63–64	0	0	0.25	2	3.5	4.5	9	21	15.75	20.75	19.5	24	22.5	10.75	4	2.25	2.25	0.5	162.5
64–65	0	0	1.25	0.75	2	6	6.5	9.75	16	18.25	23	16.75	13.75	6.75	4.75	2.25	0.25	1.5	129.5
65–66	0	0	0	0.25	1.5	1.5	3.25	5.5	9.75	7	15.5	12.75	10.5	6.25	4.25	1.75	0.25	0	80
66–67	0	0	0	0.25	1	0.75	0.5	3.5	5	3	7.25	7.75	7	3.5	2.75	1.5	0.25	0	44
67–68	0	0	0	0	0	0	0	1	2.5	1.5	2.75	3.25	2.75	1.5	1	0.5	0.25	0	17
68–69	0	0	0	0	0	0	0	0	0	1	2.5	1.25	1.25	0.5	1	0.25	0.25	0	8
69–70	0	0	0	0	0	0	0	0	0	0	0	0.25	2.25	0	2	0	0	0	4.5
70–71	0	0	0	0	0	0	0	0	0	0	0	0	1	0	0.5	0	0	0	1.5
	0.5	2.5	4.5	9.5	33	43	75.5	123.5	146.5	140.5	162.5	124	102.5	57	34	12.5	5	2.5	1079

Problems

2.1 Consider the exponential distribution

$$f(x) = \lambda e^{-\lambda x}$$

where $\lambda \geq 0$ and $x \geq 0$. Show that the distribution is properly normalized, and calculate the mean, variance and cumulative distribution $F(x)$.

2.2 Consider the sample mean as a random variable defined by

$$\bar{x} = \frac{1}{N} \sum_{i=1}^{N} x_i \tag{2.23}$$

where x_i are identical independent random variables with mean μ and variance σ^2. Show that the variance of $\bar{x}$ is equal to σ^2/N.

2.3 J.J. Thomson's experiment aimed at the measurement of the ratio between the mass and charge of the electron is presented on page 23. Using the datasets for Tube 1 and Tube 2 separately, calculate the mean and variance of the random variables W/Q and I, and the covariance and correlation coefficient between W/Q and I.

2.4 Using J.J. Thomson's experiment (page 23), verify the statement that *"It will be seen from these tables that the value of m/e is independent of the nature of the gas"* used in the experiment. You may do so by calculating the mean and standard deviation for the measurements in each gas (air, hydrogen, and carbonic acid) and testing whether the three measurements agree with each other within their standard deviations.

2.5 Calculate the sample covariance and correlation coefficient for the following set of data: $(0, 2), (2, 5), (1, 4), (3, 1)$.

2.6 Prove that the following relationship holds,

$$Var(X) = E[X^2] - \mu^2$$

where μ is the mean of the random variable X.

Chapter 3
Three Fundamental Distributions: Binomial, Gaussian, and Poisson

Abstract There are three distributions that play a fundamental role in statistics. The binomial distribution describes the number of positive outcomes in binary experiments, and it is the "mother" distribution from which the other two distributions can be obtained. The Gaussian distribution can be considered as a special case of the binomial, when the number of tries is sufficiently large. For this reason, the Gaussian distribution applies to a large number of variables, and it is referred to as the *normal* distribution. The Poisson distribution applies to counting experiments, and it can be obtained as the limit of the binomial distribution when the probability of success is small.

3.1 The Binomial Distribution

Many experiments can be considered as *binary*, meaning that they can only have two possible outcomes which we can interpret as *success* or *failure*. Even complex experiments with a larger number of possible outcomes can be described as binary, when one is simply interested about the occurrence of a specific event A, or its non-occurrence, $\overline{A}$. It is therefore of fundamental importance in statistics to determine the properties of binary experiments, and the distribution of the number of successes when the experiment is repeated for a number of times under the same experimental conditions.

3.1.1 Derivation of the Binomial Distribution

Consider a binary experiment characterized by a probability of success p a therefore a probability of failure $q = 1-p$. The probabilities p and q are determined according to the theory of probability and are assumed to be known for the experiment being considered. We seek the probability of having n successes in N tries, regardless of the order in which the successes take place. For example, consider tossing four coins, and being interested in any two of these coins showing heads up, as an indication of success of the toss. To obtain this probability, we start by counting

© Springer Science+Busines Media New York 2017

M. Bonamente, *Statistics and Analysis of Scientific Data*, Graduate Texts in Physics, DOI 10.1007/978-1-4939-6572-4_3

how many possible outcomes for the experiments are possible, and break down the derivation into three parts:

- *Probability of a specific sequence of n successes out of N tries.* Assume that successive experiments are independent, e.g., one tosses the same coin many times, each time independently of each other. The probability of having n successes and therefore $N-n$ failures occurring *in a specific sequence*, is given by

$$P(specific\ sequence\ \text{of } n \text{ successes}) = p^n \times q^{N-n}. \qquad (3.1)$$

This result can be seen by using the property of independence among the N events, of which the n successes carry a probability p, and the $(N - n)$ failures a probability q.

Example 3.1 Considering the case of four coin tosses, the probability of a given sequence, for example "heads-tails-tails-heads," is $(1/2) \times (1/2) \times (1/2) \times (1/2) = (1/2)^4$, since $p = q = 1/2$. Successive tosses are assumed to be independent.   ◇

- *Number of ordered sequences.* We start by counting how many ordered sequences exist that have n successes out of N tries. If there are no successes ($n = 0$), then there is only one possible sequence with N failures. If $n > 0$, each of the N tries can yield the "first" success, and therefore there are N possibilities for what try is the first success. Continuing on to the "second" success, there are only $N - 1$ possibilities left for what trial will be the second success, and so on. This leads to the following number of sequences containing n time-ordered successes, that is, sequences for which we keep track of the order in which the successes occurred:

$$\text{Perm}(n, N) = N \cdot (N - 1) \cdot (N - n + 1) = \frac{N!}{(N - n)!}. \qquad (3.2)$$

This is called the number of *permutations* of n successes out of N tries. This method of counting sequences can also be imagined as the placement of each success in a "success box": the first place in this box can be filled in N different ways, the second in $(N - 1)$ ways corresponding to the remaining tries, and so on.

Example 3.2 Consider the case of $n = 2$ successes out of $N = 4$ trials. According to (3.2), the number of permutations is $4!/2! = 12$. We list explicitly all 12 ordered sequences that give rise to 2 successes out of 4 tries in Table 3.1. Symbol H_1 denotes the "first success," and H_2 the "second success." Consider, for example, lines 5 and 8: both represent the same situation in which the coin 2 and 3 showed heads up, or success, and they are not really different sequences, but the separate entries in this table are the result of our method of counting time-ordered sequences.   ◇

Table 3.1 Permutations (ordered sequences) of 2 successes out of 4 tries

Sequence	Number of try 1	2	3	4	Sequence	Number of try 1	2	3	4
1	H_1	H_2	–	–	7	H_2	–	H_1	–
2	H_1	–	H_2	–	8	–	H_2	H_1	–
3	H_1	–	–	H_2	9	–	–	H_1	H_2
4	H_2	H_1	–	–	10	H_2	–	–	H_1
5	–	H_1	H_2	–	11	–	H_2	–	H_1
6	–	H_1	–	H_2	12	–	–	H_2	H_1

In reality, we are not interested in the *time order* in which the n successes occur, since it is of no consequence whether the first or the Nth, or any other, try is the "first" success. We must therefore correct for this artifact in the following.

• *Number of sequences of n successes out of N tries (regardless of order).* As it is clear from the previous example, the number of permutations is not quite the number we seek, since it is of no consequence which success happened first. According to (3.2), there are $n!$ ways of ordering n successes among themselves, or $\mathrm{Perm}(n, n) = n!$. Since all $n!$ permutations give rise to the same practical situation of n successes, we need to divide the number of (time-ordered) permutations by $n!$ in order to avoid double-counting of permutations with successes in the same trial number. It is therefore clear that, regardless of time order, the number of *combinations* of n successes out of N trials is

$$C(n, N) = \frac{\mathrm{Perm}(n, N)}{n!} = \frac{N!}{(N-n)!n!} \equiv \binom{N}{n}. \tag{3.3}$$

The number of combinations is the number we seek, i.e., the number of possible sequences of n successes in N tries.

Example 3.3 Continue to consider the case of 2 successes out of 4 trials. There are $2! = 2$ ways to order the 2 successes among themselves (either one or the other is the first success). Therefore the number of combinations of 2 successes out of 4 trials is 6, and not 12. As indicated above, in fact, each sequence had its "twin" sequence listed separately, and (3.3) correctly counts only different sequences. ◇

According to the results obtained above, what remains to be done is to use the probability of each sequence (3.1) and multiply it by the number of combinations in (3.3) to obtain the overall probability of having n successes in N trials:

$$P(n) = \binom{N}{n} p^n q^{N-n} \qquad n = 0, \ldots, N. \tag{3.4}$$

This distribution is known as the *binomial distribution* and it describes the probability of n successes in N tries of a binary experiment. It is a discrete distribution that is defined for non-negative values $n \leq N$. The factor in (3.3) is in fact the binomial coefficient and it derives its name from its use in the binomial expansion

$$(p + q)^N = \sum_{n=0}^{N} \binom{N}{n} p^n q^{N-n}. \tag{3.5}$$

3.1.2   Moments of the Binomial Distribution

The moment of mth order for a discrete random variable X of distribution $P(n)$ is given by

$$E[X^m] = \overline{n^m} = \sum_{n=0}^{N} n^m P(n). \tag{3.6}$$

We can show that the mean and the second moment of the binomial distribution are given by

$$\begin{cases} \overline{n} = pN \\ \overline{n^2} = \overline{n}^2 + pqN. \end{cases} \tag{3.7}$$

Proof Start with the mean,

$$\overline{n} = \sum_{n=0}^{N} P(n)n = \sum_{n=0}^{N} \binom{N}{n} np^n q^{N-n} = \sum_{n=0}^{N} \binom{N}{n} \left[p \frac{\partial}{\partial p} \right] p^n q^{N-n};$$

in which we have introduced a linear operator $p \dfrac{\partial}{\partial p}$ that can be conveniently applied to the entire sum,

$$\overline{n} = p \frac{\partial}{\partial p} \left[\sum_{n=0}^{N} \binom{N}{n} p^n q^{N-n} \right] = p \frac{\partial}{\partial p} (p + q)^N = pN(p + q)^{N-1} = pN.$$

The derivation for the moment $\overline{n^2}$ is similar:

$$\overline{n^2} = \sum_{n=0}^{N} P(n)n^2 = \sum_{n=0}^{N} \binom{N}{n} n^2 p^n q^{N-n} = \sum_{n=0}^{N} \binom{N}{n} \left[p \frac{\partial}{\partial p} \right]^2 q^{N-n}$$

$$= \left[p \frac{\partial}{\partial p} \right]^2 (p + q)^N = p \frac{\partial}{\partial p} \left[pN(p + q)^{N-1} \right] =$$

$$p \left[N(p + q)^{N-1} + pN(N - 1)(p + q)^{N-2} \right] =$$

$$pN + p^2 N(N - 1) = pN + (pN)^2 - p^2 N =$$

$$\overline{n^2} + p(1 - p)N = \overline{n^2} + pqN.$$

□

It follows that the variance of the binomial distribution is given by

$$\sigma^2 = E[(X - \overline{n})^2] = pqN. \tag{3.8}$$

Equations (3.7) and (3.8) describe the most important features of the binomial distribution, shown in Fig. 3.1 for the case of $N = 10$. The mean is naturally given by the product of the number of tries N and the probability of success p in each of the tries. The standard deviation σ measures the root mean square of the deviation and it is the measurement of the width of the distribution.

Example 3.4 (Probability of Overbooking)   An airline knows that 5% of the persons making reservations will not show up at the gate. On a given flight that

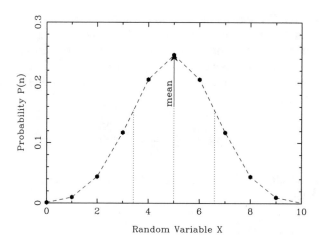

Fig. 3.1 Binomial distribution with $p = q = 0.5$ and $N = 10$. The *dotted lines* around the mean mark the $\pm\sigma$ range

can seat 50 people, 52 tickets have been sold. Calculate the probability that there will be a seat available for every passenger that will arrive at the gate.

This is a binary experiment in which $p = 0.95$ is the probability that a passenger will show. For that specific flight, $N = 52$ passenger have the choice of showing (or not). The probability that there is a seat available for each passenger is therefore given by $P = 1 - P_N(52) + P_N(51)$, which is calculated as

$$P = 1 - \binom{52}{52} p^{52} \cdot 1 - \binom{52}{51} p^{51} \cdot q = 1 - (0.95)^{52} - 52 \cdot (0.95)^{51} \cdot 0.05 = 0.741.$$

Therefore the airline is willing to take a 25.9 % chance of having an overbooked flight.   ◇

3.2   The Gaussian Distribution

The Gaussian distribution, often referred to as the *normal* distribution, can be considered as a special case of the binomial distribution in the case of a large number N of experiments performed. In this section we derive the Gaussian distribution from the binomial distribution and describe the salient features of the distribution.

3.2.1   Derivation of the Gaussian Distribution from the Binomial Distribution

The binomial distribution of (3.4) acquires a simpler form when N is large. An alternative analytic expression to the binomial distribution is a great advantage, given the numerical difficulties associated with the evaluation of the factorial of large numbers. As was evident from Fig. 3.1, the binomial distribution has a maximum at value $n = Np$. In the following we prove that the binomial distribution can be approximated as

$$P(n) \simeq \frac{1}{\sqrt{2\pi Npq}} e^{-\frac{(n-Np)^2}{2NPq}} \tag{3.9}$$

in the case in which $N \gg 1$, and for values of the variable that are close to the peak of the distribution.

Proof Expand the logarithm of the binomial probability as a Taylor series in the neighborhood of the peak value $\tilde{n}$,

$$\ln P(n) = \ln P(\tilde{n}) + \sum_{k=1}^{\infty} \frac{B_k}{k!} \Delta n^k$$

where $\Delta n = n - \tilde{n}$ is the deviation from the peak value and

$$B_k = \left. \frac{\partial \ln P(n)^k}{\partial^k n} \right|_{n=\tilde{n}}$$

is the coefficient of the Taylor series expansion. Since, by assumption, $\tilde{n}$ is a point of maximum, the first coefficient is null, $\partial \ln P(n)/\partial n|_{n=\tilde{n}} = 0$. We neglect terms of order $O(\Delta n^3)$ in the expansion, and the approximation results in

$$\ln P(n) \simeq \ln P(\tilde{n}) + \frac{1}{2} B_2 \Delta n^2,$$

where B_2 is negative, since $n = \tilde{n}$ is a point of maximum. It follows that

$$P(n) \simeq P(\tilde{n}) e^{-\frac{|B_2| \Delta n^2}{2}}.$$

Neglecting higher-order terms in Δn means that the approximation will be particularly accurate in regions where Δn is small, i.e., near the peak of the distribution. Away from the peak, the approximation will not hold with the same precision.

In the following we show that the unknown terms can be calculated as

$$\begin{cases} B_2 = -\dfrac{1}{Npq} \\ P(\tilde{n}) = \dfrac{1}{\sqrt{2\pi Npq}}. \end{cases}$$

First, we calculate the value of $|B_2|$. Start with

$$\ln P(n) = \ln \left(\frac{N!}{n!(N-n)!} p^n q^{N-n} \right) =$$

$$\ln N! - \ln n! - \ln(N-n)! + n \ln p + (N-n) \ln q.$$

At this point we need to start treating n as a continuous variable. This approximation is acceptable when $Np \gg 1$, so that values n of the random variable near the peak of the distribution are large numbers. In this case, we can approximate the derivative of the logarithm with a difference,

$$\frac{\partial \ln n!}{\partial n} = (\ln(n+1)! - \ln n!)/1 = \ln(n+1) \simeq \ln n.$$

From this it follows that the first derivative of the probability function, as expected, is zero at the peak value,

$$\frac{\partial \ln P(n)}{\partial n}\bigg|_{n=\tilde{n}} = -\ln n + \ln(N-n) + \ln p - \ln q|_{n=\tilde{n}}$$

$$= \ln\left(\frac{N-n}{n}\frac{p}{q}\right) = 0$$

so that the familiar result of $\tilde{n} = p \cdot N$ is obtained. This leads to the calculation of the second derivative,

$$B_2 = \frac{\partial^2 \ln P(n)}{\partial n^2}\bigg|_{n=\tilde{n}} = \frac{\partial}{\partial n}\ln\left(\frac{N-n}{n}\frac{p}{q}\right)\bigg|_{n=\tilde{n}}$$

$$= \frac{\partial}{\partial n}(\ln(N-n) - \ln n)\bigg|_{n=\tilde{n}} = -\frac{1}{N-n} - \frac{1}{n}\bigg|_{n=\tilde{n}}$$

$$= -\frac{1}{\tilde{n}} - \frac{1}{N-\tilde{n}} = -\frac{1}{Np} - \frac{1}{N(1-p)} = -\frac{p+q}{Npq} = -\frac{1}{Npq}.$$

Finally, the normalization constant $P(\tilde{n})$ can be calculated making use of the integral

$$\int_{-\infty}^{\infty} e^{-ax^2}dx = \sqrt{\pi/a}.$$

Enforcing the normalization condition of the probability distribution function,

$$\int_{-\infty}^{\infty} P(\tilde{n})e^{-\frac{|B_2|\Delta n^2}{2}}d\Delta n = P(\tilde{n})\sqrt{\frac{2\pi}{|B_2|}} = 1$$

we find that $P(\tilde{n}) = 1/\sqrt{2\pi Npq}$. We are therefore now in a position to obtain an approximation to the binomial distribution, valid when $n \gg 1$:

$$P(n) = \frac{1}{\sqrt{2\pi Npq}}e^{-\frac{(n-Np)^2}{2NPq}}.$$

□

Using the fact that the mean of the distribution is $\mu = Np$, and that the variance is $\sigma^2 = Npq$, the approximation takes the form

$$P(n) = \frac{1}{\sqrt{2\pi\sigma^2}} e^{-\frac{(n-\mu)^2}{2\sigma^2}} \qquad (3.10)$$

which is the standard form of the Gaussian distribution, in which n is a continuous variable. Equation (3.10) read as $P(n)$ being the probability of occurrence of the value n for a given random variable of mean μ and variance σ^2. The Gaussian distribution has the familiar "bell" shape, as shown in Fig. 3.2. When n becomes a continuous variable, which we will call x, we talk about the probability of occurrence of the variable in a given range $x, x + dx$. The Gaussian probability distribution function is thus written as

$$f(x)dx = \frac{1}{\sqrt{2\pi\sigma^2}} e^{-\frac{(x-\mu)^2}{2\sigma^2}} dx. \qquad (3.11)$$

A Gaussian of mean μ and variance σ^2 is often referred to as $N(\mu, \sigma)$. The standard Gaussian is one with zero mean and unit variance, indicated by $N(0, 1)$.

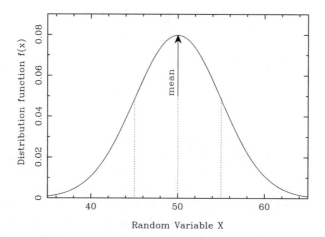

Fig. 3.2 Gaussian distribution with $\mu = 50$ and $\sigma^2 = 12.5$, corresponding to a binomial distribution of $p = q = 0.5$, and $N = 100$. The Gaussian distribution is symmetrical around the mean and therefore the mean, mode, and median coincide. The *dotted lines* around the mean mark the $\pm\sigma$ range

3.2.2   Moments and Properties of the Gaussian Distribution

The parameters μ and σ^2 are, respectively, the mean and variance of the Gaussian distribution. These results follow from the derivation of the Gaussian distribution from the binomial, and can be confirmed by direct calculation of expectations from (3.11). Central moments of odd order are zero, since the Gaussian is symmetric with respect to the mean.

Given its wide use in statistics, it is important to quantify the "effective width" of the Gaussian distribution around its mean. The *half width at half maximum*, or HWHM, is defined as the range of x between the peak and the point where $P(x) = 0.5P(\mu)$. It can be easily shown that the HWHM has a size of approximately 1.18σ, meaning that the half-maximum point is just past one standard deviation of the mean. By the same token, the *full-width at half maximum*, or FWHM, is defined as the range between the two points where $P(x) = 0.5P(\mu)$. It is twice the HWHM, or 2.35σ in size. Tables of the Gaussian distribution are provided in Appendix A.1.

The range between the points $x = \mu \pm \sigma$ is a common measure of the effective range of the random variable. The probability of a Gaussian variable to be in the range from $\mu - \sigma$ to $\mu + \sigma$ is calculated as the integral of the probability distribution function between those limits. In general, we define the integral

$$A(z) = \int_{\mu-z\sigma}^{\mu+z\sigma} f(x)dx = \frac{1}{\sqrt{2\pi}} \int_{-z}^{z} e^{-\frac{x^2}{2}} dx \qquad (3.12)$$

where $f(x)$ is the Gaussian distribution; this integral is related to the *error function*,

$$\text{erf } z = \frac{1}{\sqrt{\pi}} \int_{-z}^{z} e^{-x^2} dx. \qquad (3.13)$$

The function $A(z)$ is tabulated in Appendix A.1 The probability of the variable to be within one σ of the mean is $A(1) = 0.683$, or 68.3 %. The range of x between $\mu - \sigma$ and $\mu + \sigma$ therefore corresponds to a 68.3 % interval of probability, and it is referred to as the 1σ interval. The correspondence between the 1σ interval and the 68.3 % confidence interval applies strictly only to the Gaussian distribution, for which the value of σ is defined via the distribution function. It is common practice, however, to calculate to the 68.3 % interval (sometimes shortened to 68 %) even for those random variables that do not strictly follow a Gaussian distribution, and refer to it as the 1σ interval. The probability associated with characteristic intervals of a Gaussian variable is also reported in Table 3.2.

The cumulative distribution of a Gaussian random variable $N(0, 1)$ is defined by the following integral:

$$B(z) = \int_{-\infty}^{z} \frac{1}{\sqrt{2\pi}} e^{-\frac{x^2}{2}} dx; \qquad (3.14)$$

Table 3.2 Probability associated with characteristic intervals of a Gaussian distribution

Interval	Integrated probability
$\mu - \sigma, \mu + \sigma$ (1σ interval)	0.6827, or 68.27 %
$\mu - 2\sigma, \mu + 2\sigma$ (2σ interval)	0.9545, or 95.45 %
$\mu - 3\sigma, \mu + 3\sigma$ (3σ interval)	0.9973, or 99.73 %
$\mu - 4\sigma, \mu + 4\sigma$ (4σ interval)	0.9999, or 99.99 %

the integral can be calculated as $B(z) = 1/2 + A(z)/2$ for $z > 0$ and it is tabulated in Table A.3. For $z < 0$, the table can be used to calculate the cumulative distribution as $B(z) = 1 - B(-z)$.

3.2.3 How to Generate a Gaussian Distribution from a Standard Normal

All Gaussian distributions can be obtained from the standard $N(0, 1)$ via a simple change of variable. If X is a random variable distributed like $N(\mu, \sigma)$, and Z a standard Gaussian $N(0, 1)$, then the relationship between Z and X is given by

$$Z = \frac{X - \mu}{\sigma}. \tag{3.15}$$

The variable Z is also referred to as the *z-score* associated with the variable X. This equation means that if we can generate samples from a standard normal, we can also have samples from any other Gaussian distribution. If we call z a sample from Z, then

$$x = \sigma \cdot z + \mu \tag{3.16}$$

will be a sample drawn from X, according to the equation above. Many programming languages have a built-in function to generate samples from a standard normal, and this simple process can be used to generate samples from any other Gaussian. A more general procedure to generate a given distribution from a uniform distribution will be presented in Sect. 4.8.

3.3 The Poisson Distribution

The Poisson distribution describes the probability of occurrence of events in counting experiments, i.e., when the possible outcome is an integer number describing how many counts have been recorded. The distribution is therefore discrete and can be derived as a limiting case of the binomial distribution.

3.3.1 Derivation of the Poisson Distribution

The binomial distribution has another useful approximation in the case in which $p \ll 1$, or when the probability of success is small. In this case, the number of positive outcomes is much smaller than the number of tries, $n \ll N$, and the factorial function can be approximated as

$$N! = N(N-1)\cdots(N-n+1)\cdot(N-n)! \simeq N^n(N-n)!.$$

We are also interested in finding an approximation for the q^{N-n} term that appears in the binomial. For this we set

$$\ln q^{N-n} = \ln(1-p)^{N-n} = (N-n)\ln(1-p) \simeq -p(N-n) \simeq -pN,$$

and therefore we obtain the approximation

$$q^{N-n} \simeq e^{-pN}.$$

These two approximations can be used into (3.4) to give

$$P(n) \simeq \frac{N^n(N-n)!}{n!(N-n)!}p^n e^{-pN} = \frac{(pN)^n}{n!}e^{-pN}. \tag{3.17}$$

Since pN is the mean of the distribution, we can rewrite our approximation as

$$P(n) = \frac{\mu^n}{n!}e^{-\mu}, \tag{3.18}$$

known as the Poisson distribution. This function describes the probability of obtaining n positive outcomes, or counts, when the expected number of outcomes is μ. It can be immediately seen that the distribution is properly normalized, since

$$\sum_{n=0}^{\infty}\frac{\mu^n}{n!} = e^{\mu}.$$

A fundamental feature of this distribution is that it is described by only one parameter, the mean μ, as opposed to the Gaussian distribution that had two parameters. This clearly does not mean that the Poisson distribution has no variance—in that case, it would not be a random variable!—but that the variance can be written as function of the mean, as will be shown in the following.

3.3.2 Properties and Interpretation of the Poisson Distribution

The approximations used in the derivation of (3.18) caused the loss of any reference to the initial binomial experiment, and only the mean $\mu = Np$ is present. Using the definition of mean and variance, it is easy to prove that the mean is indeed μ, and that the variance is also equal to the mean, $Var(n) = \sigma^2 = \mu$. The fact that the mean equals the variance can be seen using the values for the binomial, $\mu = Np$ and $\sigma^2 = Npq$; since $p \ll 1$, $q \simeq 1$, and $\mu \simeq \sigma^2$. As a result, the Poisson distribution has only one parameter.

The Poisson distribution can be interpreted as the probability of occurrence of n events in the case of an experiment that detects individual counts, when the mean of the counts is μ. This makes the Poisson distribution the primary statistical tool for all experiments that can be expressed in terms of the *counting* of a specific variable associated with the experiment. Typical examples are the counting of photons or the counting of plants with a given characteristic, etc. When an experiment can be cast in terms of a counting experiment, even without a specific reference to an underlying binary experiment, then the Poisson distribution will apply. All reference to the total number of possible events (N) and the probability of occurrence of each event (p) was lost because of the approximation used throughout, i.e., $p \ll 1$, and only the mean μ remains to describe the primary property of the counting experiment, which is the mean or expectation for the number of counts.

As can be seen in Fig. 3.3, the Poisson distribution is not symmetric with respect of the mean, and the distribution becomes more symmetric for larger values of the mean. As for all discrete distributions, it is only meaningful to calculate the probability at a specific point or for a set of points, and not for an interval of points as in the case of continuous distributions. Moreover, the mean of the distribution itself can be a non-integer number, and still the outcome of the experiment described by the Poisson distribution can only take integer values.

Example 3.5 Consider an astronomical source known to produce photons, which are usually detected by a given detector in the amount of $\mu = 2.5$ in a given time interval. The probability of detecting $n = 4$ photons in a given time interval is therefore

$$P(4) = \frac{2.5^4}{4!}e^{-2.5} = 0.134$$

The reason for such apparently large probability of obtaining a measurement that differs from the expected mean is simply due to the statistical nature of the detection process. ◇

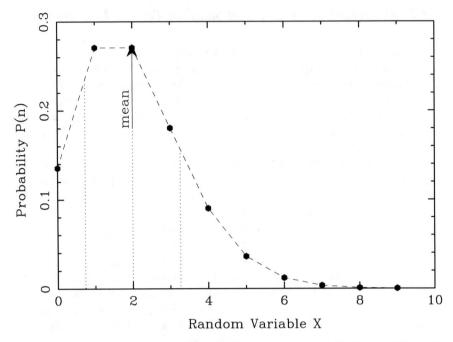

Fig. 3.3 Poisson distribution with $\mu = 2$, corresponding to a binomial distribution with $p = 0.2$ and $N = 10$. The *dotted lines* represent the mean, the $\mu - \sigma$ and $\mu + \sigma$ points

3.3.3   The Poisson Distribution and the Poisson Process

A more formal justification for the interpretation of the Poisson distribution as the distribution of counting experiments comes from the Poisson process. Although a complete treatment of this subject is beyond the scope of this book, a short description of stochastic processes will serve to strengthen the interpretation of (3.18), which is one of the foundations of statistics. More details on stochastic processes can be found, for example, in the textbook by Ross [38].

A *stochastic counting process* $\{N(t), t > 0\}$ is a sequence of random variables $N(t)$, in which t indicates time, and $N(t)$ is a random variable that indicates the number of events occurred up to time t. The stochastic process can be thought of as repeating the experiment of "counting the occurrence of a given event" at various times t; $N(t)$ is the result of the experiment. The *Poisson process with rate* λ is a particular type of stochastic process, with the following properties:

1. $N(0) = 0$, meaning that at time 0 there are no counts detected.
2. The process has *independent increments*, meaning that $N(t + s) - N(s)$ is independent of $N(t)$; this is understood with the events occurring after time t not being influenced by those occurring prior to it.

3. The process has *stationary increments*, i.e., the distribution of the number of events in an interval of time s depends only on the length of the time interval itself.
4. $P(N(h) = 1) = \lambda h + o(h)$ in which $o(h)$ is a function with the property that

$$\lim_{h \to 0} \frac{o(h)}{h} = 0.$$

5. $P(N(h) \geq 2) = o(h)$. The latter two properties mean that the probability of obtaining one count depends on the finite value λ, while it is unlikely that two or more events occur in a short time interval.

It can be shown that under these hypotheses, the number of events $N(t)$ recorded in any interval of length t is Poisson distributed,

$$P\{N(t+s) - N(s) = n\} = \frac{(\lambda t)^n}{n!} e^{-\lambda t} \tag{3.19}$$

This shows that the Poisson distribution is to be interpreted as the distribution of occurrence of n events during a time interval t, under the hypothesis that the rate of occurrence of events is λ. This interpretation is identical to the one provided above, given that $\mu = \lambda t$ is the mean of the counts in that time interval.

3.3.4 An Example on Likelihood and Posterior Probability of a Poisson Variable

The estimation of parameters of a random variable, such as the mean of the Poisson distribution, will be treated in full detail in Chap. 5. Here we present a simple application that consists of using available measurements to calculate the likelihood and to make inferences on the unknown value of the parent mean μ of a Poisson variable. The following examples illustrate how a single measurement n of a Poisson variable can be used to constrain the true mean μ, and that care must be exercised in not confusing the likelihood of a measurement with the *posterior probability*. We assume for simplicity that the mean is an integer, although in general it may be any real number.

Within the Bayesian framework, a counting experiment can be written in terms of a dataset B, consisting of the measurement n of the variable, and events A_i, representing the fact that the parent mean is $\mu = i$. It follows that the likelihood can be written as

$$P(B/A_i) = \frac{i^n}{n!} e^{-i}.$$

Example 3.6 (Calculation of Data Likelihood)   A counting experiment results in a detection of $n = 4$ units, and one wants to make a statement as to what is the probability of such measurement. Using the Poisson distribution, the probability of detecting 4 counts if, for example, $\mu = 0, 1$, or 2, is given by the likelihood

$$P(B/A_{012}) = \sum_{\mu=0}^{2} \frac{\mu^4}{4!} e^{-\mu} = 0 + \frac{1}{4!}\frac{1}{e} + \frac{2^4}{4!}\frac{1}{e^2} = 0.015 + 0.091 = 0.106,$$

or 10.6 %; this is a likelihood of the data with models that assume a specific value for the mean. Notice that if the true value of the mean is zero, there is absolutely no probability of detecting any counts. One can thus conclude that there is slightly more than a 10 % chance of detecting 4 counts, given that the source truly emits 2 or fewer counts. This is not, however, a statement of possible values of the parent mean μ.   ◇

According to Bayes' theorem, the posterior distributions are

$$P(A_i/B) = \frac{P(B/A_i)P(A_i)}{P(B)}$$

where $P(B/A_i)$ is the likelihood, corresponding to each of the three terms in the sum of the example above. In the following example, we determine posterior probabilities.

Example 3.7 (Posterior Probability of the Poisson Mean)   We want to calculate the probability of the true mean being less or equal than 2, $P(A_{012}/B)$, and start by calculating the likelihoods required to evaluate $P(B)$. We make an initial and somewhat arbitrary assumption that the mean should be $\mu \le 10$, so that only 11 likelihoods must be evaluated. This assumption is dictated simply by practical considerations, and can also be stated in terms of assuming a subjective prior knowledge that the mean is somehow known not to exceed 10. We calculate

$$P(B) \simeq \sum_{i=0}^{10} \frac{i^4}{4!} e^{-i} \times P(A_i) = 0.979 \times P(A_i)$$

Also, assuming uniform priors, we have $P(A_i) = 1/11$ and that

$$P(A_{012}/B) = \frac{P(A_i) \times \sum_{i=0}^{2} \frac{i^4}{4!} e^{-i}}{P(A_i) \times 0.979} = \frac{1}{0.979} \sum_{i=0}^{2} \frac{i^4}{4!} e^{-i} = 0.108.$$

◇

The examples presented in this section illustrate the conceptual difference between the likelihood calculation and the estimate of the posterior, though the two calculations yielded similar numerical values.

3.4   Comparison of Binomial, Gaussian, and Poisson Distributions

In this section we provide numerical calculations that compare the binomial and Gaussian functions, and also discuss under what circumstances the Poisson distribution can be approximated by a Gaussian of same mean and variance. In fact practical computations with the Poisson distribution are often hampered by the need to calculate the factorial of large numbers. In Sect. 3.2 we derived the Gaussian distribution from the binomial function, using the approximation that $Np \gg 1$. In fact we assumed that the function has values $n \gg 1$ and, since the mean of the binomial is $\mu = Np$, the value Np sets the order of magnitude for the values of the random variable that have non-negligible probability. In the left panel of Fig. 3.4 we show the binomial distribution with parameters $p = q = 0.5$, showing that for $Np = 5$ the approximation is already at the level of 1 % near the peak of the distribution.

The main limitation of the Poisson distribution (3.18) is the presence of the factorial function, which becomes very rapidly a large number as function of the integer n (for example, $20! = 2.423 \times 10^{18}$), and it may lead to overflow problems in numerical codes. For large values of n, one can use the Stirling approximation to the factorial function, which retains only the first term of the following expansion:

$$n! = \sqrt{2\pi n} \times n^n e^{-n} \left(1 + \frac{1}{12n} + \dots \right). \tag{3.20}$$

Using this approximation for values of $n \geq 10$, the right panel of Fig. 3.4 shows two Poisson distributions with mean of, respectively, 3 and 20, and the corresponding Gaussian distributions with the same mean and of variance equal to the mean, as is the case for the Poisson distribution. The difference between the Gaussian and the

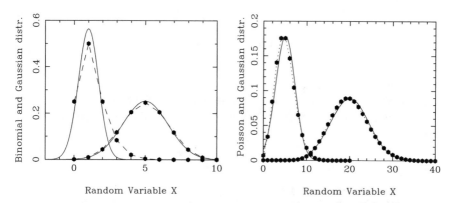

Fig. 3.4 (*Left*) Binomial distributions with $p = q = 0.5$ and, respectively, $N = 2$ and $N = 10$ as points connected by *dashed line*. Matching Gaussian distributions with same mean $\mu = Np$ and variance $\sigma^2 = Npq$ (*solid lines*). (*Right*) Gaussian distribution with $\mu = \sigma^2 = 3$ and $\mu = \sigma^2 = 20$ (*solid lines*) and Poisson distributions with same mean as points connected by a *dotted line*

Poisson distributions for a mean of $\mu = 20$ is at the percent level near the peak of the distribution. The Poisson distribution retains its characteristic asymmetry and a heavier tail at large values, and therefore deviations between the two function are larger away from the mean where, however, the absolute value of the probability becomes negligible. It can also be shown that for the value of $x = \mu$, the two distributions have the same value, when the Stirling approximation is used for the factorial function. A rule of thumb used by many is that for $x \geq 20$ the Gaussian approximation to the Poisson distribution is acceptable.

The approximation of a Poisson distribution with a Gaussian distribution is of great practical importance. Consider a counting experiment in which N counts are measured. The parent distribution of the random variable of interest is Poisson distributed and it is reasonable to assume that the best estimate of its mean is $\mu = N$ (but see Sect. 5.5.1 for a Bayesian approach that gives a slightly different answer). For values of $N > 20$ or so, the standard deviation of the parent Poisson distribution is therefore $\sigma = \sqrt{N}$. The measurement can be reported at $N \pm \sqrt{N}$, where the range of $N \pm \sqrt{N}$ corresponds to the $\mu \pm 1\sigma$ interval for a Gaussian variable.

Summary of Key Concepts for this Chapter

☐ *Binomial distribution:* It describes the probability of occurrence of n successes in N tries of a binary event,

$$P(n) = \binom{N}{n} p^n q^{N-n}$$

(mean pN and variance pqN).

☐ *Gaussian distribution*: It is an approximation of the binomial distribution when N is large,

$$f(x)dx = \frac{1}{\sqrt{2\pi\sigma^2}} e^{-\frac{(x-\mu)^2}{2\sigma^2}} dx$$

(mean μ and variance σ^2).

☐ *Poisson distribution*: It is an approximation of the binomial distribution when $p \ll 1$ that describes the probability of counting experiments,

$$P(n) = \frac{\mu^n}{n!} e^{-\mu}$$

(mean and variance have a value of μ).

Problems

3.1 Consider the Gaussian distribution

$$f(x) = \frac{1}{\sqrt{2\pi\sigma^2}} e^{-\frac{(x-\mu)^2}{2\sigma^2}}.$$

Calculate the mean and variance and show that all odd moments $E[(X - \mu)^n]$ of order $n \geq 3$ are zero.

3.2 Assume that scores from an I.Q. test follow a Gaussian distribution, and that the scores are standardized in such a way that the mean is $\mu = 100$, and the standard deviation is $\sigma = 15$.

(a) Calculate the probability that an I.Q. score is greater or equal than 145.
(b) Calculate the probability that the *mean I.Q. score* of a sample of 100 persons, chosen at random, is equal or larger than 105.

3.3 A coin is tossed ten times. Find

(a) The probability of obtaining 5 heads up and 5 tails up;
(b) The probability of having the first 5 tosses show heads up, and the final 5 tosses show tails up;
(c) The probability to have at least 7 heads up.

3.4 In a given course, it is known that 7.3 % of students fail.

(a) What is the expected number of failures in a class of 32 students?
(b) What is the probability that 5 or more students fail?

3.5 The frequency of twins in European population is about 12 in every 1000 maternities. Calculate the probability that there are no twins in 200 births, using (a) the binomial distribution, and (b) the Poisson distribution.

3.6 Given the distribution of a Poisson variable N,

$$P(n) = \frac{\mu^n}{n!} e^{-\mu}$$

show that the mean is given by μ and that the variance is also given by μ.

3.7 Consider Mendel's experiment of Table 1.1 at page 9 and refer to the "Long vs. short stem" data.

(a) Determine the parent distribution for the number of dominants.
(b) Calculate the uncertainty in the measurement of the number of plants that display the dominant character.
(c) Determine the difference between the number of measured plants with the dominant character and the expected number, in units of the standard deviation, to show that this number has an absolute value of less than one.

3.8 For Mendel's experimental data in Table 1.1 at page 9, consider the overall fraction of plants that display the dominant character, for all seven experiments combined.

(a) Determine the parent distribution of the overall fraction X of plants with dominant character and its expected value.
(b) Determine the sample mean of the fraction X;
(c) Using the parent variance of X, determine the value

$$z = \frac{x - E[X]}{\sigma}$$

which is the standardized difference between the measurement and the mean. Assuming that the binomial distribution can be approximated by a Gaussian of same mean and variance, calculate the probability of having a value of z equal or smaller (in absolute value) to the measured value.

Chapter 4
Functions of Random Variables and Error Propagation

Abstract Sometimes experiments do not directly measure the quantity of interest, but rather associated variables that can be related to the one of interest by an analytic function. It is therefore necessary to establish how we can infer properties of the interesting variable based on properties of the variables that have been measured directly. This chapter explains how to determine the probability distribution function of a variable that is function of other variables of known distribution, and how to measure its mean and variance, the latter usually referred to as *error propagation* formulas. We also establish two fundamental results of the theory of probability, the central limit theorem and the law of large numbers.

4.1 Linear Combination of Random Variables

Experimental variables are often related by a simple linear relationship. The linear combination of N random variables X_i is a variable Y defined by

$$Y = \sum_{i=1}^{N} a_i X_i \tag{4.1}$$

where a_i are constant coefficients. A typical example of a variable that is a linear combination of two variables is the signal detected by an instrument, which can be thought of as the sum of the intrinsic signal from the source plus the background. The distributions of the background and the source signals will influence the properties of the total signal detected, and it is therefore important to understand the statistical properties of this relationship in order to characterize the signal from the source.

4.1.1 General Mean and Variance Formulas

The expectation or mean of the linear combination is $E[Y] = \sum_{i=1}^{N} a_i E[X_i]$ or

$$\mu_y = \sum_{i=1}^{N} a_i \mu_i, \tag{4.2}$$

© Springer Science+Busines Media New York 2017
M. Bonamente, *Statistics and Analysis of Scientific Data*, Graduate Texts in Physics, DOI 10.1007/978-1-4939-6572-4_4

where μ_i is the mean of X_i. This property follows from the linearity of the expectation operator, and it is equivalent to a weighted mean in which the weights are given by the coefficients a_i.

In the case of the variance, the situation is more complex:

$$Var[Y] = E\left[\left(\sum_{i=1}^{N} a_i X_i - \sum_{i=1}^{N} a_i \mu_i\right)^2\right] = \sum_{i=1}^{N} a_i^2 E\left[(X_i - \mu_i)^2\right]$$

$$+ 2\sum_{i=1}^{N}\sum_{j=i+1}^{N} a_i a_j E[(X_i - \mu_i)(X_j - \mu_j)]$$

$$= \sum_{i=1}^{N} a_i^2 Var(X_i) + 2\sum_{i=1}^{N}\sum_{j=i+1}^{N} a_i a_j Cov(X_i, X_j).$$

The result can be summarized in a more compact relationship,

$$\sigma_y^2 = \sum_{i=1}^{N} a_i^2 \sigma_i^2 + 2\sum_{i=1}^{N}\sum_{j=i+1}^{N} a_i a_j \sigma_{ij}^2. \tag{4.3}$$

Equation (4.3) shows that variances add only for variables that are mutually uncorrelated, or $\sigma_{ij}^2 = 0$, but not in general. The following example illustrates the importance of a non-zero covariance between two variables, and its effect on the variance of the sum.

Example 4.1 (Variance of Anti-correlated Variables) Consider the case of the measurement of two random variables X and Y that are completely anti-correlated, $Corr(X, Y) = -1$, with mean and variance $\mu_x = 1$, $\mu_y = 1$, $\sigma_x^2 = 0.5$ and $\sigma_y^2 = 0.5$.

The mean of $Z = X + Y$ is $\mu = 1 + 1 = 2$ and the variance is $\sigma^2 = \sigma_x^2 + \sigma_y^2 - 2Cov(X, Y) = (\sigma_x - \sigma_y)^2 = 0$; this means that in this extreme case of complete anticorrelation the sum of the two random variables is actually not a random variable any more. If the covariance term had been neglected in (4.3), we would have made the error of inferring a variance of 1 for the sum.   ◇

4.1.2 Uncorrelated Variables and the $1/\sqrt{N}$ Factor

For two or more uncorrelated variables the variances add linearly, according to (4.3). Uncorrelated variables are common in statistics. For example, consider repeating the same experiment a number N of times independently, and each time measurements of a random variable X_i is made. After N experiments, one obtains N measurements from identically distributed random variables (since they resulted from the same type of experiment). The variables are independent, and therefore uncorrelated, if the experiments were performed in such a way that the outcome of one specific experiment did not affect the outcome of another.

With N uncorrelated variables X_i all of equal mean μ and variance σ^2, one is often interested in calculating the *relative uncertainty* in the variable

$$Y = \frac{1}{N} \sum_{i=1}^{N} X_i \tag{4.4}$$

which describes the sample mean of N measurements. The relative uncertainty is described by the ratio of the standard deviation and the mean,

$$\frac{\sigma_y}{\mu_y} = \frac{1}{N} \frac{\sqrt{\sigma^2 + \cdots + \sigma^2}}{\mu} = \frac{1}{\sqrt{N}} \times \frac{\sigma}{\mu} \tag{4.5}$$

where we used the property that $Var[aX] = a^2 Var[X]$ and the fact that both means and variances add linearly. The result shows that the N measurements reduced the relative error in the random variable by a factor of $1/\sqrt{N}$, as compared with a single measurement. This observation is a key factor in statistics, and it is the reason why one needs to repeat the same experiment many times in order to reduce the relative statistical error. Equation (4.5) can be recast to show that the variance in the sample mean is given by

$$\sigma_Y^2 = \frac{\sigma^2}{N} \tag{4.6}$$

where σ is the sample variance, or variance associated with one measurement. The interpretation is simple: one expects much less variance between two measurements of the sample mean, than between two individual measurements of the variable, since the statistical fluctuations of individual measurements average down with increasing sample size.

Another important observation is that, in the case of completely correlated variables, then additional measurements introduces no advantages, i.e., the relative error does not decrease with the number of measurements. This can be shown with the aid of (4.3), and is illustrated in the following example.

Example 4.2 (Variance of Correlated Variables) Consider the two measurements in Example 4.1, but now with a correlation of 1. In this case, the covariance of the sum is $\sigma^2 = \sigma_x^2 + \sigma_y^2 + 2Cov(X, Y) = (\sigma_x + \sigma_y)^2$, and therefore the relative error in the sum is

$$\frac{\sigma}{\mu} = \frac{(\sigma_x + \sigma_y)}{\mu_x + \mu_y}$$

which is the same as the relative error of each measurement. Notice that the same conclusion applies to the average of the two measurements, since the sum and the average differ only by a constant factor of $1/2$.   ◇

4.2   The Moment Generating Function

The mean and the variance provide only partial information on the random variable, and a full description would require the knowledge of all moments. The moment generating function is a convenient mathematical tool to determine the distribution function of random variables and its moments. It is also useful to prove the central limit theorem, one of the key results of statistics, since it establishes the Gaussian distribution as the normal distribution when a random variable is the sum of a large number of measurements.

The *moment generating function* of a random variable X is defined as

$$M(t) = E[e^{tX}], \tag{4.7}$$

and it has the property that all moments can be derived from it, provided they exist and are finite. Assuming a continuous random variable of probability distribution function $f(x)$, the moment generating function can be written as

$$M(t) = \int_{-\infty}^{+\infty} e^{tx} f(x) dx =$$

$$\int_{-\infty}^{+\infty} \left(1 + \frac{tx}{1} + \frac{(tx)^2}{2!} + \dots\right) f(x) dx = 1 + t\mu_1 + \frac{t^2}{2!}\mu_2 + \dots$$

and therefore all moments can be obtained as partial derivatives,

$$\mu_r = \left.\frac{\partial^r M(t)}{\partial t^r}\right|_{t=0}. \tag{4.8}$$

The most important property of the moment generating function is that there is a one-to-one correspondence between the moment generating function and the probability distribution function, i.e., the moment generating function is a sufficient description of the random variable. Some distributions do not have a moment generating function, since some of their moments may be infinite, so in principle this method cannot be used for all distributions.

4.2.1   Properties of the Moment Generating Function

A full treatment of mathematical properties of the moment generating function can be found in textbooks on theory of probability, such as [38]. Two properties of the moment generating function will be useful in the determination of the distribution function of random variables:

- If $Y = a + bX$, where a, b are constants, the moment generating function of Y is

$$M_y(t) = e^{at}M_x(bt). \tag{4.9}$$

Proof This relationship can be proved by the use of the expectation operator, according to the definition of the moment generating function:

$$E[e^{tY}] = E[e^{t(a+bX)}] = E[e^{at}e^{btX}] = e^{at}M_x(bt).$$

□

- If X and Y are independent random variables, with $M_x(t)$ and $M_y(t)$ as moment generating functions, then the moment generating function of $Z = X + Y$ is

$$M_z(t) = M_x(t)M_y(t). \tag{4.10}$$

Proof The relationship is derived immediately by

$$E[e^{tZ}] = E[e^{t(X+Y)}] = M_x(t)M_y(t).$$

□

4.2.2   The Moment Generating Function of the Gaussian and Poisson Distribution

Important cases to study are the Gaussian distribution of mean μ and variance σ^2 and the Poisson distribution of mean μ.

- The moment generating function of the Gaussian is given by

$$M(t) = e^{\mu t + \frac{1}{2}\sigma^2 t^2}. \tag{4.11}$$

Proof Start with

$$M(t) = \frac{1}{\sqrt{2\pi\sigma^2}} \int_{-\infty}^{+\infty} e^{tx} e^{-\frac{(x-\mu)^2}{2\sigma^2}} dx.$$

The exponent can be written as

$$tx - \frac{1}{2}\frac{x^2 + \mu^2 - 2x\mu}{\sigma^2} = \frac{2\sigma^2 tx - x^2 - \mu^2 + 2x\mu}{2\sigma^2}$$

$$= -\frac{(x - \mu - \sigma^2 t)^2}{2\sigma^2} + \frac{2\mu\sigma^2 t}{2\sigma^2} + \frac{\sigma^2 t}{2\sigma^2}\sigma^2 t.$$

It follows that

$$M(t) = \frac{1}{\sqrt{2\pi\sigma^2}} \int_{-\infty}^{+\infty} e^{\mu t} e^{\frac{\sigma^2 t^2}{2}} e^{-\frac{(x-\mu-\sigma^2 t)^2}{2\sigma^2}} dx$$

$$= \frac{1}{\sqrt{2\pi\sigma^2}} e^{\mu t} e^{\frac{\sigma^2 t^2}{2}} \sqrt{2\pi\sigma^2} = e^{\mu t + \frac{\sigma^2 t^2}{2}}.$$

□

- The moment generating function of the Poisson distribution is given by

$$M(t) = e^{-\mu} e^{\mu e^t}. \tag{4.12}$$

Proof The moment generating function is obtained by

$$M(t) = E[e^{tN}] = \sum_{n=0}^{\infty} e^{nt} \frac{\mu^n}{n!} e^{-\mu} = e^{-\mu} \sum_{n=0}^{\infty} \frac{(\mu e^t)^n}{n!} = e^{-\mu} e^{\mu e^t}.$$

□

Example 4.3 (Sum of Poisson Variables) The moment generating function can be used to show that the sum of two independent Poisson random variables of mean λ and μ is a Poisson random variable with mean $\lambda + \mu$. In fact that mean of the Poisson appears at the exponent of the moment generating function, and property (4.10), can be used to prove this result. The fact that the mean of two independent Poisson distributions will add is not surprising, given that the Poisson distribution relates to the counting of discrete events.   ◇

4.3   The Central Limit Theorem

The Central Limit Theorem is one of statistic's most important results, establishing that a variable obtained as the sum of a large number of independent variables has a Gaussian distribution. This result can be stated as:

Theorem 4.1 (Central Limit Theorem) *The sum of a large number of independent random variables is approximately distributed as a Gaussian. The mean of the distribution is the sum of the means of the variables and the variance of the distribution is the sum of the variances of the variables. This result holds regardless of the distribution of each individual variable.*

Proof Consider the variable Y as the sum of N variables X_i of mean μ_i and variance σ_i^2,

$$Y = \sum_{i=1}^{N} X_i, \tag{4.13}$$

with $M_i(t)$ the moment generating function of the random variable $(X_i - \mu_i)$. Since the random variables are independent, and independence is a stronger statement than uncorrelation, it follows that the mean of Y is $\mu = \sum \mu_i$, and that variances likewise add linearly, $\sigma^2 = \sum \sigma_i^2$. We want to calculate the moment generating function of the variable Z defined by

$$Z = \frac{Y - \mu}{\sigma} = \frac{1}{\sigma} \sum_{i=1}^{N} (X_i - \mu_i).$$

The variable Z has a mean of zero and unit variance. We want to show that Z can be approximated by a standard Gaussian. Using the properties of the moment generating function, the moment generating function of Z is

$$M(t) = \prod_{i=1}^{N} M_i(t/\sigma).$$

The moment generating function of each variable $(X_i - \mu_i)/\sigma$ is

$$M_i(t/\sigma) = 1 + \mu_{(x_i - \mu_i)} \frac{t}{\sigma} + \frac{\sigma_i^2}{2} \left(\frac{t}{\sigma}\right)^2 + \frac{\mu_{i,3}}{3!} \left(\frac{t}{\sigma}\right)^3 + \dots$$

where $\mu_{x_i - \mu_i} = 0$ is the mean of $X_i - \mu_i$. The quantities σ_i^2 and $\mu_{i,3}$ are, respectively, the central moments of the second and third order of X_i.

If a large number of random variables are used, $N \gg 1$, then σ^2 is large, as it is the sum of variances of the random variables, and we can ignore terms of order σ^{-3}. We therefore make the approximation

$$\ln M(t) = \sum \ln M_i \left(\frac{t}{\sigma}\right) =$$

$$\sum \ln \left(1 + \frac{\sigma_i^2}{2}\left(\frac{t}{\sigma}\right)^2\right) \simeq \sum \frac{\sigma_i^2}{2}\left(\frac{t}{\sigma}\right)^2 = \frac{1}{2}t^2.$$

This results in the approximation of the moment generating function of $(y - \mu)/\sigma$ as

$$\Rightarrow M(t) \simeq e^{\frac{t^2}{2}},$$

which shows that Z is approximately distributed as a standard Gaussian distribution, according to (4.11). Given that the random variable of interest Y is obtained by a change of variable $Z = (Y - \mu)/\sigma$, we also know that $\mu_y = \mu$ and $Var(Y) = Var(\sigma Z) = \sigma^2 Var(Z) = \sigma^2$, therefore Y is distributed as a Gaussian with mean μ and variance σ^2. $\qquad\qquad$ $\square$

The central limit theorem establishes that the Gaussian distribution is the limiting distribution approached by the sum of random variables, no matter their original shapes, when the number of variables is large. A particularly illustrative example is the one presented in the following, in which we perform the sum of a number of uniform distributions. Although the uniform distribution does not display the Gaussian-like feature of a centrally peaked distribution, with the increasing number of variables being summed, the sum rapidly approaches a Gaussian distribution.

Example 4.4 (Sum of Uniform Random Variables) We show that the sum of N independent uniform random variables between 0 and 1 tend to a Gaussian with mean $N/2$, given that each variable has a mean of $1/2$. The calculation that the sum of N uniform distribution tends to the Gaussian can be done by first calculating the moment generating function of the uniform distribution, then using the properties of the moment generating function.

We can show that the uniform distribution in the range $[0, 1]$ has $\mu_i = 1/2$, $\sigma_i^2 = 1/12$, and a moment generating function

$$M_i(t) = \frac{(e^t - 1)}{t};$$

the sum of N independent such variables therefore has $\mu = N/2$ and $\sigma^2 = N/12$. To prove that the sum is asymptotically distributed like a Gaussian with this mean and variance, we must show that

$$\lim_{N \to \infty} M(t) = e^{\frac{N}{2}t + \frac{t^2}{2}\frac{N}{12}}$$

Proof Using the property of the moment generating function of independent variables, we write

$$M(t) = M_i(t)^N = \left(\frac{1}{t}(e^t - 1)\right)^N$$

$$= \left(\frac{1 + t + t^2/2! + t^3/3!\ldots - 1}{t}\right)^N \simeq \left(1 + \frac{t}{2} + \frac{t^2}{6} + \ldots\right)^N.$$

Neglect terms of order $O(t^3)$ and higher, and work with logarithms:

$$\ln(M(t)^N) \simeq N \ln\left(1 + \frac{t}{2} + \frac{t^2}{6}\right)$$

Use the Taylor series expansion $\ln(1 + x) \simeq (x - x^2/2 + \ldots)$, to obtain

$$\ln(M_i(t)) \simeq N\left(\frac{t}{2} + \frac{t^2}{6} - \frac{1}{2}\left(\frac{t}{2} + \frac{t^2}{6}\right)^2\right) =$$

$$N(t/2 + t^2/6 - t^2/8 + O(t^3)) \simeq N(t/2 + t^2/24)$$

in which we continued neglecting terms of order $O(t^3)$. The equation above shows that the moment generating function can be approximated as

$$M(t) \simeq e^{N\left(\frac{t}{2} + \frac{t^2}{24}\right)} \tag{4.14}$$

which is in fact the moment generating function of a Gaussian with mean $N/2$ and variance $N/12$.   □

In Figure 4.1 we show the simulations of, respectively, 1000 and 100,000 samples drawn from $N = 100$ uniform and independent variables between 0 and 1. The sample distributions approximate well the limiting Gaussian with $\mu = N/2$, $\sigma = \sqrt{N/12}$. The approximation is improved when a larger number of samples are drawn, also illustrating the fact that the sample distribution approximates the parent distribution in the limit of a large number of samples collected.   ◇

Example 4.5 (Sum of Two Uniform Distributions) An analytic way to develop a practical sense of how the sum of non-Gaussian distributions progressively develops the peaked Gaussian shape can be illustrated with the sum of just two uniform distributions. We start with a uniform distribution in the range of -1 to 1, which can be shown to have

$$M(t) = 1/(2t)(e^t - e^{-t}).$$

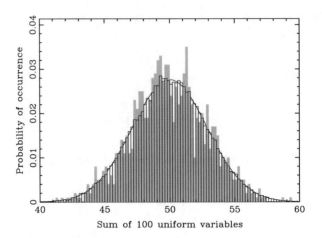

Fig. 4.1 Sample distribution functions of the sum of $N = 100$ independent uniform variables between 0 and 1, constructed from 1000 simulated measurements (*grey* histograms) and 100,000 measurements (histogram plot with *black outline*). The *solid curve* is the $N(\mu, \sigma)$ Gaussian, with $\mu = N/2$, $\sigma = \sqrt{N/12}$, the limiting distribution according to the Central Limit Theorem

The sum of two such variables will have a *triangular distribution*, given by the analytical form

$$f(x) = \begin{cases} \dfrac{1}{2} + \dfrac{x}{4} & \text{if } -2 \leq x \leq 0 \\[2mm] \dfrac{1}{2} - \dfrac{x}{4} & \text{if } 0 \leq x \leq 2. \end{cases}$$

This is an intuitive result that can be proven by showing that the moment generating function of the triangular distribution is equal to $M(t)^2$ (see Problem 4.3). The calculation follows from the definition of the moment generating function for a variable of known distribution function. The triangular distribution is the first step in the development of a peaked, Gaussian-like distribution.   ◇

4.4   The Distribution of Functions of Random Variables

The general case of a variable that is a more complex function of other variables can be studied analytically when certain conditions are met. In this book we present the method of change of variables which can be conveniently applied to one-dimensional transformations and a method based on the cumulative distribution function which can be used for multi-dimensional transformations. Additional information on this subject can be found, e.g., in the textbook by Ross [38].

4.4.1 The Method of Change of Variables

A simple method for obtaining the probability distribution function of the dependent variable $Y = Y(X)$ is by using the method of *change of variables*, which applies only if the function $Y(x)$ is strictly increasing. In this case the probability distribution of $g(y)$ of the dependent variable is related to the distribution $f(x)$ of the independent variable via

$$g(y) = f(x)\frac{dx}{dy} \qquad (4.15)$$

In the case of a decreasing function, the same method can be applied but the term dx/dy must be replaced with the absolute value, $|dx/dy|$.

Example 4.6 Consider a variable X distributed as a uniform distribution between 0 and 1, and the variable $Y = X^2$. The method automatically provides the information that the variable Y is distributed as

$$g(y) = \frac{1}{2}\frac{1}{\sqrt{y}}$$

with $0 \le y \le 1$. You can prove that the distribution is properly normalized in this domain.   ◇

The method can be naturally extended to the joint distribution of several random variables. The multi-variable version of (4.15) is

$$g(u, v) = h(x, y)\,|J| \qquad (4.16)$$

in which

$$J = \left(\frac{d(x, y)}{d(u, v)}\right) = \begin{pmatrix} \dfrac{dx}{du} & \dfrac{dx}{dv} \\ \dfrac{dy}{du} & \dfrac{dy}{dv} \end{pmatrix}$$

is the *Jacobian* of the transformation, in this case a 2 by 2 matrix, $h(x, y)$ is the joint probability distribution of the independent variables X, Y, and U, V are the new random variables related to the original ones by a transformation $U = u(X, Y)$ and $V = v(X, Y)$.

Example 4.7 (Transformation of Cartesian to Polar Coordinates) Consider two random variables X, Y distributed as standard Gaussians, and independent of one another. The joint probability distribution function is

$$h(x, y) = \frac{1}{2\pi}e^{-\frac{x^2+y^2}{2}}.$$

Consider a transformation of variables from Cartesian coordinates x, y to polar coordinates r, θ, described by

$$\begin{cases} x = r \cdot \cos(\theta) \\ y = r \cdot \sin(\theta) \end{cases}$$

The Jacobian of the transformation is

$$J = \begin{pmatrix} \cos\theta & -r\sin\theta \\ \sin\theta & r\cos\theta \end{pmatrix}$$

and its determinant is $|J| = r$. Notice that to apply the method described by (4.16) one only needs to know the inverse transformation of (x, y) as function of (r, θ). It follows that the distribution of (r, θ) is given by

$$g(r, \theta) = \frac{1}{2\pi} r e^{-\frac{r^2}{2}}$$

for $r \geq 0, 0 \leq \theta \leq 2\pi$. The distribution $r e^{-\frac{r^2}{2}}$ is called the *Rayleigh distribution*, and $1/2\pi$ can be interpreted as a uniform distribution for the angle θ between 0 and π. One important conclusion is that, since $g(r, \theta)$ can be factored out into two functions that contain separately the two variables r and θ, the two new variables are also independent.   ◇

4.4.2   A Method for Multi-dimensional Functions

We will consider the case in which the variable Z is a function of two random variables X and Y, since this is a case of common use in statistics, e.g., $X + Y$, or X/Y. We illustrate the methodology with the case of the function $Z = X + Y$, when the two variables are independent. The calculation starts with the cumulative distribution function of the random variable of interest,

$$F_Z(a) = P(Z \leq a) = \int\int_{x+y\leq a} f(x)g(y)dxdy$$

in which $f(x)$ and $g(y)$ are, respectively, the probability distribution functions of X and Y, and the limits of integration must be chosen so that the sum of the two variables is less or equal than a. The portion of parameter space such that $x + y \leq a$ includes all values $x \leq a - y$, for any given value of y, or

$$F_Z(a) = \int_{-\infty}^{+\infty} dy \int_{\infty}^{a-y} f(x)g(y)dx = \int_{-\infty}^{+\infty} g(y)dy F_x(a - y)$$

where F_x is the cumulative distribution for the variable X. It is often more convenient to express the relationship in terms of the probability distribution function, which is related to the cumulative distribution function via a derivative,

$$f_Z(a) = \frac{d}{da}F_Z(a) = \int_{-\infty}^{\infty} f(a - y)g(y)dy. \tag{4.17}$$

This relationship is called the *convolution* of the distributions $f(x)$ and $g(y)$.

Example 4.8 (Sum of Two Independent Uniform Variables) Calculate the probability distribution function of the sum of two independent uniform random variables between -1 and $+1$.

The probability distribution function of a uniform variable between -1 and $+1$ is $f(x) = 1/2$, defined for $-1 \leq x \leq 1$. The convolution gives the following integral

$$f_Z(a) = \int_{-1}^{+1} \frac{1}{2}f(a - y)dy.$$

The distribution function of the sum Z can have values $-2 \leq a \leq 2$, and the convolution must be divided into two integrals, since $f(a-y)$ is only defined between -1 and $+1$. We obtain

$$f_Z(a) = \frac{1}{4} \times \begin{cases} \int_{-1}^{a+1} dy & \text{if } -2 \leq a \leq 0 \\ \int_{a-1}^{1} dy & \text{if } 0 \leq a \leq 2. \end{cases}$$

This results in

$$f_Z(a) = \frac{1}{4} \times \begin{cases} (a + 2) & \text{if } -2 \leq a \leq 0 \\ (2 - a) & \text{if } 0 \leq a \leq 2 \end{cases}$$

which is the expected triangular distribution between -2 and $+2$.   ◇

Another useful application is for the case of $Z = X/Y$, where X and Y are again independent variables. We begin with the cumulative distribution,

$$F_Z(z) = P(Z < z) = P(X/Y < z) = P(X < zY).$$

For a given value y of the random variable Y, this probability equals $F_X(zy)$; since Y has a probability $f_Y(y)dy$ to be in the range between y and $y + dy$, we obtain

$$F_Z(z) = \int F_X(zy)f_Y(y)dy.$$

Following the same method as for the derivation of the distribution of $X + Y$, we must take the derivative of $F_Z(z)$ with respect to z to obtain:

$$f_Z(z) = \int f_X(zy) y f_Y(y) dy. \tag{4.18}$$

This is the integral than must be solved to obtain the distribution of X/Y.

4.5   The Law of Large Numbers

Consider N random variables X_i that are identically distributed, and μ is their common mean. The *Strong Law of Large Numbers* states that, under suitable conditions on the variance of the random variables, the sum of the N variables tends to the mean μ, which is a deterministic number and not a random variable. This result can be stated as

$$\lim_{n \to \infty} \frac{X_1 + \ldots + X_N}{N} = \mu, \tag{4.19}$$

and it is, together with the Central Limit Theorem, one of the most important results of the theory of probability, and of great importance for statistics. Equation (4.19) is a very strong statement because it shows that, asymptotically, the sum of random variables becomes a constant equal to the sample mean of the N variables, or N measurements. Although no indication is given towards establishing how large N should be in order to achieve this goal, it is nonetheless an important result that will be used in determining the asymptotic behavior of random variables. Additional mathematical properties of this law can be found in books of theory of probability, such as [38] or [26].

Instead of providing a formal proof of this law, we want to focus on an important consequence. Given a function $y(x)$, we would like to estimate its expected value $E[y(X)]$ from the N measurements of the variables X_i. According to the law of large numbers, we can say that

$$\lim_{n \to \infty} \frac{y(X_1) + \ldots + y(X_N)}{N} = E[y(X)]. \tag{4.20}$$

Equation (4.20) states that a large number of measurements of the variables X_i can be used to measure the expectation of $E[y(X)]$, entirely bypassing the probability distribution function of the function $y(X)$. This property is used in the following section.

4.6   The Mean of Functions of Random Variables

For a function of random variables it is often necessary or convenient to develop methods to estimate the mean and the variance without having full knowledge of its probability distribution function.

For functions of a single variable $Y = y(X)$

$$E[y(X)] = \int y(x)f(x)dx \tag{4.21}$$

where $f(x)$ is the distribution function of X. This is in fact a very intuitive result, stating that the distribution function of X is weighted by the function of interest, and it makes it straightforward to compute expectation values of variables without first having to calculate their full distribution. According to the law of large numbers, this expectation can be estimated from N measurements x_i as per (4.20),

$$\overline{y(x)} = \frac{y(x_1) + \ldots + y(x_n)}{N}. \tag{4.22}$$

An important point is that the mean of the function is *not* equal to the function of the mean, $\overline{y(x)} \neq y(\overline{x})$, as will be illustrated in the following example. Equation (4.22) says that we must have access to the *individual measurements* of the variable X, if we want to make inferences on the mean of a function of X. If, for example, we only had the mean $\overline{x}$, we cannot measure $\overline{u(x)}$. This point is relevant when one has limited access to the data, e.g., when the experimenter does not report all information on the measurements performed.

Example 4.9 (Mean of Square of a Uniform Variable)   Consider the case of a uniform variable U in the range 0–1, with mean $1/2$. If we want to evaluate the parent mean of $X = U^2$, we calculate

$$\mu = \int_0^1 u^2 du = 1/3.$$

It is important to see that the mean of U^2 is not just the square of the mean of U, and therefore the means do not transform following the same analytic expression as the random variables. You can convince yourself of this fact by assuming to draw five "fair" samples from a uniform distribution, $0.1, 0.3, 0.5, 0.7$ and 0.9—they can be considered as a dataset of measurements. Clearly their mean is $1/2$, but the mean of their squares is $1/3$ and not $1/4$, in agreement with the theoretical calculation of the parent mean.   ◇

Another example where the mean of the function does not equal to the function of the mean is reported in Problem 4.5, in which you can show that using the means of I and W/Q do not give the mean of m/e for the Thomson experiment to measure the mass to charge ratio of the electron. The problem provides a multi-dimensional extension to (4.22), since the variable m/e is a function of two variables that have been measured in pairs.

4.7 The Variance of Functions of Random Variables and Error Propagation Formulas

A random variable Z that is a function of other variables can have its variance estimated directly if the measurements of the independent variables are available, similar to the case of the estimation of the mean. Considering, for example, the case of a function $Z = z(U)$ that depends on just one variable, for which we have N measurements $u_1, \ldots, u_N$ available. With the mean estimated from (4.22), the variance can accordingly be estimated as

$$s_u^2 = \frac{(z(u_1) - \bar{z})^2 + \ldots + (z(u_N) - \bar{z})^2}{N - 1}, \tag{4.23}$$

as one would normally do, treating the numbers $z(u_1), \ldots, z(u_N)$ as samples from the dependent variable. This method can naturally be extended to more than one variable, as illustrated in the following example. When the measurements of the independent variables are available, this method is the straightforward way to estimate the variance of the function of random variables.

Example 4.10 Using the Thomson experiment described on page 23, consider the data collected for Tube 1, consisting of 11 measurements of W/Q and I, from which the variable of interest v is calculated as

$$v = 2\frac{W/Q}{I}$$

From the reported data, one obtains 11 measurements of v, from which the mean and standard deviation can be immediately calculated as $\bar{v} = 7.9 \times 10^9$, and $s_v = 2.8 \times 10^9$. ◇

There are a number of instances in which one does not have access to the original measurements of the independent variable or variables, required for an accurate estimate of the variance according to (4.23). In this care, an approximate method to estimate the variance must be used instead. This method takes the name of *error propagation*. Consider a random variable Z that is a function of a number of variables, $Z = z(U, V, \ldots)$. A method to approximate the variance of Z in terms of the variance of the independent variables U, V, etc. starts by expanding Z in a

Taylor series about the means of the independent variables, to obtain

$$z(u, v, \ldots) = z(\mu_u, \mu_v, \ldots) + (u - \mu_u) \left. \frac{\partial z}{\partial u} \right|_{\mu_u} + (v - \mu_v) \left. \frac{\partial z}{\partial v} \right|_{\mu_v}$$
$$+ \ldots + O(u - \mu_u)^2 + O(v - \mu_v)^2 + \ldots$$

Neglecting terms of the second order, the expectation of Z would be given by $E[Z] = z(\mu_u, \mu_v, \ldots)$, i.e., the mean of X would be approximated as $\mu_X = z(\mu_u, \mu_v, \ldots)$. This is true only if the function is linear, and we have shown in Sect. 4.6 that this approximation may not be sufficiently accurate in the case of nonlinear functions such as U^2. This approximation for the mean is used to estimate the variance of Z, for which we retain only terms of first order in the Taylor expansion:

$$E[(Z - E[Z])^2] \simeq E\left[\left((u - \mu_u) \left. \frac{\partial z}{\partial u} \right|_{\mu_u} + (v - \mu_v) \left. \frac{\partial z}{\partial v} \right|_{\mu_v} + \ldots \right)^2 \right]$$
$$\simeq E\left[\left((u - \mu_u) \left. \frac{\partial z}{\partial u} \right|_{\mu_u} \right)^2 \left((v - \mu_v) \left. \frac{\partial z}{\partial v} \right|_{\mu_v} \right)^2 \right.$$
$$\left. + 2(u - \mu_u) \left. \frac{\partial z}{\partial u} \right|_{\mu_u} \cdot (v - \mu_v) \left. \frac{\partial z}{\partial v} \right|_{\mu_v} + \ldots \right].$$

This formula can be rewritten as

$$\sigma_X^2 \simeq \sigma_u^2 \left. \left| \frac{\partial f}{\partial u} \right|^2 \right|_{\mu_u} + \sigma_v^2 \left. \left| \frac{\partial f}{\partial v} \right|^2 \right|_{\mu_v} + 2 \cdot \sigma_{uv}^2 \left. \frac{\partial f}{\partial u} \right|_{\mu_u} \left. \frac{\partial f}{\partial v} \right|_{\mu_v} + \ldots \tag{4.24}$$

which is usually referred to as the *error propagation formula*, and can be used for any number of independent variables. This result makes it possible to estimate the variance of a function of variable, knowing simply the variance of each of the independent variables and their covariances. The formula is especially useful for all cases in which the measured variables are independent, and all that is known is their mean and standard deviation (but not the individual measurements used to determine the mean and variance). This method must be considered as an approximation when there is only incomplete information about the measurements. Neglecting terms of the second order in the Taylor expansion can in fact lead to large errors, especially when the function has strong nonlinearities. In the following we provide a few specific formulas for functions that are of common use.

4.7.1 Sum of a Constant

Consider the case in which a constant a is added to the variable U,

$$Z = U + a$$

where a is a deterministic constant which can have either sign. It is clear that $\partial z/\partial a = 0$, $\partial z/\partial u = 1$, and therefore the addition of a constant has no effect on the uncertainty of X,

$$\sigma_z^2 = \sigma_u^2. \tag{4.25}$$

The addition or subtraction of a constant only changes the mean of the variable by the same amount, but leaves its standard deviation unchanged.

4.7.2 Weighted Sum of Two Variables

The variance of the weighted sum of two variables,

$$Z = aU + bV$$

where a, b are constants of either sign, can be calculated using $\partial z/\partial u = a$, $\partial z/\partial v = b$. We obtain

$$\sigma_z^2 = a^2\sigma_u^2 + b^2\sigma_v^2 + 2ab\sigma_{uv}^2. \tag{4.26}$$

The special case in which the two variables U, V are uncorrelated leads to the weighted sum of the variances.

Example 4.11 Consider a decaying radioactive source which is found to emit $N_1 = 50$ counts and $N_2 = 35$ counts in two time intervals of same duration, during which $B = 20$ background counts are recorded. This is an idealized situation in which we have directly available the measurement of the background counts. In the majority of real-life experiments one simply measures the sum of signal plus background, and in those cases additional considerations must be used. We want to calculate the background subtracted source counts in the two time intervals and estimate their *signal-to-noise ratio*, defined as $S/N = \mu/\sigma$. The inverse of the signal-to-noise ratio is the relative error of the variable.

Each random variable N_1, N_2, and B obeys the Poisson distribution, since it comes from a counting process. Therefore, we can estimate the following parent means and variances from the sample measurements,

$$\begin{cases} \mu_1 = 50 & \sigma_1 = \sqrt{50} = 7.1 \\ \mu_2 = 35 & \sigma_2 = \sqrt{35} = 5.9 \\ \mu_B = 20 & \sigma_B = \sqrt{20} = 4.5 \end{cases}$$

Since the source counts are given by $S_1 = N_1 - B$ and $S_2 = N_2 - B$, we can now use the approximate variance formulas *assuming* that the variables are uncorrelated, $\sigma_{S_1} = \sqrt{50 + 20} = 8.4$ and $\sigma_{S_2} = \sqrt{35 + 20} = 7.4$. The two measurements of the source counts would be reported as $S_1 = 30 \pm 8.4$ and $S_2 = 15 \pm 7.4$, from which the signal-to-noise ratios are given, respectively, as $\mu_{S1}/\sigma_{S_1} = 3.6$ and $\mu_{S2}/\sigma_{S_2} = 2.0$. ◇

4.7.3 Product and Division of Two Random Variables

Consider the product of two random variables U, V, optionally also with a constant factor a of either sign,

$$Z = aUV. \tag{4.27}$$

The partial derivatives are $\partial z/\partial u = av$, $\partial z/\partial v = au$, leading to the approximate variance of

$$\sigma_z^2 = a^2 v^2 \sigma_u^2 + a^2 u^2 \sigma_v^2 + 2auv\sigma_{uv}^2.$$

This can be rewritten as

$$\frac{\sigma_z^2}{z^2} = \frac{\sigma_u^2}{u^2} + \frac{\sigma_v^2}{v^2} + 2\frac{\sigma_{uv}^2}{uv}. \tag{4.28}$$

Similarly, the division between two random variables,

$$Z = a\frac{U}{V}, \tag{4.29}$$

leads to

$$\frac{\sigma_z^2}{z^2} = \frac{\sigma_u^2}{u^2} + \frac{\sigma_v^2}{v^2} - 2\frac{\sigma_{uv}^2}{uv}. \tag{4.30}$$

Notice the equations for product and division differ by just one sign, meaning that a positive covariance between the variables leads to a reduction in the standard deviation for the division, and an increase in the standard deviation for the product.

Example 4.12 Using the Thomson experiment of page 23, consider the data for Tube 1, and assume that the only number available are the mean and standard deviation of W/Q and I. From these two numbers we want to estimate the mean and variance of v. The measurement of the two variables are $W/Q = 13.3 \pm 8.5 \times 10^{11}$ and $I = 312.9 \pm 93.4$, from which the mean of v would have to be estimated as $\bar{v} = 8.5 \times 10^9$—compare with the value of 7.9×10^9 obtained from the individual measurements.

The estimate of the variance requires also a knowledge of the covariance between the two variables W/Q and I. In the absence of any information, we will assume that the two variables are uncorrelated, and use the error propagation formula to obtain

$$\sigma_v \simeq 2 \times \frac{13.3 \times 10^{11}}{312.9} \times \left(\left(\frac{8.5}{13.3} \right)^2 + \left(\frac{93.4}{312.9} \right)^2 \right)^{1/2} = 6 \times 10^9,$$

which is a factor of 2 larger than estimated directly from the data (see Example 4.10). Part of the discrepancy is to be attributed to the neglect of the covariance between the measurement, which can be found to be positive, and therefore would reduce the variance of v according to (4.30). Using this approximate method, we would estimate the measurement as $v = 8.5 \pm 6 \times 10^9$, instead of $7.9 \pm 2.8 \times 10^9$. ◇

4.7.4 Power of a Random Variable

A random variable may be raised to a constant power, and optionally multiplied by a constant,

$$Z = aU^b \tag{4.31}$$

where a and b are constants of either sign. In this case, $\partial z / \partial u = abu^{b-1}$ and the error propagation results in

$$\frac{\sigma_z}{z} = |b| \frac{\sigma_u}{u}. \tag{4.32}$$

This results states that the relative error in Z is b times the relative error in U.

4.7.5   *Exponential of a Random Variable*

Consider the function

$$Z = ae^{bU}, \tag{4.33}$$

where a and b are constants of either sign. The partial derivative is $\partial z/\partial u = abe^{bu}$, and we obtain

$$\frac{\sigma_z}{z} = |b|\sigma_u. \tag{4.34}$$

4.7.6   *Logarithm of a Random Variable*

For the function

$$Z = a\ln(bU), \tag{4.35}$$

where a is a constant of either sign, and $b > 0$. The partial derivative is $\partial z/\partial u = a/U$, leading to

$$\sigma_z = |a|\frac{\sigma_u}{u}. \tag{4.36}$$

A similar result applies for a base-10 logarithm,

$$Z = a\log(bU), \tag{4.37}$$

where a is a constant of either sign, and $b > 0$. The partial derivative is $\partial z/\partial u = a/(U\ln(10))$, leading to

$$\sigma_z = |a|\frac{\sigma_u}{u\ln(10)}. \tag{4.38}$$

Similar error propagation formulas can be obtained for virtually any analytic function for which derivatives can be calculated. Some common formulas are reported for convenience in Table 4.1, where the terms z, u, and v refer to the random variables evaluated at their estimated mean value.

Example 4.13 With reference to Example 4.11, we want to give a quantitative answer to the following question: what is the *probability* that during the second time interval the radioactive source was actually detected? In principle a fluctuation of the number of background counts could give rise to all detected counts.

Table 4.1 Common error propagation formulas

Function	Error propagation formula	Notes
$Z = U + a$	$\sigma_z^2 = \sigma_u^2$	a is a constant
$Z = aU + bV$	$\sigma_z^2 = a^2\sigma_u^2 + b^2\sigma_b^2 + 2ab\sigma_{uv}^2$	a, b are constants
$Z = aUV$	$\dfrac{\sigma_z^2}{z^2} = \dfrac{\sigma_u^2}{u^2} + \dfrac{\sigma_v^2}{v^2} + 2\dfrac{\sigma_{uv}^2}{uv}$	a is a constant
$Z = a\dfrac{U}{V}$	$\dfrac{\sigma_z^2}{z^2} = \dfrac{\sigma_u^2}{u^2} + \dfrac{\sigma_v^2}{v^2} - 2\dfrac{\sigma_{uv}^2}{uv}$	a is a constant
$Z = aU^b$	$\dfrac{\sigma_z}{z} = b\dfrac{\sigma_u}{u}$	a, b are constants
$Z = ae^{bU}$	$\dfrac{\sigma_z}{z} = |b|\sigma_u$	a, b are constants
$Z = a\ln(bU)$	$\sigma_z = |a|\dfrac{\sigma_u}{u}$	a, b are constants, $b > 0$
$Z = a\log(bU)$	$\sigma_z = |a|\dfrac{\sigma_u}{u\ln(10)}$	a, b are constants, $b > 0$

A solution to this question can be provided by stating the problem in a Bayesian way:

$$P(\text{detection}) = P(S_2 > 0/\text{data})$$

where the phrase "data" refers also to the available measurement of the background where $S_2 = N_2 - B$ is the number of source counts. This could be elaborated by stating that the data were used to estimate a mean of 15 and a standard deviation of 7.4 for S_2, and therefore we want to calculate the probability to exceed zero for such random variable. We can use the Central Limit Theorem to say that the sum of two random variables—each approximately distributed as a Gaussian since the number of counts is sufficiently large—is Gaussian, and the probability of a positive detection of the radioactive source therefore becomes equivalent to the probability of a Gaussian-distributed variable to have values larger than approximately $\mu - 2\sigma$. According to Table A.3, this probability is approximately 97.7 %. We can therefore conclude that source were detected in the second time period with such confidence.   ◇

4.8   The Quantile Function and Simulation of Random Variables

In data analysis one often needs to simulate a random variable, that is, drawing random samples from a parent distribution. The simplest such case is the generation of a random number between two limits, which is equivalent to drawing samples from a uniform distribution. In particular, several Monte Carlo methods including the Markov chain Monte Carlo method discussed in Chap. 16 will require random variables with different distributions. Most computer languages and programs do have available a uniform distribution, and thus it is useful to learn how to simulate any distribution based on the availability of a simulator for a uniform variable.

Given a variable X with a distribution $f(x)$ and a cumulative distribution function $F(x)$, we start by defining the *quantile function* $F^{-1}(p)$ as

$$F^{-1}(p) = \min\{x \varepsilon \mathcal{R}, p \le F(x)\} \tag{4.39}$$

with the meaning that x is the minimum value of the variable at which the cumulative distribution function reaches the value $0 \le p \le 1$. The word "minimum" in the definition of the quantile function is necessary to account for those distributions that have steps—or discontinuities—in their cumulative distribution, but in the more common case of a strictly increasing cumulative distribution, the quantile function is simply defined by the relationship $p = F(x)$. This equation can be solved for x, to obtain the quantile function $x = F^{-1}(p)$.

Example 4.14 (Quantile Function of a Uniform Distribution) For a uniform variable in the range 0–1, the quantile function has a particularly simple form. In fact, $F(x) = x$, and the quantile function defined by the equation $p = F(x)$ yields $x = p$, and therefore

$$x = F^{-1}(p) = p. \tag{4.40}$$

Therefore the analytical form of both the cumulative distribution and the quantile function is identical for the uniform variable in 0–1, meaning that, e.g., the value 0.75 of the random variable is the $p = 0.75$, or 75 % quantile of the distribution. ◇

The basic property of the quantile function can be stated mathematically as

$$p \le F(x) \Leftrightarrow x \le F^{-1}(p) \tag{4.41}$$

meaning that the value of $F^{-1}(p)$ is the value x at which the probability of having $X \ge x$ is p.

Example 4.15 (Quantile Function of an Exponential Distribution) Consider a random variable distributed like an exponential,

$$f(x) = \lambda e^{-\lambda x},$$

with $x \ge 0$. Its cumulative distribution function is

$$F(x) = 1 - e^{-\lambda x}.$$

The quantile function is obtained from,

$$p = F(x) = 1 - e^{-\lambda x},$$

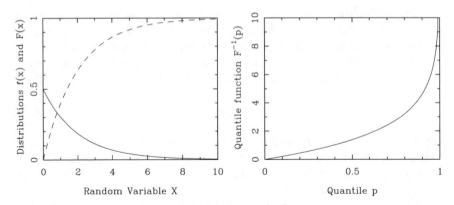

Fig. 4.2 Distribution function $f(x)$, cumulative distribution $F(x)$, and quantile function $F^{-1}(p)$ of an exponential variable with $\lambda = 1/2$

leading to $x = \ln(1 - p)/(-\lambda)$, and therefore the quantile function is

$$x = F^{-1}(p) = \frac{ln(1 - p)}{-\lambda}.$$

Figure 4.2 shows the cumulative distribution and the quantile function for the exponential distribution.   ◇

4.8.1 General Method to Simulate a Variable

The method to simulate a random variable is summarized in the following equation,

$$X = F^{-1}(U), \tag{4.42}$$

which states that any random variable X can be expressed in terms of the uniform variable U between 0 and 1, F is the cumulative distribution of the variable X, and F^{-1} is the quantile function. If a closed analytic form for F is available for that distribution, this equation results in a simple method to simulate the random variable.

Proof We have already seen that for the uniform variable the quantile function is $F^{-1}(U) = U$, i.e., it is the uniform random variable itself. The proof therefore simply consists of showing that, assuming (4.42), then the cumulative distribution of X is indeed $F(X)$, or $P(X \leq x) = F(x)$. This can be shown by writing

$$P(X \leq x) = P(F^{-1}(U) \leq x) = P(U \leq F(x)) = F(x),$$

in which the second equality follows from the definition of the quantile function, and the last equality follows from the fact that $P(U \leq u) = u$, for u a number between 0 and 1, for a uniform variable. □

Example 4.16 (Simulation of an Exponential Variable) Consider a random variable distributed like an exponential, $f(x) = \lambda e^{-\lambda x}$, $x \geq 0$. Given the calculations developed in the example above, the exponential variable can be simulated as

$$X = \frac{\ln(1-U)}{-\lambda}.$$

Notice that, although this relationship is between random variables, its practical use is to draw random samples u from U, and a random sample x from X is obtained by simply using the equation

$$x = \frac{\ln(1-u)}{-\lambda}.$$

Therefore, for a large sample of values u, the above equation returns a random sample of values for the exponential variable X. ◇

Example 4.17 (Simulation of the Square of Uniform Variable) It can be proven that the simulation of the square of a uniform random variable $Y = U^2$ is indeed achieved by squaring samples from a uniform distribution, a very intuitive result.

In fact, we start with the distribution of Y as $g(y) = \frac{1}{2} y^{-1/2}$. Since its cumulative distribution is given by $G(y) = \sqrt{y}$, the quantile function is defined by $p = \sqrt{y}$, or $y = p^2$ and therefore the quantile function for U^2 is

$$y = G^{-1}(p) = p^2.$$

This result, according to (4.42), defines U^2, or the square of a uniform distribution, as the function that needs to be simulated to draw fair samples from Y. ◇

4.8.2 Simulation of a Gaussian Variable

This method of simulation of random variables relies on the knowledge of $F(x)$ and the fact that such a function is analytic and invertible. In the case of the Gaussian distribution, the cumulative distribution function is a special function,

$$F(x) = \frac{1}{2\pi} \int_{-\infty}^{x} e^{-\frac{x^2}{2}} dx$$

which cannot be inverted analytically. Therefore, this method cannot be applied. This complication must be overcome, given the importance of Gaussian distribution in probability and statistics. Fortunately, a relatively simple method is available that permits the simulation of two Gaussian distributions from two uniform random variables.

In Sect. 4.4 we showed that the transformation from Cartesian to polar coordinates results in two random variables R, Θ that are distributed, respectively, like a Rayleigh and a uniform distribution:

$$
\begin{cases}
h(r) = re^{-\frac{r^2}{2}} & r \geq 0 \\
i(\theta) = \dfrac{1}{2\pi} & 0 \leq \theta \leq 2\pi.
\end{cases}
\tag{4.43}
$$

Since these two distributions have an analytic form for their cumulative distributions, R and Θ can be easily simulated. We can then use the transformation given by (4.7) to simulate a pair of independent standard Gaussians. We start with the Rayleigh distribution, for which the cumulative distribution function is

$$
H(r) = 1 - e^{-\frac{r^2}{2}}.
$$

The quantile function is given by

$$
p = 1 - e^{-\frac{r^2}{2}},
$$

and from this we obtain

$$
r = \sqrt{-2\ln(1 - p)} = H^{-1}(p)
$$

and therefore $R = \sqrt{-2\ln(1 - U)}$ simulates a Rayleigh distribution, given the uniform variable U. For the uniform variable Θ, it is clear that the cumulative distribution is given by

$$
I(\theta) = \begin{cases}
\theta/(2\pi) & 0 \leq \theta \leq 2\pi \\
0 & \text{otherwise;}
\end{cases}
$$

the quantile function is $\theta = 2\pi p = I^{-1}(p)$, and therefore $\Theta = 2\pi V$ simulates a uniform distribution between 0 and 2π, with V the uniform distribution between 0 and 1.

Therefore, with the use of two uniform distributions U, V, we can use R and Θ to simulate a Rayleigh and a uniform angular distribution

$$
\begin{cases}
R = \sqrt{-2\ln(1 - U)} \\
\Theta = 2\pi V.
\end{cases}
\tag{4.44}
$$

Then, using the Cartesian-Polar coordinate transformation, we arrive at the formulas needed to simulate a pair of Gaussians X and Y:

$$\begin{cases} X = R\cos(\Theta) = \sqrt{-2\ln(1-U)} \cdot \cos(2\pi V) \\ Y = R\sin(\Theta) = \sqrt{-2\ln(1-U)} \cdot \sin(2\pi V) \end{cases} \tag{4.45}$$

Equations (4.45) can be easily implemented by having available two simultaneous and independent uniform variables between 0 and 1.

Summary of Key Concepts for this Chapter

☐ Linear combination of variables: The formulas for the mean and variance of the linear combination of variables are

$$\begin{cases} \mu = \sum a_i \mu_i \\ \sigma^2 = \sum_{i=1}^{N} a_i^2 \sigma_i^2 + 2 \sum_{i=1}^{N} \sum_{j=i+1}^{N} a_i a_j \sigma_{ij}^2 \end{cases}$$

☐ *Variance of uncorrelated variables*: When variables are uncorrelated the variances add linearly. The variance of the mean of N independent measurements is $\sigma_{\bar{Y}}^2 = \sigma^2/N$.

☐ *Moment generating function*: It is a mathematical function that enables the calculation of moments of a distribution, $M(t) = E[e^{tX}]$.

☐ *Central Limit theorem*: The sum of a large number of independent variables is distributed like a Gaussian of mean equal to the sum of the means and variance equal to the sum of the variances.

☐ *Method of change of variables*: A method to obtain the distribution function of a variable Y that is a function of another variable X, $g(y) = f(x)dx/dy$.

☐ *Law of Large Numbers*: The sum of a large number of random variables with mean μ tends to a constant number equal to μ.

☐ *Error propagation formula*: It is an approximation for the variance of a function of random variables. For a function $x = f(u, v)$ of two uncorrelated variables U and V, the variance of X is given by

$$\sigma_x^2 = \sigma_u^2 \left| \frac{\partial f}{\partial u} \right|^2 + \sigma_v^2 \left| \frac{\partial f}{\partial v} \right|^2$$

☐ *Quantile function*: It is the function $x = F^{-1}(p)$ used to find the value x of a variable that corresponds to a given quantile p.

☐ *Simulation of a Gaussian*: Two Gaussians can be obtained from two uniform random variables U, V via

$$\begin{cases} X = \sqrt{-2\ln(1-U)} \cos(2\pi V) \\ Y = \sqrt{-2\ln(1-U)} \sin(2\pi V) \end{cases}$$

Problems

4.1 Consider the data from Thomson's experiment of Tube 1, from page 23.

(a) Calculate the mean and standard deviation of the measurements of v.
(b) Use the results from Problem 2.3, in which the mean and standard deviation of W/Q and I were calculated, to calculate the approximate values of mean and standard deviation of v using the relevant error propagation formula, assuming no correlation between the two measurements.

This problem illustrates that the error propagation formulas may give different results than direct measurement of the mean and variance of a variable, when the individual measurements are available.

4.2 Calculate the mean, variance, and moment generating function $M(t)$ for a uniform random variable in the range 0–1.

4.3 Consider two uniform independent random variables X, Y in the range -1 to 1.

(a) Determine the distribution function, mean and variance, and the moment generating function of the variables.
(b) We speculate that the sum of the two random variables is distributed like a "triangular" distribution between the range -2 to 2, with distribution function

$$f(x) = \begin{cases} \dfrac{1}{2} + \dfrac{x}{4} & \text{if } -2 \le x \le 0 \\ \dfrac{1}{2} - \dfrac{x}{4} & \text{if } 0 \le x \le 2 \end{cases}$$

Using the moment generation function, prove that the variable $Z = X + Y$ is distributed like the triangular distribution above.

4.4 Using a computer language of your choice, simulate the sum of $N = 100$ uniform variables in the range 0–1, and show that the sampling distribution of the sum of the variables is approximately described by a Gaussian distribution with mean equal to the mean of the N uniform variables and variance equal to the sum of the variances. Use 1,000 and 100,000 samples for each variable.

4.5 Consider the J.J. Thomson experiment of page 23.

(a) Calculate the sample mean and the standard deviation of m/e for Tube 1.
(b) Calculate the approximate mean and standard deviation of m/e from the mean and standard deviation of W/Q and I, according to the equation

$$\frac{m}{e} = \frac{I^2}{2} \frac{Q}{W};$$

Assume that W/Q and I are uncorrelated.

4.6 Use the data provided in Example 4.11. Calculate the probability of a positive detection of source counts S in the first time period (where there are $N_1 = 50$ total counts and $B = 20$ background counts), and the probability that the source emitted ≥ 10 source counts. You will need to assume that the measured variable can be approximated by a Gaussian distribution.

4.7 Consider the data in the Thomson experiment for Tube 1 and the fact that the variables W/Q and I are related to the variable v via the relationship

$$v = \frac{2W}{QI}.$$

Calculate the sample mean and variance of v from the direct measurements of this variable, and then using the measurements of W/Q and I and the error propagation formulas. By comparison of the two estimates of the variance, determine if there is a positive or negative correlation between W/Q and I.

4.8 Provide a general expression for the error propagation formula when three independent random variables are present, to generalize (4.24) that is valid for two variables.

Chapter 5
Maximum Likelihood and Other Methods to Estimate Variables

Abstract In this chapter we study the problem of estimating parameters of the distribution function of a random variable when N observations of the variable are available. We discuss methods that establish what *sample* quantities must be calculated to estimate the corresponding *parent* quantities. This establishes a firm theoretical framework that justifies the definition of the sample variance as an unbiased estimator of the parent variance, and the sample mean as an estimator of the parent mean. One of these methods, the maximum likelihood method, will later be used in more complex applications that involve the fit of two-dimensional data and the estimation of fit parameters. The concepts introduced in this chapter constitute the core of the statistical techniques for the analysis of scientific data.

5.1 The Maximum Likelihood Method for Gaussian Variables

Consider a random variable X distributed like a Gaussian. The probability of making a measurement between x_i and $x_i + dx$ is given by

$$f(x_i)dx = \frac{1}{\sqrt{2\pi\sigma^2}}e^{-\frac{(x_i-\mu)^2}{2\sigma^2}}dx.$$

This probability describes the *likelihood* of collecting the data point x_i given that the distribution has a fixed value of μ and σ. Assume now that N measurements of the random variable have been made. The goal is to estimate the *most likely* values of the true—yet unknown—values of μ and σ, the two parameters that determine the distribution of the random variable. The method of analysis that follows this principle is called the *maximum likelihood method*. The method is based on the postulate that the values of the unknown parameters are those that yield a maximum probability of observing the measured data. Assuming that the measurements are made independently of one another, the quantity

$$P = \prod_{i=1}^{N} P(x_i) = \prod_{i=1}^{N} \frac{1}{\sqrt{2\pi\sigma^2}}e^{-\frac{(x_i - \mu)^2}{2\sigma^2}} \tag{5.1}$$

© Springer Science+Busines Media New York 2017
M. Bonamente, *Statistics and Analysis of Scientific Data*, Graduate Texts
in Physics, DOI 10.1007/978-1-4939-6572-4_5

is the probability of making N independent measurements in intervals of unit length around the values x_i, which can be viewed as the probability of measuring the dataset composed of the given N measurements.

The method of maximum likelihood consists therefore of finding the parameters of the distribution that maximize the probability in (5.1). This is simply achieved by finding the point at which the first derivative of the probability P with respect to the relevant parameter of interest vanishes, to find the extremum of the function. It can be easily proven that the second derivative with respect to the two parameters is negative at the point of extremum, and therefore this is a point of maximum for the likelihood function.

5.1.1 Estimate of the Mean

To find the maximum-likelihood estimate of the mean of the Gaussian distribution we proceed with the calculation of the first derivative of $\ln P$, instead of P, with respect to the mean μ. Given that the logarithm is a monotonic function of the argument, maximization of $\ln P$ is equivalent to that of P, and the logarithm has the advantage of ease of computation. We obtain

$$\frac{\partial}{\partial \mu} \sum_{i=1}^{N} \frac{(x_i - \mu)^2}{2\sigma^2} = 0.$$

The solution is the maximum-likelihood estimator of the mean, which we define as μ_{ML}, and is given by

$$\mu_{ML} = \frac{1}{N} \sum_{i=1}^{N} x_i = \bar{x}. \tag{5.2}$$

This result was to be expected: the maximum likelihood method shows that the "best" estimate of the mean is simply the sample average of the measurements.

The quantity μ_{ML} is a quantity that, despite the Greek letter normally reserved for parent quantities, is a function of the measurements. Although it appears obvious that the sample average is the correct estimator of the true mean, it is necessary to prove this statement by calculating its expectation. It is clear that the expectation of the sample average is in fact

$$E[\bar{x}] = \frac{1}{N} E[x_1 + \ldots + x_N] = \mu,$$

This calculation is used to conclude that the sample mean is an *unbiased* estimator of the true mean.

5.1.2 Estimate of the Variance

Following the same method used to estimate the mean, we can also take the first derivative of $\ln P$ with respect to σ^2, to obtain

$$\frac{\partial}{\partial \sigma^2}\left(N \ln \frac{1}{\sqrt{2\pi\sigma^2}}\right) + \frac{\partial}{\partial \sigma^2}\left(\sum -\frac{1}{2}\frac{(x_i - \mu)^2}{\sigma^2}\right) = 0$$

from which we obtain

$$N\left(-\frac{1}{2}\frac{1}{\sigma^2}\right) + \sum -\frac{1}{2}(x_i - \mu)^2\left(-\frac{1}{\sigma^4}\right) = 0$$

and finally the result that the maximum likelihood estimator of the variance is

$$\sigma_{ML}^2 = \frac{1}{N}\sum_{i=1}^{N}(x_i - \mu)^2. \tag{5.3}$$

It is necessary to notice that in the maximum likelihood estimate of the variance we have implicitly assumed that the mean μ was known, while in reality we can only estimate it as the sample mean, from the same data used also to estimate the variance. To account for the fact that μ is not known, we replace it with $\bar{x}$ in (5.3), and call

$$s_{ML}^2 = \frac{1}{N}\sum_{i=1}^{N}(x_i - \bar{x})^2 \tag{5.4}$$

the *maximum likelihood sample variance estimator*, which differs from the sample variance defined in (2.11) by a factor of $(N-1)/N$. The fact that $\bar{x}$ replaced μ in its definition leads to the following expectation:

$$E[s_{ML}^2] = \frac{N-1}{N}\sigma_{ML}^2. \tag{5.5}$$

Proof Calculation of the expectation is obtained as

$$E[s_{ML}^2] = E[\frac{1}{N}\sum(x_i - \bar{x})^2] = \frac{1}{N}E[\sum(x_i - \mu + \mu - \bar{x})^2]$$

$$= \frac{1}{N}E[\sum(x_i - \mu)^2 + \sum(\mu - \bar{x})^2 + 2(\mu - \bar{x})\sum(x_i - \mu)]$$

The term $E[\sum(\mu - \bar{x})^2]$ is the variance of the sample mean, which we know from Sect. 4.1 to be equal to σ^2/N. The last term in the equation is $\sum(x_i - \mu) = N(\bar{x} - \mu)$, therefore:

$$E[s^2_{ML}] = \frac{1}{N}\left(E[\sum(x_i - \mu)^2] + NE[(\mu - \bar{x})^2] + 2NE[(\mu - \bar{x})(\bar{x} - \mu)]\right)$$

$$= \frac{1}{N}(N\sigma^2 + N\sigma^2/N - 2NE[(\mu - \bar{x})^2])$$

leading to the result that

$$E[s^2_{ML}] = \frac{1}{N}\left(N\sigma^2 + N\sigma^2/N - 2N\sigma^2/N\right) = \frac{N-1}{N}\sigma^2.$$

In this proof we used the notation $\sigma^2 = \sigma^2_{ML}$ □

This result is at first somewhat surprising, since there is an extra factor $(N-1)/N$ that makes $E[s^2_{ML}]$ different from the maximum likelihood estimator of σ^2. This is actually due to the fact that, in estimating the variance, the mean needed to be estimated as well and was not known beforehand. The unbiased estimator of the variance is therefore

$$s^2 = s^2_{ML} \times \frac{N}{N-1} = \frac{1}{N-1}\sum_{i=1}^{N}(x_i - \bar{x})^2 \tag{5.6}$$

for which we have shown that $E[s^2] = \sigma^2$. This is the reason for the definition of the sample variance according to (5.6), and not (5.4).

It is important to pay attention to the fact that (5.6) defines a statistic for which we could also find, in addition to its expectation, also its variance, similar to what was done for the sample mean. In Chap. 7 we will study how to determine the probability distribution function of certain statistics of common use, including the distribution of the sample variance s^2.

Example 5.1 We have already made use of the sample mean and the sample variance as estimators for the parent quantities in the analysis of the data from Thomson's experiment (page 23). The estimates we obtained are unbiased if the assumptions of the maximum likelihood method are satisfied, namely that I and W/Q are Gaussian distributed.   ◇

5.1.3   Estimate of Mean for Non-uniform Uncertainties

In the previous sections we assumed a set of measurements x_i of the same random variable, i.e., the parent mean μ and variance σ^2 were the same. It is often the case with real datasets that observations are made from variables with the same mean, but with different variance. This could be the case when certain measurements are more precise than others, and therefore they feature the same mean (since they are drawn from the same process), but the standard error varies with the precision of the instrument, or because some measurements were performed for a longer period of time. In this case, each measurement x_i is assigned a different standard deviation σ_i, which represents the precision with which that measurement was made.

Example 5.2   A detector is used to measure the rate of arrival of a certain species of particles. One measurement consists of 100 counts in 10 s, another of 180 particles in 20 s, and one of 33 particles in 3 s. The measured count rates would be reported as, respectively, 10.0, 9.0, and 11.0 counts per second. Given that this is a counting experiment, the Poisson distribution applies to each of the measurements. Moreover, since the number of counts is sufficiently large, it is reasonable to approximate the Poisson distribution with a Gaussian, with variance equal to the mean. Therefore the variance of the counts is 100, 180, and 33, and the variance of the count rate can be calculated by the property that $Var[X/t] = Var[X]/t^2$, where t is the known time of each measurement. It follows that the standard deviation σ of the count rates is, respectively, 1.0, 0.67, and 1.91 for the three measurements. The three measurements would be reported as 10.0 ± 1.0, 9.0 ± 0.67, and 11.0 ± 1.91, with the last measurement being clearly of lower precision because of the shorter period of observation. $\diamond$

Our goal is therefore now focused on the maximum likelihood estimate of the parent mean μ, which is the same for all measurements. This is achieved by using (5.1) in which the parent mean of each measurement is μ, and the parent variance of each measurement is σ_i^2. Following the same procedure as in the case of equal standard deviations for each measurement, we start with the probability P of making the N measurements,

$$P = \prod_{i=1}^{N} P(x_i) = \prod_{i=1}^{N} \frac{1}{\sqrt{2\pi\sigma_i^2}} e^{-\frac{(x_i-\mu)^2}{2\sigma_i^2}} . \tag{5.7}$$

Setting the derivative of $\ln P$ with respect to μ—the common mean to all measurements—equal to zero, we obtain that the maximum likelihood estimates is

$$\mu_{ML} = \frac{\sum_{i=1}^{N}(x_i/\sigma_i^2)}{\sum_{i=1}^{N}(1/\sigma_i^2)} \tag{5.8}$$

This is the weighted mean of the measurements, where the weights are the inverse of the variance of each measurement, $1/\sigma_i^2$.

The variance in this weighted sample mean can be calculated using the expectation of the weighted mean, by assuming that the σ_i^2 are constant numbers:

$$Var(\mu_{ML}) = \frac{1}{\left(\sum_{i=1}^{N}(1/\sigma_i^2)\right)^2} \times \sum_{i=1}^{N} \frac{Var(x_i)}{\sigma_i^4} = \frac{\sum_{i=1}^{N}(1/\sigma_i^2)}{\left(\sum_{i=1}^{N}(1/\sigma_i^2)\right)^2},$$

which results in

$$\sigma_\mu^2 = \frac{1}{\sum_{i=1}^{N}(1/\sigma_i^2)}. \tag{5.9}$$

The variance of the weighted mean on (5.9) becomes the usual σ^2/N if all variances σ_i^2 are identical.

Example 5.3 Continuing Example 5.2 of the count rate of particle arrivals, we use (5.8) and (5.9) to calculate a weighted mean and standard deviation of 9.44 and 0.53. Since the interest is just in the overall mean of the rate, the more direct means to obtain this number is by counting a total of 313 counts in 33 s, for an overall measurement of the count rate of 9.48 ± 0.54, which is virtually identical to that obtained using the weighted mean and its variance.   ◇

It is common, as in the example above, to assume that the parent variance σ_i^2 is equal to the value estimated from the measurements themselves. This approximation is necessary, unless the actual precision of the measurement is known beforehand by some other means, for example because the apparatus used for the experiment has been calibrated by prior measurements.

5.2   The Maximum Likelihood Method for Other Distributions

The method of maximum likelihood can also be applied when the measurements do not follow a Gaussian distribution.

A typical case is that of N measurements n_i, $i = 1, \ldots, N$, from a Poisson variable N of parameter λ, applicable to all situations in which the measurements are derived from a counting experiment. In this case, the maximum likelihood method can be used to estimate λ, which is the mean of the random variable, and the only parameter of the Poisson distribution.

The Poisson distribution is discrete in nature and the probability of making N independent measurements is simply given by

$$P = \prod_{i=1}^{N} \frac{\lambda^{n_i}}{n_i!} e^{-\lambda}.$$

It is convenient to work with logarithms,

$$\ln P = \sum_{i=1}^{N} \ln e^{-\lambda} + \sum_{i=1}^{N} \ln \frac{1}{n_i!} + \sum_{i=1}^{N} \ln \lambda^{n_i} = A - N\lambda + \ln \lambda \sum_{i=1}^{N} n_i$$

in which A is a term that doesn't depend on λ. The condition that the probability must be maximum requires $\partial P / \partial \lambda = 0$. This condition results in

$$\frac{1}{\lambda} \sum_{i=1}^{N} x_i - N = 0,$$

and therefore we obtain that the maximum likelihood estimator of the λ parameter of the Poisson distribution is

$$\lambda_{ML} = \frac{1}{N} \sum_{i=1}^{N} n_i.$$

This result was to be expected, since λ is the mean of the Poisson distribution, and the linear average of N measurements is an unbiased estimate of the mean of a random variable, according to the Law of Large Numbers.

The maximum likelihood method can in general be used for any type of distribution, although often the calculations can be mathematically challenging if the distribution is not a Gaussian.

5.3   Method of Moments

The method of moments takes a more practical approach to the estimate of the parameters of a distribution function. Consider a random variable X for which we have N measurements and whose probability distribution function f(x) depends on M unknown parameters, for example $\theta_1 = \mu$ and $\theta_2 = \sigma^2$ for a Gaussian ($M = 2$), or $\theta_1 = \lambda$ for an exponential ($M = 1$), etc. The idea is to develop a method that yields as many equations as there are free parameters and solve for the parameters of the distribution. The method starts with the determination of arbitrary functions $a_j(x), j = 1, \ldots M$, that make the distribution function integrable:

$$E[a_j(X)] = \int_{-\infty}^{\infty} a_j(x) f(x) dx = g_j(\theta) \tag{5.10}$$

where $g_j(\theta)$ is an analytic function of the parameters of the distribution. Although we have assumed that the random variable is continuous, the method can also be applied to discrete distributions. According to the law of large numbers, the left-hand side of (5.10) can be approximated by the sample mean of the function of the N measurements, and therefore we obtain a linear system of M equations:

$$\frac{1}{N}(a_j(x_1) + \ldots + a_j(x_N)) = g_j(\theta) \tag{5.11}$$

which can be solved for the parameters $\boldsymbol{\theta}$ as function of the N measurements x_i.

As an illustration of the method, consider the case in which the parent distribution is a Gaussian of parameters μ, σ^2. First, we need to decide which functions $a_1(x)$ and $a_2(x)$ to choose. A simple and logical choice is to use $a_1(x) = x$ and $a_2(x) = x^2$; this choice is what gives the name of "moments," since the right-hand side of (5.10) will be, respectively, the first and second order moment. Therefore we obtain the two equations

$$\begin{cases} E[a_1(X)] = \dfrac{1}{N}(X_1 + \ldots + X_N) = \mu \\[2mm] E[a_2(X)] = \dfrac{1}{N}(X_1^2 + \ldots + X_N^2) = \sigma^2 + \mu^2. \end{cases} \tag{5.12}$$

The estimator for mean and variance are therefore

$$\begin{cases} \mu_{MM} = \dfrac{1}{N}(X_1 + \ldots + X_N) \\[2mm] \sigma_{MM}^2 = \dfrac{1}{N}(X_1^2 + \ldots + X_N^2) - \left(\dfrac{1}{N}(X_1 + \ldots + X_N)\right)^2 = \dfrac{1}{N}\sum(x_i - \mu_{MM})^2 \end{cases} \tag{5.13}$$

which, in this case, are identical to the estimates obtained from the likelihood method. This method is often easier computationally than the method of maximum likelihood, since it does not require the maximization of a function, but just a careful choice of the integrating functions $a_j(x)$. Also, notice that in this application we did not make explicit use of the assumption that the distribution is a Gaussian, since the same results will apply to *any* distribution function with mean μ and variance σ^2. Equation (5.13) can therefore be used in a variety of situations in which the distribution function has parameters that are related to the mean and variance, even if they are not identical to them, as in the case of the Gaussian. The method of moments therefore returns unbiased estimates for the mean and variance of every distribution in the case of a large number of measurements.

Example 5.4 Consider the five measurements presented in Example 4.9: 0.1, 0.3, 0.5, 0.7, and 0.9, and assume that they are known to be drawn from a uniform distribution between 0 and a. The method of moments can be used to estimate the parameter a of the distribution from the measurements. The probability distribution function is $f(x) = 1/a$ between 0 and a, and null otherwise. Using the integrating function $a_1(x) = x$, the method of moments proceeds with the calculation of the first moment of the distribution,

$$E[X] = \int_0^a xf(x)dx = \frac{a}{2}$$

Therefore, using (5.12), we can estimate the only parameter a of the distribution function as

$$a = 2 \times \frac{1}{N} \sum_{i=1}^{N} x_i$$

where $N = 5$, for the result of $a = 1$. The result confirms that the five measurements are compatible with a parent mean of $1/2$.   ◇

5.4   Quantiles and Confidence Intervals

The parameters of the distribution function can be used to determine the range of values that include a given probability, for example, 68.3 %, or 90 %, or 99 %, etc. This range, called *confidence interval*, can be conveniently described by the cumulative distribution function $F(x)$.

Define the α-quantile x_α, where α is a number between 0 and 1, as the value of the variable such that $x \le x_\alpha$ with probability α:

$$\alpha \text{ quantile } x_\alpha : P(x \le x_\alpha = \alpha) \text{ or } F(x_\alpha) = \alpha. \tag{5.14}$$

For example, consider the cumulative distribution shown in Fig. 5.1 (right panel): the $\alpha = 0.05$ quantile is the number of the variable x where the lower horizontal dashed line intersects the cumulative distribution $F(x)$, $x_\alpha \simeq 0.2$, and the $\beta = 0.95$ quantile is the number of the variable x where the upper dashed line intersects $F(x)$, $x_\beta \simeq 6$. Therefore the range x_α to x_β, or 0.2–6, corresponds to the $(\beta - \alpha) = 90\%$ confidence interval, i.e., there is 90 % probability that a measurement of the variable falls in that range. These confidence intervals are called *central* because they are centered at the mean (or median) of the distribution, and are the most commonly used type of confidence intervals.

Confidence intervals can be constructed at any confidence level desired, depending on applications and on the value of probability that the analyzer wishes to include in that interval. It is common to use 68 % confidence intervals because this is the probability between $\pm\sigma$ of the mean for a Gaussian variable (see Sect. 5.4.1). Normally a confidence interval or limit at a significance lower than 68 % is not considered interesting, since there is a significant probability that the random variable will be outside of this range.

One-sided confidence intervals that extends down to $-\infty$, or to the lowest value allowed for that random variable, is called an *upper limit*, and intervals that extend

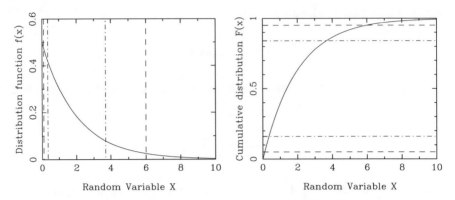

Fig. 5.1 (*Left*) Distribution function of an exponential variable with central 68 and 90 % confidence interval marked by, respectively, *dot-dashed* and *dotted lines*. (*Right*) The confidence interval are obtained as the intersection of *dot-dashed* and *dotted lines* with the cumulative distribution (*solid line*)

to $+\infty$, or to the highest allowed value, is called a *lower limit*. A lower limit describes a situation in which a large number is detected, for example counts from a Poisson experiment, and we want to describe how small the value of the variable can be, and still be consistent with the data. An upper limit is used for a situation in which a small number is detected, to describe how high can the variable be and still be consistent with the data. Lower and upper limits depend on the value of the probability that we want to use; for example, using a value for α that is closer to 0 results in a lower limit that progressively becomes $\lambda_{lo} = -\infty$ (or lowest allowed value), which is not a very interesting statement. If β is progressively closer to 1, the upper limit will tend to $\lambda_{up} = \infty$.

5.4.1   Confidence Intervals for a Gaussian Variable

When the variable is described by a Gaussian distribution function we can use integral tables (Table A.2) to determine confidence intervals that enclose a given probability. It is usually meaningful to have central confidence intervals, i.e., intervals centered at the mean of the distribution and extending by equal amounts on either side of the mean. For central confidence intervals, the relationship between the probability p enclosed by a given interval (say $p = 0.9$ or 90 % confidence interval) and the size $\Delta x = 2(z \times \sigma)$ of the interval is given by

$$p = \int_{\mu - z\sigma}^{\mu + z\sigma} f(x)dx, \tag{5.15}$$

where $f(x)$ is a Gaussian of mean μ and variance σ^2. The number z represents the number of standard deviations allowed by the interval in each direction (positive and negative relative to the mean). The most common central confidence intervals for a Gaussian distribution are reported in Table 5.1. For example, for a mean μ and variance σ^2 and the interval from $\mu - \sigma$ to $\mu + \sigma$ is a 68.3 % confidence interval, the interval from $\mu - 1.65\sigma$ to $\mu + 1.65\sigma$ is a 90 % confidence interval. In principle, one could have confidence intervals that are not centered on the mean of the distribution—such intervals would still be valid confidence intervals. It can be shown that central confidence intervals are the smallest possible, for a given confidence level.

Example 5.5 Using the data for the J.J. Thomson experiment on the measurement of the electron's mass-to-charge ratio, we can calculate the 90 % confidence interval on m/e for Tube 1 and Tube 2. For Tube 1, we estimated the mean as $\mu_1 = 0.42$ and the standard error as $\sigma_1 = 0.07$, and for Tube 2 $\mu_2 = 0.53$ and $\sigma = 0.08$. Since the random variable is assumed to be Gaussian, the 90 % confidence interval corresponds to the range between $\mu - 1.65\sigma$ and $\mu + 1.65\sigma$; therefore for the Thomson measurements of Tube 1 and Tube 2, the 90 % central confidence intervals are, respectively, 0.30–0.54 and 0.40–0.66. ◇

Upper and lower limits can be easily calculated using the estimates of μ and σ for a Gaussian variable. They are obtained numerically from the following relationships,

$$p = \int_{-\infty}^{\lambda_{up}} f(x)dx = F(\lambda_{up}) \qquad \text{upper limit } \lambda_{up}$$

$$p = \int_{\lambda_{lo}}^{\infty} f(x)dx = 1 - F(\lambda_{lo}) \qquad \text{lower limit } \lambda_{lo}$$

(5.16)

making use of Tables A.2 and A.3. The quantites $F(\lambda_{up})$ and $F(\lambda_{lo})$ are the values of the cumulative distribution of the Gaussian, showing that λ_{up} is the p-quantile and λ_{lo} is the $(1 - p)$-quantile of the distribution. Useful upper and lower limits for the Gaussian distribution are reported in Table 5.2. Upper limits are typically of interest when the measurements result in a low value of the mean of the variable. In this case we usually want to know how high the variable can be and still be

Table 5.1 Common confidence intervals for a Gaussian distribution

Interval	Range	Enclosed probability (%)
50 % confidence interval	$\mu - 0.68\sigma, \mu + 0.68\sigma$	50
1-σ interval	$\mu - \sigma, \mu + \sigma$	68.3
90 % confidence interval	$\mu - 1.65\sigma, \mu + 1.65\sigma$	90
2-σ interval	$\mu - 2\sigma, \mu + 2\sigma$	95.5
3-σ interval	$\mu - 3\sigma, \mu + 3\sigma$	99.7

Table 5.2 Common upper and lower limits for a Gaussian distribution

Upper limit	Range	Enclosed probability (%)	Lower limit	Range	Enclosed probability (%)
50 % confidence	$\leq \mu$	50	50 % confidence	$\geq \mu$	50
90 % confidence	$\leq \mu + 1.28\sigma$	90	90 % confidence	$\geq \mu - 1.28\sigma$	90
95 % confidence	$\leq \mu + 1.65\sigma$	95	95 % confidence	$\geq \mu - 1.65\sigma$	95
99 % confidence	$\leq \mu + 2.33\sigma$	99	99 % confidence	$\geq \mu - 2.33\sigma$	99
1-σ	$\leq \mu + \sigma$	84.1	1-σ	$\geq \mu - \sigma$	84.1
2-σ	$\leq \mu + 2\sigma$	97.7	2-σ	$\geq \mu - 2\sigma$	97.7
3-σ	$\leq \mu + 3\sigma$	99.9	3-σ	$\geq \mu - 3\sigma$	99.9

consistent with the measurement, at a given confidence level. For example, in the case of the measurement of Tube 1 for the Thomson experiment, the variable m/e was measured to be 0.42 ± 0.07. In this case, it is interesting to ask the question of how high can m/e be and still be consistent with the measurement at a given confidence level.

Example 5.6 Using the data for the J.J. Thomson experiment on the measurement of the electron's mass-to-charge ratio, we can calculate the 90 % upper limits to m/e for Tube 1 and Tube 2. For Tube 1, we estimated the mean as $\mu_1 = 0.42$ and the standard error as $\sigma_1 = 0.07$, and for Tube 2 $\mu_2 = 0.53$ and $\sigma = 0.08$.

To determine the upper limit $m/e_{UL,90}$ of the ratio, we calculate the probability of occurrence of $m/e \leq m/e_{UL,90}$:

$$P(m/e \leq m/e_{UL,90}) = 0.90$$

Since the random variable is assumed to be Gaussian, the value $x \simeq \mu + 1.28\sigma$ corresponds to the 90 percentile of the distribution (see Table 5.2). The two 90 % upper limits are, respectively, 0.51 and 0.63.   ◇

A common application of upper limits is when an experiment has failed to detect the variable of interest. In this case we have a *non-detection* and we want to place upper limits based on the measurements made. This problem is addressed by considering the parent distribution of the variable that we did not detect, for example a Gaussian of zero mean and given variance. We determine the upper limit as the value of the variable that exceeds the mean by 1, 2 or 3 σ, corresponding to the probability levels shown in Table 5.2. A 3-σ upper limit, for example, is the value of the variable that has only a 0.1 % chance of being observed based on the parent distribution for the non-detection, and therefore we are 99.9 % confident that the true value of the variable is lower than this upper limit.

Example 5.7 (Gaussian Upper Limit to the Non-detection of a Source) A measurement of $n = 8$ counts in a given time interval is made in the presence of a source of unknown intensity. The instrument used for the measurement has a background level with a mean of 9.8 ± 0.4 counts, as estimated from an independent experiment of long duration. Given that the measurement is below the expected background level, it is evident that there is no positive detection of the source. The hypothesis that the source has zero emission can be described by a distribution function with a mean of approximately 9.8 counts and, since this is a counting experiment, the probability distribution of counts should be Poisson. We are willing to approximate the distribution with a Gaussian of same mean, and variance equal to the mean, or $\sigma \simeq 3.1$, to describe the distribution of counts one expects from an experiment of the given duration as the one that yielded $n = 8$ counts.

A 99 % upper limit to the number of counts that can be recorded by this instrument, in the given time interval, can be calculated according to Table 5.2 as

$$\mu + 2.33\sigma = 9.8 + 2.33 \times 3.1 \simeq 17.$$

This means that we are 99 % confident that the true value of the source plus background counts is less than 17. A complementary way to interpret this number is that the experimenter can be 99 % sure that the measurement *cannot* be due to just the background if there is a detection of ≥ 17 total counts. A conservative analyst might also want to include the possibility that the Gaussian distribution has a slightly higher mean, since the level of the background is not known exactly, and conservatively assume that perhaps 18 counts are required to establish that the source does have a positive level of emission. After subtraction of the assumed background level, we can conclude that the 99 % upper limit to the source's true emission level in the time interval is 8.2 counts. This example was adapted from the analysis of an astronomical source that resulted in a non-detection [4].　　　◇

5.4.2 Confidence Intervals for the Mean of a Poisson Variable

The Poisson distribution does not have the simple analytical properties of the Gaussian distribution. For this distribution it is convenient to follow a different method to determine its confidence intervals.

Consider the case of a single measurement of a Poisson variable of unknown mean λ for which n_{obs} was recorded. We want to make inferences on the parent mean based on this information. Also, we assume that the measurement includes a uniform and known background λ_B. The measurement is therefore drawn from a random variable

$$X = N_S + N_B \tag{5.17}$$

in which $N_B = \lambda_B$ is assumed to be a constant, i.e., the background is known exactly (this generalization can be bypassed by simply setting $\lambda_B = 0$). The probability distribution function of X (the total source plus background counts) is

$$f(n) = \frac{(\lambda + \lambda_B)^n}{n!} e^{-(\lambda + \lambda_B)}, \tag{5.18}$$

where n is an integer number describing possible values of X. Equation (5.18) is true even if the background is not known exactly, since the sum of two Poisson variables is also a Poisson variable with mean equal to the sum of the means. It is evident that, given the only measurement available, the estimate of the source mean is

$$\hat{\lambda} = n_{obs} - \lambda_B.$$

This estimate is the starting point to determine a confidence interval for the parent mean. We define the lower limit λ_{lo} as the value of the source mean that results in the observation of $n \geq n_{obs}$ with a probability α:

$$\alpha = \sum_{n=n_{obs}}^{\infty} \frac{(\lambda_{lo} + \lambda_B)^n}{n!} e^{-(\lambda_{lo} + \nu_B)} = 1 - \sum_{n=0}^{n_{obs}-1} \frac{(\lambda_{lo} + \lambda_B)^n}{n!} e^{-(\lambda_{lo} + \lambda_B)}. \tag{5.19}$$

The mean λ_{lo} corresponds to the situation shown in the left panel of Fig. 5.2: assuming that the actual mean is as low as λ_{lo}, there is only a small probability α (say 5 %) to make a measurement above or equal to what was actually measured. Thus, we can say that there is only a very small chance (α) that the actual mean could have been as low (or lower) than λ_{lo}. The quantity λ_{lo} is the *lower limit with confidence* $(1 - \alpha)$, i.e., we are $(1 - \alpha)$, say 95 %, confident that the mean is higher than this value.

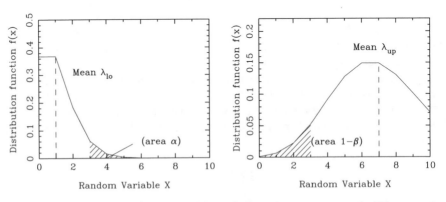

Fig. 5.2 This illustration of the upper and lower limits to the measurement of a Poisson mean assumes a measurement of $n_{obs} = 3$. On the *left*, the lower limit to the parent mean is such that there is a probability of α to measure n_{obs} or higher (*hatched area*); on the *right*, the upper limit leaves a probability of $(1 - \beta)$ that a measurement is n_{obs} or lower

By the same logic we also define λ_{up} as the parent value of the source mean that results in the observation of $n \leq n_{obs}$ with a probability $(1 - \beta)$, or

$$1 - \beta = \sum_{n=0}^{n_{obs}} \frac{(\lambda_{up} + \lambda_B)^n}{n!} e^{-(\lambda_{up} + \lambda_B)}. \tag{5.20}$$

This is illustrated in the right panel of Fig. 5.2, where the number $(1 - \beta)$ is intended as a small number, of same magnitude as α. Assuming that the mean is as high as λ_{up}, there is a small probability of $1 - \beta$ to make a measurement equal or lower than the actual measurement. Therefore we say that there is only a small probability that the true mean could be as high or higher than λ_{up}. The number λ_{up} is the *upper limit with confidence* β, that is, we are β (say 95 %) confident that the mean is lower than this value.

If we combine the two limits, the probability that the true mean is above λ_{up} or below λ_{lo} is just $(1 - \beta) + \alpha$, say 10 %, and therefore the interval λ_{lo} to λ_{up} includes a probability of

$$P(\lambda_{lo} \leq \lambda \leq \lambda_{up}) = 1 - (1 - \beta) - \alpha = \beta - \alpha,$$

i.e., this is a $(\beta - \alpha)$, say 90 %, confidence interval.

The upper and lower limits defined by (5.19) and (5.20) can be approximated analytically using a relationship that relates the Poisson sum with an analytic distribution function:

$$\sum_{x=0}^{n_{obs}-1} \frac{e^{-\lambda} \lambda^x}{x!} = 1 - P_{\chi^2}(\chi^2, \nu) \tag{5.21}$$

where $P_{\chi^2}(\chi^2, \nu)$ is the cumulative distribution of the χ^2 probability distribution function defined in Sect. 7.2, with parameters $\chi^2 = 2\lambda$ and $\nu = 2n_{obs}$,

$$P_{\chi^2}(\chi^2, \nu) = \int_{-\infty}^{\chi^2} f_{\chi^2}(x, \nu) dx.$$

The approximation is due to Gehrels [16], and makes use of mathematical relationships that can be found in the handbook of Abramowitz and Stegun [1]. The result is that the upper and lower limits can be simply approximated once we specify the number of counts n_{obs} and the probability level of the upper or lower limit. The probability level is described by the number S, which is the equivalent number of Gaussian σ that corresponds to the confidence level chosen (for example, 84 %

confidence interval corresponds to $S = 1$, etc. see Table 5.3)

$$\begin{cases} \lambda_{up} = n_{obs} + \dfrac{S^2 + 3}{4} + S\sqrt{n_{obs} + \dfrac{3}{4}} \\ \lambda_{lo} = n_{obs}\left(1 - \dfrac{1}{9n_{obs}} - \dfrac{S}{3\sqrt{n_{obs}}}\right)^3. \end{cases} \tag{5.22}$$

The S parameter is also a quantile of a standard Gaussian distribution, enclosing a probability as illustrated in Table 5.3.

Proof Use of (5.21) into (5.19) and (5.20) gives a relationship between the function P_{χ^2} and the probability levels α and β,

$$\begin{cases} P_{\chi^2}(2\lambda_{lo}, 2n_{obs}) = \alpha \\ P_{\chi^2}(2\lambda_{up}, 2n_{obs} + 2) = \beta. \end{cases} \tag{5.23}$$

We use the simplest approximation for the function P_{χ^2} described in [16], one that is guaranteed to give limits that are accurate within 10 % of the true values. The approximation makes use of the following definitions: for any probability $a \leq 1$, y_a is the a-quantile of a standard normal distribution, or $G(y_a) = a$,

$$G(y_a) = \frac{1}{\sqrt{2\pi}} \int_{-\infty}^{y_a} e^{-t^2/2} dt.$$

If $P_{\chi^2}(\chi_a^2, \nu) = a$, then the simplest approximation between χ_a^2 and y_a given by Gehrels [16] is

$$\chi_a^2 \simeq \frac{1}{2}\left(y_a + \sqrt{2\nu - 1}\right)^2. \tag{5.24}$$

Consider the upper limit in (5.23). We can solve for λ_{up} by using (5.24) with $2\lambda_{up} = \chi_a^2$, $\nu = 2n_{obs} + 2$ and $S = y_a$, since y_a is the a-quantile of a standard

Table 5.3 Poisson parameters S and corresponding probabilities

Upper or lower limit	Range	Probability (%)	Poisson S parameter
90 % confidence	$\leq \mu + 1.28\sigma$	90	1.28
95 % confidence	$\leq \mu + 1.65\sigma$	95	1.65
99 % confidence	$\leq \mu + 2.33\sigma$	99	2.33
1-σ	$\leq \mu + \sigma$	84.1	1.0
2-σ	$\leq \mu + 2\sigma$	97.7	2.0
3-σ	$\leq \mu + 3\sigma$	99.9	3.0

normal distribution, thus equivalent to S. It follows that

$$2\lambda_{up} = \frac{1}{2}\left(S + \sqrt{4n_{obs} + 3}\right)^2$$

and from this the top part of (5.22) after a simple algebraic manipulation.
A similar result applies for the lower limit.   □

Equation (5.22) is tabulated in Tables A.5 and A.6 for several interesting values
of n_{obs} and S. A few cases of common use are also shown in Table 5.4.

Example 5.8 (Poisson Upper Limit to Non-detection with No Background) An
interesting situation that can be solved analytically is that corresponding to the
situation in which there was a complete *non-detection* of a source, $n_{obs} = 0$.
Naturally, it is not meaningful to look for a lower limit to the Poisson mean, but it is
quite interesting to solve (5.20) in search for an upper limit with a given confidence
β. In this case of $n = 0$ the equation simplifies to

$$1 - \beta = e^{-(\lambda_{up} + \lambda_B)} \quad \Rightarrow \quad \lambda_{up} = -\lambda_B - \ln\beta.$$

For $\beta = 0.84$ and zero background ($\lambda_B = 0$) this corresponds to an upper limit of
$\lambda_{up} = -\ln 0.16 = 1.83$. This example can also be used to test the accuracy of the
approximation given by (5.22). Using $n_{obs} = 0$, we obtain

$$\lambda_{up} = \frac{1}{4}(1 + \sqrt{3})^2 = 1.87$$

which is in fact just 2 % higher than the exact result. An example of upper limits in
the presence of a non-zero background is presented in Problem 5.8.   ◇

Table 5.4 Selected Upper and Lower limits for a Poisson variable using the Gehrels approximation (see Tables A.5 and A.6 for a complete list of values)

n_{obs}	Poisson parameter S or confidence level		
	$S = 1$	$S = 2$	$S = 3$
	(1-σ, or 84.1 %)	(2-σ, or 97.7 %)	(3-σ, or 99.9 %)
Upper limit			
0	1.87	3.48	5.60
1	3.32	5.40	7.97
...			
Lower limit			
...			
9	6.06	4.04	2.52
10	6.90	4.71	3.04
...			

5.5   Bayesian Methods for the Poisson Mean

The Bayesian method consists of determining the *posterior probability* $P(\lambda/obs)$, having calculated the likelihood as

$$P(n_{obs}/\lambda) = \frac{(\lambda + \lambda_B)^{n_{obs}}}{n_{obs}!}e^{-(\lambda+\lambda_B)}. \tag{5.25}$$

We use Bayes' theorem,

$$P(\lambda/obs) = \frac{P(n_{obs}/\lambda)\pi(\lambda)}{\int_0^\infty P(n_{obs}/\lambda')\pi(\lambda'), d\lambda'} \tag{5.26}$$

in which we needed to introduce a *prior probability* distribution $\pi(\lambda)$ in order to calculate the posterior probability. The use of a prior distribution is what constitutes the Bayesian approach. The simplest assumption is that of a uniform prior, $\pi(\lambda) = C$, over an arbitrarily large range of $\lambda \geq 0$, but other choices are possible according to the information available on the Poisson mean prior to the measurements. In this section we derive the Bayesian expectation of the Poisson mean and upper and lower limits and describe the differences with the classical method.

5.5.1   Bayesian Expectation of the Poisson Mean

The posterior distribution of the Poisson mean λ (5.26) can be used to calculate the Bayesian expectation for the mean can be calculated as the integral of (5.26) over the entire range allowed to the mean,

$$E[\lambda/obs] = \frac{\int_0^\infty P(\lambda/obs)\lambda d\lambda}{\int_0^\infty P(\lambda/obs)d\lambda}. \tag{5.27}$$

The answer will in general depend on the choice of the prior distribution $\pi(\lambda)$. Assuming a constant prior, the expectation becomes

$$E[\lambda/obs] = \int_0^\infty \frac{e^{-\lambda}\lambda^{n_{obs}+1}}{n!}d\lambda = n_{obs} + 1, \tag{5.28}$$

where we made use of the integral

$$\int_0^\infty e^{-\lambda}\lambda^n d\lambda = n!$$

The interesting result is therefore that a measurement of n_{obs} counts implies a Bayesian expectation of $E[\lambda] = n_{obs} + 1$, i.e., one count more than the observation. Therefore even a non-detection results in an expectation for the mean of the parent distribution of 1, and not 0. This somewhat surprising result can be understood by considering the fact that even a parent mean of 1 results in a likelihood of $1/e$ (i.e., a relatively large number) of obtaining zero counts as a result of a random fluctuations. Moreover, the Poisson distribution is skewed, with a heavier tail at large values of λ. This calculation is due to Emslie [12].

5.5.2 Bayesian Upper and Lower Limits for a Poisson Variable

Using a uniform prior, we use the Bayesian approach (5.26) to calculate the upper limit to the source mean with confidence β (say, 95 %). This is obtained by integrating (5.26) from the lower limit of 0 to the upper limit λ_{up},

$$\beta = \frac{\int_0^{\lambda_{up}} P(n_{obs}/\lambda)d\lambda}{\int_0^\infty P(n_{obs}/\lambda)d\lambda} = \frac{\int_0^{\lambda_{up}} (\lambda + \lambda_B)^{n_{obs}} e^{-(\lambda+\lambda_B)} d\lambda}{\int_0^\infty (\lambda + \lambda_B)^{n_{obs}} e^{-(\lambda+\lambda_B)} d\lambda} \tag{5.29}$$

Similarly, the lower limit can be estimated according to

$$\alpha = \frac{\int_0^{\lambda_{lo}} P(n_{obs}/\lambda)d\lambda_S}{\int_0^\infty P(n_{obs}/\lambda)d\lambda} = \frac{\int_0^{\lambda_{lo}} (\lambda + \lambda_B)^{n_{obs}} e^{-(\lambda+\lambda_B)} d\lambda}{\int_0^\infty (\lambda + \lambda_B)^{n_{obs}} e^{-(\lambda+\lambda_B)} d\lambda} \tag{5.30}$$

where α is a small probability, say 5 %. Since n_{obs} is always an integer, these integrals can be evaluated analytically.

The difference between the classical upper limits described by (5.19) and (5.20) and the Bayesian limits of (5.29) and (5.30) is summarized by the different variable of integration (or summation) in the relevant equations. For the classical limits we use the Poisson probability to make n_{obs} measurements for a true mean of λ. We then estimate the upper or lower limits as the values of the mean that gives a probability of, respectively, $1 - \beta$ and α, to observe $n \leq n_{obs}$ events. In this case, the probability is evaluated as a sum over the number of counts, for a fixed value of the parent mean.

In the case of the Bayesian limits, on the other hand, we first calculate the posterior distribution of λ, and then require that the range between 0 and the limits λ_{up} and λ_{lo} includes, respectively, a $1 - \beta$ and α probability, evaluated as an integral over the mean for a fixed value of the detected counts. In general, the two methods will give different results.

Example 5.9 (Bayesian Upper Limit to a Non-detection) The case of non-detection, $n_{obs} = 0$, is especially simple and interesting, since the background drops out of the equation, resulting in $\beta = 1 - e^{-\lambda_{up}}$, which gives

$$\lambda_{up} = -\ln(1 - \beta) \text{ (case of } n_{obs} = 0, \text{ Bayesian upper limit)} \tag{5.31}$$

The Bayesian upper limit is therefore equal to the classical limit, when there is no background. When there is background, the two estimate will differ.[1]   ◇

Summary of Key Concepts for this Chapter

☐ *Maximum Likelihood (ML) method*: A method to estimate parameters of a distribution under the assumption that the best-fit parameters maximize the likelihood of the measurements.

☐ *ML estimates of mean and variance*: For a Gaussian variable, the unbiased ML estimates are

$$\begin{cases} \mu_{ML} = \bar{x} \\ s^2 = \dfrac{1}{N-1} \sum (x_i - \bar{x})^2 \end{cases}$$

☐ *Estimates of mean with non-uniform uncertainties*: They are given by

$$\begin{cases} \mu_{ML} = \dfrac{\sum x_i/\sigma_i^2}{\sum 1/\sigma_i^2} \\ \sigma_\mu^2 = \dfrac{1}{\sum 1/\sigma_i^2} \end{cases}$$

☐ *Confidence intervals*: Range of the variable that contains a given probability of occurrence (e.g., $\pm 1\sigma$ range contains 68 % of probability for a Gaussian variable).

☐ *Upper and lower limits*: An upper (lower) limit is the value below (above) which there is a given probability (e.g., 90 %) to observe the variable.

Problems

5.1 Using the definition of weighted sample mean as in (5.8), derive its variance and show that it is given by (5.9).

5.2 Using the data from Mendel's experiment (Table 1.1), calculate the standard deviation in the measurement of each of the seven fractions of dominants, and the weighted mean and standard deviation of the seven fractions.

Compare your result from a direct calculation of the overall fraction of dominants, obtained by grouping all dominants from the seven experiments together.

---

[1]Additional considerations on the measurements of the mean of a Poisson variable, and the case of upper and lower limits, can be found in [10].

5.3 The Mendel experiment of Table 1.1 can be described as n number n of measurements of n_i, the number of plants that display the dominant character, out of a total of N_i plants. The experiment is described by a binomial distribution with probability $p = 0.75$ for the plant to display the dominant character.

Using the properties of the binomial distribution, show analytically that the weighted average of the measurements of the fraction $f_i = n_i/N_i$ is equal to the value calculated directly as

$$\mu = \frac{\sum_{i=1}^n n_i}{\sum_{i=1}^n N_i}$$

5.4 Consider a decaying radioactive source observed in a time interval of duration $T = 15\,s$; N is the number of total counts, and B is the number of background counts (assumed to be measured independently of the total counts):

$$\begin{cases} N = 19 & \text{counts} \\ B = 14 & \text{counts} \end{cases}.$$

The goal is to determine the probability of detection of source counts $S = N - B$ in the time interval T.

(a) Calculate this probability directly via:

$$\text{Prob(detection)} = \text{Prob}(S > 0/\text{data})$$

in which S is treated as a random variable, with Gaussian distribution of mean and variance calculated according to the error propagation formulas. Justify why the Gaussian approximation *may* be appropriate for the variable S.

(b) Use the same method as in (a), but assuming that the background B is known without error (e.g., as if it was observed for such along time interval that its error becomes negligible).

(c) Assume that the background is a variable with mean of 14 counts in a 15 s interval, and that it can be observed for an interval of time $T \gg 15\,s$. Find what interval of time T makes the error σ_{B15} of the background over a time interval of 15-s have a value $\sigma_{B15}/B_{15} = 0.01$, e.g., negligible.

5.5 For the Thomson experiment of Table 2.1 (tube 1) and Table 2.2 (tube 2), calculate:

(a) The 90 % central confidence intervals for the variable v;
(b) The 90 % upper and lower limits, assuming that the variable is Gaussian.

5.6 Consider a Poisson variable X of mean μ.

(a) We want to set 90 % confidence upper limits to the value of the parent mean λ, assuming that one measurement of the variable yielded the result of $N = 1$. Following the classical approach, find the equation that determines the exact

90 % upper limit to the mean λ_{up}. Recall that the classical 90 % confidence upper limit is defined as the value of the Poisson mean that yields a $P(X \leq N) = \beta$, where $1 - \beta = 0.9$.

(b) Using the Bayesian approach, which consists of defining the $1 - \beta = 0.9$ upper limit via

$$\beta = \frac{\int_0^{\lambda_{up}} P(n_{obs}/\mu)d\mu}{\int_0^{\infty} P(n_{obs}/\mu)d\mu} \tag{5.32}$$

where $n_{obs} = N$; find the equation that determines the 90 % upper limit to the mean λ_{up}.

5.7 The data provided in Table 2.3 from Pearson's experiment on biometric data describes the cumulative distribution function of heights from a sample of 1,079 couples. Calculate the 2σ upper limit to the fraction of couples in which both mother and father are taller than 68 in.

5.8 Use the data presented in Example 5.7, in which there is a non-detection of a source in the presence of a background of $\lambda_B \simeq 9.8$. Determine the Poisson upper limit to the source count at the 99 % confidence level and compare this upper limit with that obtained in the case of a zero background level.

Chapter 6
Mean, Median, and Average Values of Variables

Abstract The data analyst often faces the question of what is the "best" value to report from N measurements of a random variable. In this chapter we investigate the use of the linear average, the weighted average, the median and a logarithmic average that may be applicable when the variable has a log-normal distribution. The latter may be useful when a variable has errors that are proportional to their measurements, avoiding the inherent bias arising in the weighted average from measurements with small values and small errors. We also introduce a relative-error weighted average that can be used as an approximation for the logarithmic mean for log-normal distributions.

6.1 Linear and Weighted Average

In the previous chapter (see Sect. 5.1.3) we have shown that the weighted mean is the most likely value of the mean of the random variable. Therefore, the weighted mean is a commonly accepted quantity to report as the best estimate for the value of a measured quantity. If the measurements have the same standard deviation, then the weighted mean becomes the linear average; in general, the linear and weighted means differ unless all measurement errors are identical.

The difference between linear average and weighted mean can be illustrated with an example. Consider the $N = 25$ measurements shown in Table 6.1, which reports the measurement of the energy of certain astronomical sources made at a given radius [5]. This dataset is illustrative of the general situation of the measurement of a quantity (in this example, the ratio between the two measurements) in the presence of different measurement error. The weighted mean is 0.90 ± 0.02, while the linear average is 1.01 (see Problem 6.1). The difference is clearly due to the presence of a few measurements with a low value of the ratio that carry higher weight because of the small measurement error (for example, source 15).

Which of the two values is more representative? This question can be addressed by making the following observations. The measurement error reported in the table reflects the presence of such sources of uncertainty as Poisson fluctuations in the detection of photons from the celestial sources. The same type of uncertainty would also apply to other experiments, in particular those based on the counting of events.

© Springer Science+Busines Media New York 2017
M. Bonamente, *Statistics and Analysis of Scientific Data*, Graduate Texts
in Physics, DOI 10.1007/978-1-4939-6572-4_6

Table 6.1 Dataset with measurement of energy for $N = 25$ different sources and their ratio

Source	Radius	Energy		Ratio
		Method #1	Method #2	
1	$221.1\pm^{11.0}_{12.3}$	$8.30\pm^{0.76}_{0.88}$	$9.67\pm^{1.14}_{1.12}$	$0.86\pm^{0.08}_{0.07}$
2	$268.5\pm^{22.1}_{20.7}$	$4.92\pm^{0.77}_{0.70}$	$4.19\pm^{0.82}_{0.70}$	$1.17\pm^{0.16}_{0.15}$
3	$138.4\pm^{12.7}_{11.9}$	$3.03\pm^{0.53}_{0.49}$	$2.61\pm^{0.59}_{0.49}$	$1.16\pm^{0.20}_{0.18}$
4	$714.3\pm^{23.5}_{34.5}$	$49.61\pm^{3.15}_{3.19}$	$60.62\pm^{4.84}_{6.13}$	$0.82\pm^{0.06}_{0.05}$
5	$182.3\pm^{18.5}_{15.1}$	$2.75\pm^{0.49}_{0.43}$	$3.30\pm^{0.81}_{0.61}$	$0.83\pm^{0.14}_{0.14}$
6	$72.1\pm^{5.5}_{5.7}$	$1.01\pm^{0.23}_{0.20}$	$0.86\pm^{0.14}_{0.13}$	$1.17\pm^{0.24}_{0.21}$
7	$120.3\pm^{8.6}_{7.5}$	$5.04\pm^{0.66}_{0.57}$	$3.80\pm^{0.72}_{0.57}$	$1.33\pm^{0.16}_{0.15}$
8	$196.2\pm^{15.1}_{15.5}$	$5.18\pm^{0.73}_{0.70}$	$6.00\pm^{1.17}_{1.11}$	$0.86\pm^{0.14}_{0.11}$
9	$265.7\pm^{8.7}_{8.6}$	$12.17\pm^{1.22}_{1.17}$	$10.56\pm^{0.93}_{0.95}$	$1.14\pm^{0.13}_{0.10}$
10	$200.0\pm^{9.6}_{10.7}$	$7.74\pm^{0.57}_{0.58}$	$6.26\pm^{0.78}_{0.83}$	$1.24\pm^{0.14}_{0.11}$
11	$78.8\pm^{5.6}_{5.1}$	$1.08\pm^{0.16}_{0.15}$	$0.73\pm^{0.11}_{0.10}$	$1.49\pm^{0.26}_{0.24}$
12	$454.4\pm^{20.3}_{20.3}$	$17.10\pm^{2.64}_{2.03}$	$23.12\pm^{2.36}_{2.32}$	$0.75\pm^{0.07}_{0.06}$
13	$109.4\pm^{8.3}_{8.3}$	$3.31\pm^{0.34}_{0.34}$	$3.06\pm^{0.54}_{0.52}$	$1.09\pm^{0.18}_{0.15}$
14	$156.5\pm^{11.5}_{10.2}$	$2.36\pm^{0.61}_{0.58}$	$2.31\pm^{0.36}_{0.31}$	$1.02\pm^{0.26}_{0.23}$
15	$218.0\pm^{6.6}_{5.9}$	$14.02\pm^{0.75}_{0.75}$	$21.59\pm^{1.82}_{1.82}$	$0.65\pm^{0.04}_{0.04}$
16	$370.7\pm^{7.6}_{8.0}$	$31.41\pm^{1.56}_{1.56}$	$29.67\pm^{1.56}_{1.57}$	$1.06\pm^{0.06}_{0.06}$
17	$189.1\pm^{16.4}_{15.4}$	$2.15\pm^{0.45}_{0.39}$	$2.52\pm^{0.57}_{0.51}$	$0.86\pm^{0.22}_{0.18}$
18	$150.5\pm^{4.2}_{4.6}$	$3.39\pm^{0.57}_{0.50}$	$4.75\pm^{0.44}_{0.46}$	$0.72\pm^{0.11}_{0.11}$
19	$326.7\pm^{12.1}_{9.9}$	$15.73\pm^{1.43}_{1.30}$	$18.03\pm^{1.54}_{1.26}$	$0.87\pm^{0.06}_{0.06}$
20	$189.1\pm^{9.9}_{9.1}$	$5.04\pm^{0.65}_{0.55}$	$4.61\pm^{0.61}_{0.50}$	$1.09\pm^{0.12}_{0.12}$
21	$147.7\pm^{8.0}_{11.1}$	$2.53\pm^{0.29}_{0.30}$	$2.76\pm^{0.37}_{0.48}$	$0.93\pm^{0.12}_{0.10}$
22	$504.6\pm^{12.5}_{11.2}$	$44.97\pm^{2.99}_{2.74}$	$43.93\pm^{3.08}_{2.59}$	$1.02\pm^{0.05}_{0.05}$
23	$170.5\pm^{8.6}_{8.1}$	$3.89\pm^{0.30}_{0.29}$	$3.93\pm^{0.49}_{0.42}$	$0.98\pm^{0.10}_{0.09}$
24	$297.6\pm^{13.1}_{13.6}$	$10.78\pm^{1.04}_{1.02}$	$10.48\pm^{1.34}_{1.22}$	$1.04\pm^{0.10}_{0.11}$
25	$256.2\pm^{13.4}_{14.4}$	$7.27\pm^{0.81}_{0.77}$	$7.37\pm^{0.97}_{0.95}$	$0.99\pm^{0.09}_{0.09}$

This type of uncertainty is usually referred to as *statistical error*. Many experiments and measurements are also subject to other sources of uncertainty that may not be explicitly reported in the dataset. For example, the measurement of events recorded by a detector is affected by the calibration of the detector, and a systematic offset in the calibration would affect the numbers recorded. In the case of the data of Table 6.1, the uncertainty due to the calibration of the detector is likely to affect by the same amount of all measurements, regardless of the precision indicated by the statistical error. This type of uncertainty is typically referred to as *systematic error*, and the inclusion of such additional source of uncertainty would modify the value of the weighted mean. As an example of this effect, if we add an error of ±0.1 to all values of the ratio of Table 6.1, the weighted mean becomes 0.95 ± 0.04 (see Problem 6.2). It is clear that the addition of a constant error for each measurement causes a de-weighting of datapoints with small statistical errors, and in the limit of a large systematic error the weighted mean becomes the linear average. Therefore, the linear average can be used when the data analyst wants to weigh equally all datapoints, regardless of the precision indicated by the statistical errors. Systematic errors are discussed in more detail in Chap. 11.

6.2 The Median

Another quantity that can be calculated from the N measurements is the median, defined in Sect. 2.3.1 as the value of the variable that is greater than 50 % of the measurements, and also lower than 50 % of the measurements. In the case of the measurement of the ratios in Table 6.1, this is simply obtained by ordering the 25 measurements in ascending order, and using the 13th measurement as an approximation for the median. The value obtained in this case is 1.02, quite close to the value of the linear average, since both statistics do not take into account the measurement errors.

One useful feature of the median is that it is not very sensitive to "outliers" in the distribution. For example, if one of the measurements was erroneously reported as 0.07 ± 0.01 (instead of 0.72 ± 0.11, such as source 18 in the Table), both linear and weighted averages would be affected by the error, but the median would not. The median may therefore be an appropriate value to report in cases where the analyst suspects the presence of outliers in the dataset.

6.3 The Logarithmic Average and Fractional or Multiplicative Errors

The quantity "Ratio" in Table 6.1 can be used to illustrate a type of variables that may require a special attention when calculating their averages. Consider a variable whose errors are proportional to their measured values. In this case, a weighted average will be skewed towards *lower* values because of the smaller errors in those measurements. The question we want address is whether it is appropriate to use a weighted average of these measurements or whether one should use a different approach.

To illustrate this situation, let's use two measurements such as $x_1 = 1.2 \pm 0.24$ and $x_2 = 0.80 \pm 0.16$. Both measurements have a relative error of 20 %, the linear average is 1.00 and the weighted average is 0.923. The base-10 logarithm of these measurements are $\log x_1 = 0.0792$ and $\log x_2 = -0.0969$, with the same error. In fact, using the error propagation method (Sect. 4.7.6), the error in the logarithm is proportional to the fractional error according to

$$\sigma_{\log x} = \frac{\sigma_x}{x} \frac{1}{\ln 10}. \tag{6.1}$$

For our measurements, this equation gives a value of $\sigma_{\log x} = 0.087$ for both measurements. The weighted average of these logarithms is therefore the linear average $\overline{\log x} = -0.0088$, leading to an average of $\bar{x} = 0.980$. This value is much closer to the linear average of 1.00 than to the weighted average.

Errors that are exactly proportional to the measurement, or

$$\sigma_x = x\sigma_r \tag{6.2}$$

may be called *fractional* or *multiplicative errors*. The quantity σ_r is the relative error and it remains constant for purely multiplicative errors. In most cases, including that of Table 6.1, the relative error σ_x/x varies among the measurements, and therefore (6.2) applies only as an approximation. In the following we investigate when it is in fact advisable to use the logarithm of measurements, instead of the measurements themselves, to obtain a more accurate determination of the mean of a variable that has multiplicative errors.

6.3.1   The Weighted Logarithmic Average

The maximum likelihood method applied to the logarithm of measurements of a variable X can be used to estimate the mean and the error of $\log X$. The weighted logarithmic average of N measurements x_i is defined as

$$\overline{\log x} = \frac{\sum_{i=1}^{N} \dfrac{\log x_i}{\sigma_{\log x_i}^2}}{\sum_{i=1}^{N} \dfrac{1}{\sigma_{\log x_i}^2}} \tag{6.3}$$

where $\sigma_{\log x_i}^2$ is the variance of the logarithm of the measurements, which can be obtained from (6.1). The uncertainty in the weighted logarithmic average is given by

$$\sigma_{\log x}^2 = \frac{1}{\sum_{i=1}^{N} \dfrac{1}{\sigma_{\log x_i}^2}}. \tag{6.4}$$

The use of this logarithmic average is justified when the variable X has a log-normal distribution, i.e., when $\log X$ has a Gaussian distribution, rather than the variable X itself. An example of a log-normal variable is illustrated in Fig. 6.1. In this case, the maximum likelihood method estimator of the mean of $\log X$ is the logarithmic mean of (6.3). Clearly, a variable can only be log-normal when the variable has positive values, such as the ratio of two positive quantities. The

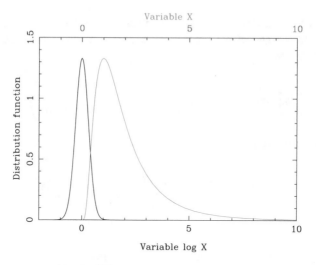

Fig. 6.1 Log-normal distribution with mean $\mu = 0$ and standard deviation $\sigma = 0.3$ (*black line*) and linear plot of the same distribution (*red line*). A heavier right-hand tail in the linear plot may be indicative of a log-normal distribution

determination of the log-normal shape can be made if one has available random samples from its distribution.

In the limit of measurements with the same fractional error and small deviations from the mean μ, the weighted logarithmic average is equivalent to the linear average.

Proof This can be shown by proving that

$$\overline{\log x} = \log \bar{x}$$

where $\bar{x}$ is the ordinary linear average. Notice that $\log x$ in (6.3) is a base-10 logarithm. In this proof we make use of the base-e logarithm ($\ln x$), the two are related by

$$\log x = \ln x / \ln 10.$$

Consider N measurements x_i in the neighborhood of the mean μ of the random variable, $x_i = \mu + \Delta x_i$. A Taylor series expansion yields

$$\ln x_i = \ln \mu (1 + \frac{\Delta x_i}{\mu}) = \ln \mu + \frac{\Delta x_i}{\mu} - \frac{(\Delta x_i/\mu)^2}{2} + \ldots$$

If the deviation is $\Delta x_i \ll \mu$, one can neglect terms of the second order and higher. The average of the logarithms of the N measurements can thus be approximated as

$$\overline{\log x} = \frac{1}{N} \sum_{i=1}^{N} \log x_i \simeq \frac{1}{\ln 10} \left(\ln \mu + \frac{1}{N} \sum_{i=1}^{N} \frac{\Delta x_i}{\mu} \right).$$

On the other hand, the logarithm of the mean $\bar{x}$ is

$$\log \bar{x} = \log \frac{1}{N} \sum_{i=1}^{N} x_i = \log \left(\frac{1}{N} \sum_{i=1}^{N} \mu(1 + \frac{\Delta x_i}{\mu}) \right).$$

This leads to

$$\log \bar{x} = \frac{1}{\ln 10} \left(\ln \mu + \ln \left(1 + \frac{1}{N} \sum_{i=1}^{N} \frac{\Delta x_i}{\mu} \right) \right) \simeq$$

$$\frac{1}{\ln 10} \left(\ln \mu + \frac{1}{N} \sum_{i=1}^{N} \frac{\Delta x_i}{\mu} \right) = \overline{\log x}$$

where we retained only the first-order term in the Taylor series expansion of the logarithm since $\sum \Delta x_i / \mu \ll N$.   □

As discussed earlier in this section, the logarithmic average is an appropriate quantity for log-normal distributed variables. The results of this section show that this average is closer to the linear average of the measurements than the standard weighted average, when measurement errors are positively correlated to the measurements themselves.

Example 6.1 The data of Table 6.1 can be used to calculate the logarithmic average of the column "Ratio" according to (6.3) and (6.4) as $\overline{\log x} = -0.023 \pm 0.018$. These quantities can be converted easily to linear quantities taking into account the error propagation formula $\sigma_{\log x} = \sigma/(x \ln 10)$, to obtain a value of 0.95 ± 0.04.

Notice how the logarithmic mean has a value that is somewhat between that of the linear average $\bar{x} = 1.01$ and the traditional weighted average of 0.90 ± 0.02. It should not be surprising that the logarithmic mean is not exactly equal to the linear average. In fact, the measurements of Table 6.1 have different relative errors. Only in the case of identical relative errors for all measurements we expect that the two averages have the same value.   ◇

6.3.2 The Relative-Error Weighted Average

Although transforming measurements to their logarithms is a simple procedure, we also want to investigate another type of average that deals directly with the measurements without the need to calculate their logarithms.

We introduce the *relative-error weighted average* as

$$\overline{x_{RE}} = \frac{\sum_{i=1}^{N} x_i/(\sigma_i/x_i)^2}{\sum_{i=1}^{N} 1/(\sigma_i/x_i)^2}. \tag{6.5}$$

The only difference with the weighted mean defined in Sect. 5.1.3 is the use of the extra factor of x_i in the error term, so that σ_i/x_i is the relative error of each measurement.

The reason to introduce this new average is that, for log-normal variables, this relative-error weighted mean is equivalent to the logarithmic mean of (6.3). This can be proven by showing that $\overline{\ln x} = \ln \overline{x_{RE}}$.

Proof Start with the logarithm of the relative-error weighted average,

$$\ln \overline{x_{RE}} = \ln\left(\frac{\sum_{i=1}^{N} x_i/(\sigma_i/x_i)^2}{\sum_{i=1}^{N} 1/(\sigma_i/x_i)^2}\right) = \ln\left(\frac{\sum_{i=1}^{N} x_i/\sigma_{\log x_i}^2}{\sum_{i=1}^{N} 1/\sigma_{\log x_i}^2}\right).$$

From this, expand the measurement term $x_i = \mu + \Delta x_i$, where μ is the parent mean of the variable X,

$$\ln\left(\mu + \frac{\sum_{i=1}^{N} \Delta x_i/\sigma_{\log x_i}^2}{\sum_{i=1}^{N} 1/\sigma_{\log x_i}^2}\right) = \ln \mu + \ln\left(1 + \frac{\sum_{i=1}^{N} \Delta x_i/(\mu\sigma_{\log x_i}^2)}{\sum_{i=1}^{N} 1/\sigma_{\log x_i}^2}\right).$$

If $\Delta x_i \ll \mu$, then

$$\ln \overline{x_{RE}} = \ln \mu + \frac{\sum_{i=1}^{N} \Delta x_i/(\mu\sigma_{\log x_i}^2)}{\sum_{i=1}^{N} 1/\sigma_{\log x_i}^2}$$

leading to

$$\log \overline{x_{RE}} = \log \mu + \frac{1}{\ln 10}\frac{\sum_{i=1}^{N} \Delta x_i/(\mu\sigma_{\log x_i}^2)}{\sum_{i=1}^{N} 1/\sigma_{\log x_i}^2}.$$

The logarithmic average can also be expanded making use of

$$\sum_{i=1}^{N} \frac{\log x_i}{\sigma_{\log x_i}^2} = \sum_{i=1}^{N} \frac{\log \mu + \log(1 + \Delta x_i)/\mu}{\sigma_{\log x_i}^2} \simeq \sum_{i=1}^{N} \left(\frac{\log \mu}{\sigma_{\log x_i}^2} + \frac{\Delta x_i/\mu}{\sigma_{\log x_i}^2 \ln 10}\right).$$

This leads to

$$\overline{\log x} = \log \mu + \frac{1}{\ln 10} \frac{\sum_{i=1}^{N} \Delta x_i / (\mu \sigma_{\log x_i}^2)}{\sum_{i=1}^{N} 1/\sigma_{\log x_i}^2} = \log \overline{x_{RE}}.$$

□

The use of the relative-error weighted average should be viewed as an ad hoc method to obtain an average value that is consistent with the logarithmic average, especially in the limit measurements with equal relative errors. The statistical uncertainty in this error-weighted average can be simply assigned as the error in the traditional weighted average (5.8). In fact, the statistical error should be determined by the "physical" uncertainties in the measurements, as is the case for the variance in (5.8). It would be tempting to use the inverse of the denominator of (6.5) as the variance; however, the result would be biased by our somewhat arbitrary choice of weighing the measurements by the relative errors, instead of the error themselves.

Example 6.2 Continuing with the values of "Ratio" in Table 6.1, the error-weighted average is calculated as $\overline{x_{RE}} = 0.96$. The error in the traditional weighted average was 0.02, therefore we may report the result as 0.96 ± 0.02. Comparison with the values of 0.95 ± 0.04 for the logarithmic average shows the general agreement between these two values.

◇

Summary of Key Concepts for this Chapter

□ *Linear average*: The mean $\bar{x}$ of N measurements.
□ *Median*: The 50 % quantile, or the number below and above which there are 50 % of the variable's values.
□ *Logarithmic average*: In some cases (e.g., when errors are proportional to the measured values) it is meaningful to calulate the weighted average of the logarithm of the variable,

$$\begin{cases} \overline{\log x} = \dfrac{\sum \log x_i / \sigma_{logx_i}^2}{\sum 1/\sigma_{logx_i}^2} \\ \sigma_{\log x}^2 = \dfrac{1}{\sigma_{logx_i}^2} \end{cases}$$

where $\sigma_{logx_i} = \sigma_i / (x_i \ln 2)$.
□ *Relative-error weighted average*: An approximation of the logarithmic average that does not require logarithms,

$$\overline{x_{RE}} = \frac{\sum x_i / (\sigma_i / x_i)^2}{\sum 1/(\sigma_i / x_i)^2}.$$

Problems

6.1 Calculate the linear average and the weighted mean of the quantity "Ratio" in Table 6.1.

6.2 Consider the 25 measurements of "Ratio" in Table 6.1. Assume that an additional uncertainty of ± 0.1 is to be added linearly to the statistical error of each measurement reported in the table. Show that the addition of this source of uncertainty results in a weighted mean of 0.95 ± 0.04.

6.3 Given two measurements x_1 and x_2 with values in the neighborhood of 1.0, show that the logarithm of the average of the measurements is approximately equal to the average of the logarithms of the measurements.

6.4 Given two measurements x_1 and x_2 with values in the neighborhood of a positive number A, show that the logarithm of the average of the measurements is approximately equal to the average of the logarithms of the measurements.

6.5 For the data in Table 6.1, calculate the linear average, weighted average and median of each quantity (Radius, Energy Method 1, Energy Method 2 and Ratio). You may assume that the error of each measurements is the average of the asymmetric errors of each measurement reported in the table.

6.6 Table 6.1 contains the measurement of the thermal energy of certain sources using two independent methods labeled as method #1 and method #2. For each source, the measurement is made at a given radius, which varies from source to source. The error bars indicate the 68 %, or 1σ, confidence intervals; the fact that most are asymmetric indicate that the measurements do not follow exactly a Gaussian distribution. Calculate the weighted mean of the ratios between the two measurements and its standard deviation, assuming that the errors are Gaussian and equal to the average of the asymmetric errors, as it is often done in this type of situation.

Chapter 7
Hypothesis Testing and Statistics

Abstract Every quantity that is estimated from the data, such as the mean or the variance of a Gaussian variable, is subject to statistical fluctuations of the measurements. For this reason they are referred to as a *statistics*. If a different sample of measurements is collected, statistical fluctuations will certainly give rise to a different set of measurements, even if the experiments are performed under the same conditions. The use of different data samples to measure the same statistic results in the determination of the *sampling distribution* of the statistic, to describe what is the expected range of values for that quantity. In this chapter we derive the distribution of a few fundamental statistics that play a central role in data analysis, such as the χ^2 statistic. The distribution of each statistic can be used for a variety of tests, including the acceptance or rejection of the fit to a model.

7.1 Statistics and Hypothesis Testing

In this book we have already studied several quantities that are estimated from the data, such as the sample mean and the sample variance. These quantities are subject to random statistical fluctuations that occur during the measurement and collection process and they are often referred to as random variables or *statistics*. For example, a familiar statistic is the sample mean of a variable X. Under the hypothesis that the variable X follows a Gaussian distribution of mean μ and variance σ^2, the sample mean of N measurements is Gaussian-distributed with mean μ and variance equal to σ^2/N (see Sect. 4.1.2). This means that different samples of size N will in general give rise to different sample means and that ones expects a variance of order σ^2/N among the various samples. This knowledge lets us establish whether a given sample mean is consistent with this theoretical expectation.

Hypothesis testing is the process that establishes whether the measurement of a given statistic, such as the sample mean, is consistent with its theoretical distribution. Before describing this process in detail, we illustrate with the following example the type of statistical statement that can be made from a given measurement and the knowledge of its parent distribution.

Example 7.1 Consider the case of the measurement of the ratio m/e from Tube 1 of Thomson's experiment, and arbitrarily assume (this assumption will be relaxed

© Springer Science+Busines Media New York 2017
M. Bonamente, *Statistics and Analysis of Scientific Data*, Graduate Texts in Physics, DOI 10.1007/978-1-4939-6572-4_7

in more realistic applications) that the parent mean is known to be equal to $\mu = 0.475$, and that the parent variance is $\sigma = 0.075$. We want to make quantitative statements regarding the possibility that the measurements are drawn from the parent distribution.

For example, we can make the following statements concerning the measurement $m/e = 0.42$: since $m/e = \mu - 0.73\sigma$, there is a probability of 24% that a measurement of 0.42 or lower is recorded. This statement addresses the fact that, despite the measurement fell short of the parent mean, there is still a significant (24%) chance that any given measurement will be that low, or even lower. We can also make this statement: the measurement is within the $1 - \sigma$ central confidence interval, which encompasses 68% of the probability. This statement looks at the distance of the measurement from the mean, regardless of its sign.

Before we can say: the measurement is consistent with the parent distribution, we need to quantify the meaning of the word *consistent*.   ◇

The process of hypothesis testing requires a considerable amount of care in the definition the hypothesis to test and in drawing conclusions. The method can be divided into the following four steps.

1. Begin with the definition of a hypothesis to test. For the measurements of a variable X, a possible hypothesis is that the measurements are consistent with a parent mean of $\mu = 0$ and a variance of $\sigma^2 = 1$. For a fit of a dataset to a linear model (Chap. 8) we may want to test whether the linear model is a constant, i.e., whether the parent value of the slope coefficient is $b = 0$. This initial step in the process identifies a so-called *null hypothesis* that we want to test with the available data.

2. The next step is to determine the statistic to use for the null hypothesis. In the example of the measurements of a variable X, the statistic we can calculate from the data is the sample mean. For the fit to the linear model, we will learn that the χ^2_{min} is the statistic to use for a Gaussian dataset. The choice of statistic means that we are in a position to use the theoretical distribution function for that statistic to tell whether the actual measurements are consistent with its expected distribution, according to the null hypothesis.

3. Next we need to determine a probability or *confidence level* for the agreement between the statistic and its expected distribution under the null hypothesis. This level of confidence p, say $p = 0.9$ or 90%, defines a range of values for the statistics that are consistent with its expected distribution. We will refer to this range as the *acceptable region* for the statistic. For example, a standard Gaussian of zero mean and unit variance has 90% of its values in the range from -1.65 to $+1.65$. For a confidence level of $p = 0.9$, the analyst would require that the measurement must fall within this range. The choice of probability p is somewhat arbitrary: some analysts may choose 90%, some may require 99.99%, some may even be satisfied with 68%, which is the probability associated with $\pm 1\sigma$ for a Gaussian distribution. Values of the statistics outside of the acceptable range define the *rejection region*. For the standard Gaussian, the rejection region at $p = 0.9$ consists of values ≥ 1.65 and values ≤ -1.65, i.e., the rejection region

is two-sided, as obtained from

$$P\{|S| \geq \overline{S}\} = 1 - \int_{-\overline{S}}^{\overline{S}} f(s)ds = 1 - p \tag{7.1}$$

where $f(s)$ is the probability distribution of the statistic (in this example the standard Gaussian) and $\overline{S}$ is the critical value of the statistic at the level of confidence p. For two-sided rejection regions such as this, where large values of the absolute value of the statistic S are not acceptable, the null hypothesis can be summarized as

$$H_0 = \{\text{The statistic has values } |S| \leq \overline{S}\}$$

Here we have assumed that the acceptable region is centered at 0, but other choices are also possible.

   In other cases of interest, such as for the χ^2 distribution, the rejection region is one-sided. The critical value at confidence level p for the statistic can be found from

$$P\{S \geq \overline{S}\} = \int_{\overline{S}}^{\infty} f(s)ds = 1 - p \tag{7.2}$$

where $f(s)$ is the probability distribution function of the statistic S. For one-sided rejection regions where large values of the statistic are not acceptable the null hypothesis can now be summarized as

$$H_0 = \{\text{The statistic has values } S \leq \overline{S}\}.$$

Clearly p and $\overline{S}$ are related: the larger the value of the probability p, the larger the value of $\overline{S}$, according to (7.1) and (7.2). Larger values of p, such as $p = 0.9999$, increase the size of the acceptable region and reduce the size of the rejection region.

   In principle, other choices for the acceptable and rejection regions are possible, such as multiple intervals or intervals that are not centered at zero. The corresponding critical value(s) of the statistic can be calculated using expression similar to the two reported above. The majority of cases for the rejection region are, however, either a one-sided interval extending to infinity or a two-sided region centered at zero.

4. Finally we are in a position to make a quantitative and definitive statement regarding the null hypothesis. Since we have partitioned the range of the statistic into an acceptable region and a rejection region, only two cases are possible:

   • *Case 1*: The measured value of the statistic S falls into the rejection region. This means that the distribution function of the statistic of interest, under the null hypothesis, does not allow the measured value at the confidence level p.

In this case the null hypothesis *must be rejected* at the stated confidence level p. The rejection of the null hypothesis means that the data should be tested for alternative hypotheses and the procedure can be repeated.

- *Case 2*: The measured value of the statistic S is within the acceptable region. This means that there is a reasonable probability that the measured value of the statistic is consistent with the null hypothesis. In that case the null hypothesis *cannot be rejected*, i.e., the null hypothesis could be true. In this case one can state that the null hypothesis or the underlying model is consistent with the data. Sometimes this situation can be referred to as the null hypothesis being *acceptable*. This is, however, not the same as stating that the null hypothesis *is* the correct hypothesis and that the null hypothesis is accepted. In fact, there could be other hypotheses that could be acceptable and one cannot be certain that the null hypothesis tested represents the parent model for the data.

Example 7.2 Consider $N = 5$ independent measurements of a random variable X, namely $x_i = (10, 12, 15, 11, 13)$. We would like to test the hypothesis that the measurements are drawn from a Gaussian random variable with $\mu = 13$ and $\sigma^2 = 2$).

Next we need to determine the test statistic that we want to use. Since there are N independent measurements of the same variable, we can consider the sum of all measurements as the statistic of interest,

$$Y = \sum_{i=1}^{5} X_i,$$

which is distributed like a Gaussian $N(N \cdot \mu, N \cdot \sigma^2) = N(65, 10)$. We could have chosen the average of the measurements instead. It can be proven that the results of the hypothesis testing are equivalent for the two statistics.

The next step requires the choice of a confidence level for our hypothesis. Assume that we are comfortable with a value of $p = 95\%$ level. This means that the rejection region includes values that are $\pm 1.96\sigma$ (or ± 6.2 units) away from the parent mean of $\mu = 65$, as shown by the cross-hatched are in Fig. 7.1.

Next, we calculate the value of the statistic as $Y = 61$, and realize that the measured value *does not* fall within the region of rejection. We conclude that the data are consistent with the hypothesis that the measurements are drawn from the parent Gaussian at the 95\% probability level (or 1.96σ level).

Assume next that another analyst is satisfied with a $p = 68\%$ probability, instead of 95\%. This means that the region of rejection will be $\pm 1.0\sigma = 1.0 \cdot \sqrt{10} = 3.2$ away from the mean. In this case, the rejection region becomes the hatched area in Fig. 7.1, and the measured value of the test statistic Y falls in the rejection region. In this case, we conclude that *the hypothesis must be rejected at the 68\% probability level (or at the 1σ level)*. $\diamond$

The example above illustrates the importance of the choice of the confidence level p—the same null hypothesis can be acceptable or must be rejected depending

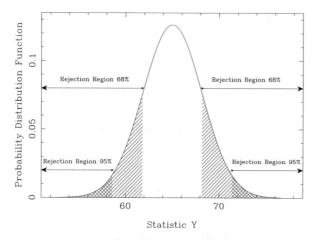

Fig. 7.1 Rejection regions at $p = 0.95$ and $p = 0.68$ confidence level for the test of the Gaussian origin of measurements $x_i = (10, 12, 15, 11, 13)$. The null hypothesis is that the sum of the measurements are drawn from a random variable $Y \sim N(\mu = 65, \sigma^2 = 10)$

on its value. To avoid this ambiguity, some analysts prefer to take a post-facto approach to the choice of p. In this example, the measured value of the sample mean corresponds to an absolute value of the deviation of 1.26σ from the parent mean. Such deviation corresponds to a probability of approximately 79 % to exceed the parent mean. It is therefore possible to report this result with the statement that the data are consistent with the parent model at the 79 % confidence level. In general, for a two-dimensional rejection region, the measurement S_{data} corresponds to a level of confidence p via

$$P\{S \geq |S_{\text{data}}|\} = 1 - \int_{-S_{\text{data}}}^{S_{\text{data}}} f(s)ds = 1 - p, \tag{7.3}$$

where $f(s)$ is the probability distribution of the test statistic under the null hypothesis (an equivalent expression applies to a one-sided rejection region). This equation can be used to make the statement that the measurement of S_{data} is consistent with the model at the p confidence level.

It is necessary to discuss further the meaning of the word "acceptable" with regard to the null hypothesis. The fact that the measurements were within 1-σ of a given mean does not imply that the parent distribution of the null hypothesis *is* the correct one; in fact, there could be other parent distributions that are equally well "acceptable." Therefore, any null hypothesis can only be conclusively disproved (if the measurements were beyond, say, 3- or 5-σ of the parent mean, depending on the choice of probability p), but *never conclusively proven* to be the correct one, since this would imply exhausting and discarding all possible alternative hypotheses. The process of hypothesis testing is therefore slanted towards trying to disprove the null hypothesis, possibly in favor of alternative hypotheses. The rejection of the null hypothesis is the only outcome of the hypothesis testing process that is conclusive, in that it requires to discard the hypothesis.

7.2   The χ^2 Distribution

Consider N random variables X_i, each distributed like a Gaussian with mean μ_i, variance σ_i^2, and independent of one other. For each variable X_i, the associated z-score

$$Z_i = \frac{X_i - \mu_i}{\sigma_i}$$

is a standard Gaussian of zero mean and unit variance. We are interested in finding the distribution function of the random variable given by the sum of the square of all the deviations,

$$Z = \sum_{i=1}^{N} Z_i^2. \tag{7.4}$$

This quantity will be called a χ^2-distributed variable.

The reason for our interest in this distribution will become apparent from the use of the maximum likelihood method in fitting two-dimensional data (Chap. 8). In fact, the sum of the squares of the deviations of the measurements from their mean,

$$\chi^2 = \sum_{i=1}^{N} \left(\frac{x_i - \mu_i}{\sigma_i} \right)^2,$$

represents a measure of how well the measurements follow the expected values μ_i.

7.2.1   The Probability Distribution Function

The theoretical distribution of Z is obtained by making use of the Gaussian distribution for its components. To derive the distribution function of Z, we first prove that the moment generating function of the square of each Gaussian Z_i is given by

$$M_{Z_i^2}(t) = \sqrt{\frac{1}{1 - 2t}}. \tag{7.5}$$

This result enables the comparison with the moment generating function of another distribution and the determination of the distribution function of Z.

Proof The moment generating function of the square of a standard Gaussian Z_i is given by

$$M_{Z_i^2}(t) = E[e^{Z_i^2 t}] = \int_{-\infty}^{+\infty} e^{x^2 t} \frac{1}{\sqrt{2\pi}} e^{-\frac{x^2}{2}} dx = \frac{1}{\sqrt{2\pi}} \int_{-\infty}^{+\infty} e^{-x^2\left(\frac{1}{2}-t\right)} dx$$

We use the fact that $\int_{-\infty}^{+\infty} e^{-y^2} dy = \sqrt{\pi}$; thus, change variable $y^2 = x^2(1/2 - t)$, and use $2x dx(1/2 - t) = 2y dy$:

$$dx = \frac{y}{x} \frac{dy}{(1/2 - t)} = \frac{\sqrt{1/2 - t}}{1/2 - t} dy = \frac{dy}{\sqrt{1/2 - t}}.$$

This results in the following moment generating function for Y^2:

$$M_{Z_i^2}(t) = \int_{-\infty}^{+\infty} \frac{e^{-y^2}}{\sqrt{\pi}} \frac{dy}{\sqrt{2(1/2 - t)}} = \sqrt{\frac{1}{1 - 2t}}. \tag{7.6}$$

$\square$

We make use of the property that $M_{x+y}(t) = M_x(t) \cdot M_y(t)$ for independent variables (4.10). Since the variables X_i are independent of one another, so are the variables Z_i^2. Therefore, the moment generating function of Z is given by

$$M_Z(t) = \left(M_{Z_i^2}(t)\right)^N = \left(\sqrt{\frac{1}{1 - 2t}}\right)^{N/2}.$$

To connect this result with the distribution function for Z, we need to introduce the *gamma distribution*:

$$f_\gamma(r, \alpha) = \frac{\alpha(\alpha x)^{r-1} e^{-\alpha x}}{\Gamma(r)} \tag{7.7}$$

where α, r are positive numbers, and $x \geq 0$. Its name derives from the following relationship with the Gamma function:

$$\Gamma(r) = \int_0^\infty e^{-x} x^{r-1} dx. \tag{7.8}$$

For integer arguments, $\Gamma(n) = (n-1)!$ It can be shown that the mean of the gamma distribution is $\mu = r/\alpha$, and the variance is $\sigma^2 = r/\alpha^2$. From property (7.8), it is also clear that the gamma distribution in (7.7) is properly normalized.

Next, we show that the moment generating function of a gamma distribution is a generalization of the moment generating function of the square of a standard normal distribution,

$$M_g(t) = \frac{1}{\left(1 - \dfrac{t}{\alpha}\right)^r}. \tag{7.9}$$

Proof The moment generating function of a gamma distribution is calculated as

$$M_g(t) = E[e^{tG}] = \int_0^\infty e^{tz} f_\gamma(r, \alpha)dz = \int_0^\infty \frac{\alpha^r}{\Gamma(r)} z^{r-1} e^{-z(\alpha - t)}dz$$

$$= \frac{\alpha^r}{\Gamma(r)} \int_0^\infty (\alpha - t)(\alpha - t)^{r-1} z^{r-1} e^{-z(\alpha - t)}dz.$$

The change of variable $x = z(\alpha - t)$, $dx = dz(\alpha - t)$ enables us to use the normalization property of the gamma distribution,

$$M_g(t) = \frac{\alpha^r}{(\alpha - t)^r} \int_0^\infty \frac{x^{r-1}}{\Gamma(r)} e^{-x}dx = \frac{\alpha^r}{(\alpha - t)^r} = \frac{1}{\left(1 - \dfrac{t}{\alpha}\right)^r}. \tag{7.10}$$

$\square$

The results shown in (7.5) and (7.9) prove that the moment generating functions for the Z and gamma distributions are related to one another. This relationship can be used to conclude that the random variable Z is distributed like a gamma distribution with parameters $r = N/2$ and $\alpha = 1/2$. The random variable Z is usually referred to as a χ^2 variable with N *degrees of freedom*, and has a probability distribution function

$$f_{\chi^2}(z, N) = f_Z(z) = \left(\frac{1}{2}\right)^{N/2} \frac{1}{\Gamma(N/2)} e^{-z/2} z^{N/2-1}. \tag{7.11}$$

An example of χ^2 distribution is shown in Fig. 7.2. The distribution is unimodal, although not symmetric with respect to the mean.

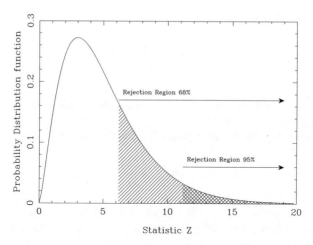

Fig. 7.2 The Z statistic is a χ^2 distribution with 5 degrees of freedom. The *hatched area* is the 68 % rejection region, and the *cross-hatched area* the 95 % region

7.2.2 Moments and Other Properties

Since the mean and variance of a gamma distribution with parameters r, α, are $\mu = \dfrac{r}{\alpha}$ and $\sigma^2 = \dfrac{r}{\alpha^2}$, the χ^2 distribution has the following moments:

$$\begin{cases} \mu = N \\ \sigma^2 = 2N. \end{cases} \tag{7.12}$$

This result shows that the expectation of a χ^2 variable is equal to the number of degrees of freedom. It is common to use the *reduced χ^2 square* variable defined by

$$\chi^2_{red} = \frac{\chi^2}{N}. \tag{7.13}$$

The mean or expectation of the reduced χ^2 and the variance are therefore given by

$$\begin{cases} \mu = 1 \\ \sigma^2 = \dfrac{2}{N}. \end{cases} \quad (\text{reduced } \chi^2) \tag{7.14}$$

As a result, the ratio between the standard deviation and the mean for the reduced χ^2, a measure of the spread of the distribution, decreases with the number of degrees of freedom,

$$\frac{\sigma}{\mu} = \sqrt{\frac{2}{N}}.$$

As the numbers of degrees of freedom increase, the values of the reduced χ^2 are more closely distributed around 1.

As derived earlier, the moment generating function of the χ^2 distribution

$$M_{\chi^2}(t) = E[e^{tZ}] = \frac{1}{(1 - 2t)^{N/2}}. \tag{7.15}$$

This form of the moment generating function highlights the property that, if two independent χ^2 distributions have, respectively, N and M degrees of freedom, then the sum of the two variables will also be a χ^2 variable, and it will have $N+M$ degrees of freedom. In fact, the generating function of the sum of independent variables is the product of the two functions, and the exponents in (7.15) will add.

7.2.3 Hypothesis Testing

The null hypothesis for a χ^2 distribution is that all measurements are consistent with the parent Gaussians. Under this hypothesis, we have derived the probability distribution function $f_{\chi^2}(z, N)$, where N is the number of degrees of freedom of the distribution. If the N measurements are consistent with their parent distributions, one expects a value of approximately $\chi^2 \simeq N$, i.e., each of the N measurements contributes approximately a value of one to the χ^2. Large values of χ^2 clearly indicate that some of the measurements are not consistent with the parent Gaussian, i.e., some of the measurements x_i differ by several standard deviations from the expected mean, either in defect or in excess. Likewise, values of $\chi^2 \ll N$ are also not expected. Consider, for example, the extreme case of N measurements all identical to the parent mean, resulting in $\chi^2 = 0$. Statistical fluctuations of the random variables make it extremely unlikely that all N measurements match the mean. Clearly such an extreme case of perfect agreement between the data and the parent model is suspicious and the data should be checked for possible errors in the collection or analysis.

Despite the fact that very small value of χ^2 is unlikely, it is customary to test for the agreement between a measurement of χ^2 and its theoretical distribution using a one-sided rejection region consisting of values of χ^2 exceeding a critical value. This means that the acceptable region is for values of χ^2 that are between zero and the critical value. Critical values of the χ^2 distribution for a confidence level p can be calculated via

$$P(Z \geq \chi^2_{crit}) = \int_{\chi^2_{crit}}^{\infty} f_{\chi^2}(z, n)dz = 1 - p \qquad (7.16)$$

and are tabulated in Table A.7.

Example 7.3 Assume the $N = 5$ measurements of a variable X, $(10, 12, 15, 11, 13)$, presented in Example 7.2. We want to test the hypothesis that these were independent measurements of a Gaussian variable X of mean $\mu = 13$ and variance $\sigma^2 = 2$. Under this assumption, we could use the χ^2 statistic to try and falsify the null hypothesis that the data are drawn from the given Gaussian. The procedure for a quantitative answer to this hypothesis is that of deciding a level of probability p, then to calculate the value of the statistic,

$$\chi^2 = 1/2 \cdot ((10-13)^2 + (12-13)^2 + (15-13)^2 + (11-13)^2 + (13-13)^2) = 9.$$

In Fig. 7.2 we show the rejection regions for a probability $p = 0.95$ and $p = 0.68$, which are determined according to the tabulation of the integral of the χ^2 distribution with $N = 5$ degrees of freedom: $\chi^2_{crit} = 6.1$ marks the beginning of the 70 % rejection region, and $\chi^2_{crit} = 11.1$ that of the 95 % rejection region. The hypothesis is therefore rejected at the 68 % probability level, but cannot be rejected at the 95 % confidence level.

Moreover, we calculate from Table A.7

$$P(\chi^2 \geq 9) = \int_9^{\infty} f_{\chi^2}(z, 5)dz \simeq 0.10.$$

We therefore conclude that there is a 10 % probability of observing such value of χ^2, or higher, under the hypothesis that the measurements were made from a Gaussian distribution of such mean and variance (see Fig. 7.2). Notice that the results obtained using the χ^2 distribution are similar to those obtained with the test that made use of the sum of the five measurements. ◇

7.3 The Sampling Distribution of the Variance

The distribution function of the sample variance, or *sampling distribution of the variance*, is useful to compare a given measurement of the sample variance s^2 with the parent variance σ^2. We consider N measurements of X that are distributed like a Gaussian of mean μ, variance σ^2 and independent of each other. The variable S^2 defined by

$$S^2 = (N-1)s^2 = \sum_{i=1}^{N}(X_i - \bar{X})^2 \qquad (7.17)$$

is proportional to the sample variance s^2. We seek a distribution function for S^2/σ^2 that enables a comparison of the measured sample variance with the parent variance σ^2.

In determining the sampling distribution of the variance we do not want to assume that the mean of the parent Gaussian is known, as we did in the previous section for the determination of the χ^2 distribution. This is important, since in a typical experiment we do not know a priori the parent mean of the distribution, but we can easily calculate the sample mean. One complication in the use of (7.17) is therefore that $\bar{X}$ is itself a random variable, and not an exactly known quantity. This fact must be taken into account when calculating the expectation of S^2. A measurement of S^2 is equal to

$$S^2 = \sum_{i=1}^{N}(x_i - \mu + \mu - \bar{x})^2 = \sum_{i=1}^{N}(x_i - \mu)^2 - N(\mu - \bar{x})^2. \qquad (7.18)$$

Dividing both terms by σ^2, we obtain the following result:

$$\frac{\sum_{i=1}^{N}(x_i - \mu)^2}{\sigma^2} = \frac{S^2}{\sigma^2} + \frac{(\bar{x} - \mu)^2}{\sigma^2/N}. \qquad (7.19)$$

According to the result in Sect. 7.2, the left-hand side term is distributed like a χ^2 variable with N degrees of freedom, since the parent mean μ and variance σ^2 appear in the sum of squares. For the same reason, the second term in the right-hand side is also distributed like a χ^2 variable with 1 degree of freedom, since we have already determined that the sample mean $\bar{X}$ is distributed like a Gaussian with mean μ and with variance σ^2/N. Although it may not be apparent at first sight, it can be proven that the two terms on the right-hand side are two independent random variables. If we can establish the independence between these two variables, then it must be true that the first variable in the right-hand side, S^2/σ^2, is also distributed like a χ^2 distribution with $N-1$ degrees of freedom. This follows from the fact that the sum of two independent χ^2 variables is also a χ^2 variable featuring the sum of the degrees of freedom of the two variables, as shown in Sect. 7.2.

Proof The proof of the independence between S^2/σ^2 and $\frac{(\bar{X}-\mu)^2}{\sigma^2\mu}$, and the fact that both are distributed like χ^2 distributions with, respectively, $N-1$ and 1 degrees of freedom, can be obtained by making a suitable change of variables from the original N standard normal variables that appear in the left-hand side of (7.19),

$$Z_i = \frac{X_i - \mu}{\sigma},$$

to a new set of N variables Y_i. The desired transformation is one that has the property

$$Z_1^2 + \ldots + Z_N^2 = Y_1^2 + \ldots + Y_N^2.$$

This is called an orthonormal (linear) transformation, and in matrix form it can be expressed by a transformation matrix A, of dimensions $N \times N$, such that a row vector $z = (Z_1, \ldots, Z_N)$ is transformed into another vector y by way of the product $y = zA$. For such a transformation, the dot product between two vectors is expressed as $yy^T = zAA^Tz^T$. Since for an orthonormal transformation the relationship $AA^T = I$ holds, where I is the $N \times N$ identity matrix, then the dot product remains constant upon this transformation. An orthonormal transformation, expressed in extended form as

$$\begin{cases} Y_1 = a_1Z_1 + \ldots + a_NZ_N \\ Y_2 = b_1Z_1 + \ldots + b_NZ_N \\ \ldots \end{cases}$$

is obtained when, for each row vector, $\sum a_i^2 = 1$; and, for any pair of row vectors, $\sum a_ib_i = 0$, so that the Y_i's are independent of one another.

Any such orthonormal transformation, when applied to N independent variables that are standard Gaussians, $Z_i \sim N(0,1)$, as is the case in this application, is such that the transformed variables Y_i are also independent standard Gaussians. In fact, the joint probability distribution function of the Z_i's can be written as

$$f(z) = \frac{1}{(2\pi)^{N/2}} e^{-\frac{z_1^2 + \ldots + z_N^2}{2}};$$

and, since the transformed variables have the same dot product, $z_1^2 + \ldots + z_N^2 = y_1^2 + \ldots + y_N^2$, the N variables Y_i have the same joint distribution function, proving that they are also independent standard Gaussians.

We want to use these general properties of orthonormal transformations to find a transformation that will enable a proof of the independence between S^2/σ^2 and $(\overline{X} - \mu)^2/\sigma_\mu^2$. The first variable is defined by the following linear combination,

$$Y_1 = \frac{Z_1}{\sqrt{N}} + \ldots + \frac{Z_N}{\sqrt{N}}$$

in such a way that the following relationships hold:

$$
\begin{cases}
Y_1^2 = \dfrac{(\bar{X} - \mu)^2}{\sigma^2/N} \\[2mm]
\displaystyle\sum_{i=1}^{N} Z_i^2 = \dfrac{1}{\sigma^2} \sum_{i=1}^{N}(X_i - \mu)^2, \text{ or} \\[2mm]
\displaystyle\sum_{i=1}^{N} Z_i^2 = Y_1^2 + \sum_{i=2}^{N} Y_i^2.
\end{cases}
$$

The other $N - 1$ variables $Y_2, \ldots, Y_N$ can be chosen arbitrarily, provided they satisfy the requirements of orthonormality. Since $\sum_{i=1}^{N} Z_i^2 - Y_1^2 = S^2/\sigma^2$, we can conclude that

$$
\frac{S^2}{\sigma^2} = \sum_{i=2}^{N} Y_i^2
$$

proving that S^2/σ^2 is distributed like a χ^2 distribution with $N - 1$ degrees of freedom, as the sum of squares on $N - 1$ independent standard Gaussians, and that S^2/σ^2 is independent of the sampling distribution of the mean, Y_1^2, since the variables Y_i are independent of each other. This proof is due to Bulmer [7], who used a derivation done earlier by Helmert [20]. □

We are therefore able to conclude that the ratio S^2/σ^2 is distributed like a χ^2 variable with $N - 1$ degrees of freedom,

$$
\frac{S^2}{\sigma^2} \sim \chi^2(N - 1). \tag{7.20}
$$

The difference between the χ^2 distribution (7.11) and the distribution of the sample variance (7.20) is that in the latter case the mean of the parent distribution is not assumed to be known, but it is calculated from the data. This is in fact the more common situation, and therefore when N measurements are obtained, the quantity

$$
\frac{S^2}{\sigma^2} = \sum_{i=1}^{N} \left(\frac{x_i - \bar{x}}{\sigma} \right)^2
$$

is distributed like a χ^2 distribution with just $N - 1$ degrees of freedom, not N. This reduction in the number of degrees of freedom can be expressed by saying that one degree of freedom is being used to estimate the mean.

Example 7.4 Assume $N = 10$ measurements of a given quantity (10, 12, 15, 11, 13, 16, 12, 10, 18, 13). We want to answer the following question: Are these measurements consistent with being drawn from the same Gaussian random variable with $\sigma^2 = 2$? If the measurements are in fact derived from the *same* variable, then the probability of measuring the actual value of s^2 for the sample variance will be consistent with its theoretical distribution that was just derived in (7.20).

The value of the sample variance is obtained by $\bar{x} = 13$ as $S^2 = 62$. Therefore, the measurement $s^2/\sigma^2 = 62/2 = 36$ must be compared with the χ^2 distribution with $N - 1 = 9$ degrees of freedom. The measurement is equivalent to a reduced χ^2 value of 4, which is inconsistent with a χ^2 distribution with 9 degrees of freedom at more than the 99 % confidence level. We therefore conclude that the hypothesis must be rejected with this confidence level.

It is necessary to point out that, in this calculation, we assumed that the parent variance was known. In the following section we will provide another test that can be used to compare two measurements of the variance that does not require knowledge of the parent variance. That is in fact the more common experimental situation and it requires a detailed study. ◇

7.4 The F Statistic

The distribution of the sample variance discussed above in Sect. 7.3 shows that if the actual variance σ is not known, then it is impossible to make a quantitative comparison of the sample variance with the parent distribution. Alternatively, one can compare two different measurements of the variance, and ask the associated question of whether the ratio between the two measurements is reasonable. In this case the parent variance σ^2 drop out of the equation and the parent variance is not required to compare two measurements of the sample variance.

For this purpose, consider two independent random variables Z_1 and Z_2, respectively, distributed like a χ^2 distribution with f_1 and f_2 degrees of freedom. We define the random variable F as

$$F = \frac{Z_1/f_1}{Z_2/f_2}. \tag{7.21}$$

The variable F is equivalent to the ratio of two reduced χ^2, and therefore is expected to have values close to unity.

7.4.1   The Probability Distribution Function

We show that the probability distribution function of the random variable F is given by

$$f_F(z) = \frac{\Gamma\left(\dfrac{f_1 + f_2}{2}\right)}{\Gamma\left(\dfrac{f_1}{2}\right)\Gamma\left(\dfrac{f_2}{2}\right)} \left(\frac{f_1}{f_2}\right)^{\frac{f_1}{2}} \frac{z^{\frac{f_1}{2}-1}}{\left(1 + z\dfrac{f_1}{f_2}\right)^{\frac{f_1+f_2}{2}}}. \tag{7.22}$$

Proof The proof makes use of the methods described in Sects. 4.4.1 and 4.4.2. First we derive the distribution functions of the numerator and denominator of (7.21), and then we calculate the distribution function for the ratio of two variables with known distribution.

Given that $Z_1 \sim \chi^2(f_1)$ and $Z_2 \sim \chi^2(f_2)$, the distribution functions of $X' = Z_1/f_1$ and $Y' = Z_2/f_2$ are found using change of variables; for X',

$$f_{X'}(x') = f(z)\frac{dz}{dx'} = f(z)f_1,$$

where $f(z)$ is the distribution of Z_1. This results in

$$f_{X'}(x') = \frac{z^{f_1/2-1}e^{-z/2}}{\Gamma(f_1/2)2^{f_1/2}}f_1 = \frac{(x'f_1)^{f_1/2-1}e^{-(x'f_1)/2}}{\Gamma(f_1/2)2^{f_1/2}}f_1;$$

same transformation applies to Y'. Now we can use (4.18),

$$
\begin{aligned}
f_F(z) &= \int_0^\infty f_{X'}(z\zeta)\zeta f_{Y'}(\zeta)d\zeta \\
&= \int_0^\infty \frac{(z\zeta f_1)^{f_1/2-1}e^{-(z\zeta f_1)/2}}{\Gamma(f_1/2)2^{f_1/2}}f_1\zeta \frac{(\zeta f_2)^{f_2/2-1}e^{-(\zeta f_2)/2}}{\Gamma(f_2/2)2^{f_2/2}}d\zeta \\
&= \frac{f_1 f_2 z^{f_1/2-1}f_1^{f_1/2-1}f_2^{f_2/2-1}}{\Gamma(f_1/2)2^{(f_1+f_2)/2}\Gamma(f_2/2)}\int_0^\infty \zeta^{f_1/2-1+f_2/2-1+1}e^{-1/2\zeta(zf_1+f_2)}d\zeta \\
&= \frac{z^{f_1/2-1}f_1^{f_1/2}f_2^{f_2/2}}{\Gamma(f_1/2)\Gamma(f_2/2)2^{(f_1+f_2)/2}}\int_0^\infty \zeta^{(f_1+f_2)/2-1}e^{-1/2\zeta(zf_1+f_2)}d\zeta
\end{aligned}
$$

After another change of variables, $t = \zeta(zf_1 + f_2)/2$, $dt = d\zeta(zf_1 + f_2)/2$, the integral becomes

$$\int_0^\infty \left(\frac{2t}{zf_1 + f_2}\right)^{(f_1+f_2)/2-1} e^{-t} \frac{dt}{\left(\dfrac{zf_1 + f_2}{2}\right)}$$

$$= \frac{2^{(f_1+f_2)/2}}{(zf_1 + f_2)^{1+(f_1+f_2)/2-1}} \int_0^\infty t^{(f_1+f_2)/2-1} e^{-t} dt$$

$$= \frac{2^{(f_1+f_2)/2}}{(zf_1 + f_2)^{(f_1+f_2)/2}} \Gamma\left(\frac{f_1 + f_2}{2}\right).$$

Therefore the distribution of Z is given by

$$f_F(z) = \frac{z^{f_1/2-1} f_1^{f_1/2} f_2^{f_2/2}}{\Gamma(f_1/2)\Gamma(f_2/2) 2^{(f_1+f_2)/2}} \frac{2^{(f_1+f_2)/2} \Gamma\left(\dfrac{f_1 + f_2}{2}\right)}{(zf_1 + f_2)^{(f_1+f_2)/2}}$$

$$= \frac{f_1^{f_1/2} f_2^{f_2/2} \Gamma\left(\dfrac{f_1 + f_2}{2}\right)}{\Gamma(f_1/2)\Gamma(f_2/2)} \frac{z^{f_1/2-1}}{(1 + z\dfrac{f_1}{f_2})^{(f_1+f_2)/2} f_2^{(f_1+f_2)/2}}$$

$$= \left(\frac{f_1}{f_2}\right)^{f_1/2} \frac{\Gamma\left(\dfrac{f_1 + f_2}{2}\right)}{\Gamma\left(\dfrac{f_1}{2}\right)\Gamma\left(\dfrac{f_2}{2}\right)} \frac{z^{f_1/2-1}}{\left(1 + z\dfrac{f_1}{f_2}\right)^{(f_1+f_2)/2}}.$$

□

The distribution of F is known as the F *distribution*. It is named after Fisher [13], who was the first to study it.

7.4.2 Moments and Other Properties

The mean and higher-order moments of the F distribution can be calculated by making use of the Beta function,

$$B(x, y) = \int_0^\infty \frac{f^{x-1}}{(1 + t)^{x+y}} dt = \frac{\Gamma(x)\Gamma(y)}{\Gamma(x + y)}, \tag{7.23}$$

to find that

$$
\begin{cases}
\mu = \dfrac{f_2}{f_2 - 2} \, (f_2 > 2) \\[2mm]
\sigma^2 = \dfrac{2 f_2^2 (f_1 + f_2 - 2)}{f_1 (f_2 - 2)^2 (f_2 - 4)} \, (f_2 > 4)
\end{cases}
\tag{7.24}
$$

The mean is approximately 1, provided that f_2 is not too small.

It is possible to find an approximation to the F distribution when either f_1 or f_2 is a large number:

$$
\begin{cases}
\lim\limits_{f_2 \to \infty} f_F(z, f_1, f_2) = f_{\chi^2}(x, f_1) \ \ \text{where } x = f_1 z \\[2mm]
\lim\limits_{f_1 \to \infty} f_F(z, f_1, f_2) = f_{\chi^2}(x, f_2) \ \ \text{where } x = f_2/z.
\end{cases}
\tag{7.25}
$$

The approximation, discussed, for example, in [1], is very convenient, since it overcomes the problems with the evaluation of the Gamma function for large numbers.

7.4.3   Hypothesis Testing

The F statistic is a ratio

$$
F = \frac{\chi_1^2/f_1}{\chi_2^2/f_2}
\tag{7.26}
$$

between two independent χ^2 measurements of, respectively, f_1 and f_2 degrees of freedom. A typical application of the F test is the comparison of two χ^2 statistics from independent datasets using the parent Gaussians as models for the data. The null hypothesis is that both sets of measurements follow the respective Gaussian distribution. In this case, the measured ratio F will follow the F distribution. This implies that the measured value of F should not be too large under the null hypothesis that both measurements follow the parent models.

It is customary to do hypothesis testing of an F distribution using a one-sided rejection region above a critical value. The critical value at confidence level p is calculated via

$$
P(F > F_{crit}) = \int_{F_{crit}}^{\infty} f_F(z) dz = 1 - p.
\tag{7.27}
$$

Critical values are tabulated in Table A.8 for the case of fixed $f_1 = 1$, and Tables A.9, A.10, A.11, A.12, A.13, A.14, and A.15 for various values of p, and as function of f_1 and f_2. The values of F_{crit} calculated from (7.27) indicate how high a value of the

F statistic can be, and still be consistent with the hypothesis that the two quantities at the numerator and denominator are χ^2-distributed variables.

The approximations for the F distribution in (7.25) can be used to calculate critical values when one of the degrees of freedom is very large. For example, the critical value of F at 90 % confidence, $p = 0.90$, for $f_1 = 100$ and $f_2 \to \infty$ (e.g., Table A.13) is calculated from Table A.7 as $\overline{F} = 1.185$. Note that Table A.7 reports the value of the reduced χ^2, or z in the notation of the top equation in (7.25).

Example 7.5 Consider the data set composed of the ten measurements

$$(10, 12, 15, 11, 13, 16, 12, 10, 18, 13).$$

We assume that the measurements follow a Gaussian distribution of mean of $\mu = 13$ and variance σ^2. The goal is to compare the calculation of the χ^2 of the first five measurements with the last five to address whether both subsets are equally likely to be described by the same Gaussian.

We obtain $\chi_1^2 = 18/\sigma^2$ and $\chi_2^2 = 44/\sigma^2$, respectively, for the first and the second set of five measurements. Both variables, under the null hypothesis that the measurements follow the reference Gaussian, are distributed like χ^2 with 5 degrees of freedom (since both mean and variance are assumed to be known). We therefore can calculate an F statistic of $F = 44/18 = 2.44$. For simplicity, we have placed the initial five measurements at the denominator.

In the process of calculating the F statistic, the variances σ^2 cancel, and therefore the null hypothesis is that of a mean of $\mu = 13$ and same variance for both sets, regardless of its value. In Fig. 7.3 we plot the F distribution for $f_1 = 5$ and $f_2 = 5$

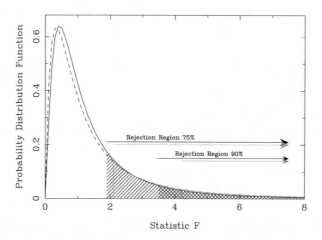

Fig. 7.3 *Solid curve* is the F distribution with $\nu_1 = 5$, $\nu_2 = 5$ degrees of freedom; the *hatched area* is the 75 % rejection region, and the *cross-hatched area* is the 90 % rejection region. For comparison, the F distribution with $\nu_1 = 4$, $\nu_2 = 4$ degrees of freedom is shown as the *dashed line*, and the two rejection regions are outlined in *green and red*, respectively. The rejection region for the F distribution with $\nu_1 = 4$, $\nu_2 = 4$ degrees of freedom is shifted to higher values, relative to that with $\nu_1 = 5$, $\nu_2 = 5$ degrees of freedom, because of its heavier tail

as the solid line, and its 75 and 90 % rejection regions, marked, respectively, by the critical values $\overline{F} = 1.89$ and 3.45, as hatched and cross-hatched areas. The measurements are therefore consistent with the null hypothesis at the 90 % level, but the null hypothesis must be discarded at the 75 % confidence level. Clearly the first set of five numbers follows the parent Gaussian more closely than the second set. Yet, there is a reasonable chance ($\geq 10\%$) that both sets follow the Gaussian.

If the parent variance was given, say $\sigma^2 = 4$, we could have tested both subsets independently for the hypothesis that they follow a Gaussian of mean $\mu = 13$ and variance $\sigma^2 = 4$ using the χ^2 distribution. The two measurements are $\chi_1^2 = 4.5$ and $\chi_2^2 = 11$ for 5 degrees of freedom. Assuming a confidence level of $p = 0.9$, the critical value of the χ^2 distribution is $\chi_{crit}^2 = 9.2$. At this confidence level, we would reject the null hypothesis for the second measurement.   ◇

The ratio between two measurements of the sample variance follows the F distribution. For two independent sets of, respectively, N and M measurements, the sample variances s_1^2 and s_2^2 are related to the parent variances σ_1^2 and σ_2^2 of the Gaussian models via

$$F = \frac{Z_1/f_1}{Z_2/f_2} = \frac{\dfrac{S_1^2}{\sigma_1^2 f_1}}{\dfrac{S_2^2}{\sigma_2^2 f_2}}, \tag{7.28}$$

where

$$\begin{cases} S_1^2 = (N-1)s_1^2 = \sum_{i=1}^{N}(x_i - \bar{x})^2 \\ S_2^2 = (M-1)s_2^2 = \sum_{j=1}^{M}(y_j - \bar{y})^2. \end{cases} \tag{7.29}$$

The quantities $Z_1 = S_1^2/\sigma_1^2$ and $Z_2 = S_2^2/\sigma_2^2$ are χ^2-distributed variables with, respectively, $f_1 = N - 1$ and $f_2 = M - 1$ degrees of freedom. The statistic F can be used to test whether both measurements of the variance are equally likely to have come from the respective models.

The interesting case is clearly when the two variances are equal, $\sigma_1^2 = \sigma_2^2$, so that the value of the variance drops out of the equation and the F statistic becomes

$$F = \frac{S_1^2/f_1}{S_2^2/f_2} \; (\sigma_1^2 = \sigma_2^2). \tag{7.30}$$

In this case, the null hypothesis becomes that the two samples are Gaussian distributed, regardless of values for the mean and the variance. The statistic therefore measure if the variances or variability of the data in the two measurements are consistent with one another or if one measurement has a sample variance that is significantly larger than the other. If the value of F exceeds the critical value, then the null hypothesis must be rejected and the conclusion is that the measurement with the largest value of Z/f, which is placed at the numerator, is not as likely to have come from the parent model as the other set. This type of analysis will have specific applications to model fitting in Chap. 13.

Example 7.6 Using the same data as in the previous example, we can calculate the sample variance using the sample mean for each of the two 5-measurement sets. We calculate a sample mean of $\bar{x}_1 = 12.2$ and $\bar{x}_2 = 13.8$, for a value of $S_1^2 = 14.8$ and $S_2^2 = 40.8$, for a ratio of $F = 2.76$. Given that the sample mean was estimated from the data, the null hypothesis is that both sets are drawn from the same Gaussian distribution, without specification of the value of either variance or mean, and each measurement of S^2/σ^2 is distributed now like a χ^2 variable with just 4 degrees of freedom (and not 5). The value of the F statistic must therefore be compared with an F distribution with $f_1 = 4$ and $f_2 = 4$ degrees of freedom, reported in Fig. 7.3 as a dashed line. The 75 and 90 % rejection regions, marked, respectively, by the critical values $\overline{F} = 2.06$ and 4.1, are outlined in green and red, respectively. The measurements are therefore consistent at the 90 % confidence level, but not at the 75 % level.

We conclude that there is at least a 10 % probability that the two measurements of the variance are consistent with one another. At the $p = 0.9$ level we therefore cannot reject the null hypothesis.   ◇

7.5   The Sampling Distribution of the Mean and the Student's *t* Distribution

In many experimental situations we want to compare the sample mean obtained from the data to a parent mean based on theoretical considerations. Other times we want to compare two sample means to one another. The question we answer in this section is how the sample mean is expected to vary when estimated from independent samples of size N.

7.5.1   *Comparison of Sample Mean with Parent Mean*

For measurements of a Gaussian variable of mean μ and variance σ^2, the sample mean $\bar{x}$ is distributed as a Gaussian of mean μ and variance σ^2/N. Therefore, if both the mean and the variance of the parent distribution are known, the sample mean $\bar{X}$ is such that

$$\frac{\bar{X} - \mu}{\sigma/\sqrt{N}} \sim N(0, 1).$$

A simple comparison between the z-score of the sample mean to the $N(0, 1)$ Gaussian therefore addresses the consistency between the measurement and the model.

Example 7.7 Continue with the example of the five measurements of a random variable $(10, 12, 15, 11, 13)$, assumed to be distributed like a Gaussian of $\mu = 13$ and $\sigma^2 = 2$. Assuming knowledge of the parent mean and variance, the z-score of the sample mean is

$$\frac{\bar{x} - \mu}{\sigma_\mu} = \frac{12.2 - 13}{\sqrt{2/5}} = -1.27.$$

According to Table A.2, there is a probability of about 20 % to exceed the absolute value of this measurement according to the parent distribution $N(0, 1)$. Therefore the null hypothesis that the measurements are distributed like a Gaussian of $\mu = 13$ and $\sigma^2 = 2$ cannot be rejected at the 90 % confidence level. Notice that this is the same probability as obtained by using the sum of the five measurements, instead of the average. This was to be expected, since the mean differs from the sum by a constant value, and therefore the two statistics are equivalent.   ◇

A more common situation is when the mean μ of the parent distribution is known but the parent variance is unknown. In those cases the parent variance can only be estimated from the data themselves via the sample variance s^2 and one needs to allow for such uncertainty when estimating the distribution of the sample mean. This additional uncertainty leads to a deviation of the distribution function from the simple Gaussian shape. We therefore seek to find the distribution of

$$T = \frac{\bar{x} - \mu}{s/\sqrt{n}} \tag{7.31}$$

in which we define the sample variance in such a way that it is an unbiased estimator of the parent variance,

$$s^2 = \frac{1}{N-1} \sum (x_i - \bar{x})^2 = \frac{S^2}{N-1}.$$

The variable T can be written as

$$T = \frac{\bar{x} - \mu}{s/\sqrt{N}} = \left[\frac{\bar{x} - \mu}{\sigma/\sqrt{N}}\right] / (s/\sigma) = \left[\frac{\bar{x} - \mu}{\sigma/\sqrt{N}}\right] / \left[\frac{S^2}{(N-1)\sigma^2}\right]^{1/2}, \tag{7.32}$$

in which S^2 is the sum of the squares of the deviations from the sample mean. As shown in previous sections,

$$\begin{cases} \dfrac{\bar{x} - \mu}{\sigma/\sqrt{n}} & \sim N(0, 1) \\[2mm] \dfrac{S^2}{\sigma^2} & \sim \chi^2(N - 1). \end{cases}$$

We therefore need to determine the distribution function of the ratio of these two variables. We will show that a random variable T defined by the ratio

$$T = \frac{X}{\sqrt{Z/f}}, \tag{7.33}$$

in which $X \sim N(0, 1)$ and $Z \sim \chi^2(f)$ (a χ^2 distribution with f degrees of freedom) is said to be distributed like a *t distribution* with f degrees of freedom:

$$f_T(t) = \frac{1}{\sqrt{f\pi}} \frac{\Gamma((f + 1)/2)}{\Gamma(f/2)} \times \left(1 + \frac{t^2}{f}\right)^{-\frac{f + 1}{2}}. \tag{7.34}$$

Proof The proof of (7.34) follows the same method as that of the F distribution. First, we can derive the distribution function of $Y = \sqrt{Z/f}$ using the usual method of change of variables,

$$g(y) = h(z)\frac{dz}{dy} = h(z)2\sqrt{fZ}$$

where

$$h(z) = \frac{z^{f/2-1}e^{-z/2}}{2^{f/2}\Gamma(f/2)}.$$

Therefore the distribution of Y is given by substituting $z = fy^2$ into the first equation,

$$g(y) = \frac{f^{(f-1)/2}y^{f-1}e^{-fy^2/2}\sqrt{f}}{2^{f/2-1}\Gamma(f/2)}. \tag{7.35}$$

The distribution function of the numerator of (7.33) is simply

$$f(x) = \frac{1}{\sqrt{2\pi}}e^{-x^2/2},$$

and therefore the distribution of T is given by applying (4.18),

$$f_T(t) = \int_0^\infty \frac{1}{\sqrt{2\pi}} e^{-(ty)^2/2} y \frac{f^{(f-1)/2} y^{f-1} e^{-fy^2/2} \sqrt{f}}{2^{f/2-1} \Gamma(f/2)} dy. \tag{7.36}$$

The integral can be shown to be equal to (7.34) following a few steps of integration as in the case of the F distribution.   □

This distribution is symmetric and has a mean of zero, and it goes under the name of *Student's t distribution*. This distribution was studied first by Gosset in 1908 [18], who published a paper on the subject under the pseudonym of "Student."

The random variable T defined in (7.31) therefore is distributed like a t variable with $N-1$ degrees of freedom. It is important to notice the difference between the sample distribution of the mean in the case in which the variance is known, which is $N(0,1)$, and the t distribution. In particular, the latter depends on the number of measurements, while the former does not. One expects that, in the limit of a large number of measurements, the t distribution tends to the standard normal (see Problem 7.10). The t distribution has in fact broader wings than the standard Gaussian, and in the limit of an infinite number of degrees of freedom, the two distributions are identical; an example of the comparison between the two distributions is shown in Fig. 7.4. The t distribution has heavier tails than the Gaussian distribution, indicative of the additional uncertainty associated with the fact that the variance is estimated from the data and not known a priori.

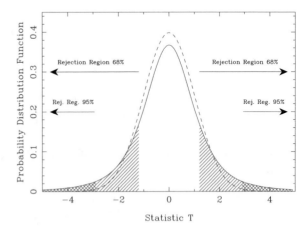

Fig. 7.4 Student's t distribution with $f = 4$ degrees of freedom. The *dashed curve* is the $N(0,1)$ Gaussian, to which the t-distribution tends for a large number of degrees of freedom. The *hatched area* is the 68 % rejection region (compare to the $\pm 1\sigma$ region for the $N(0,1)$ distribution) and the *cross-hatched area* is the 95 % region (compare to $\pm 1.95\sigma$ for the $N(0,1)$ distribution)

7.5.1.1   Hypothesis Testing

Hypothesis testing with the t distribution typically uses a two-sided rejection region. After obtaining a measurement of the t variable from a given dataset, we are usually interested in knowing how far the measurement can be from the expected mean of 0 and still be consistent with the parent distribution. The critical value for a confidence level p is calculated via

$$P(|t| \leq T_{crit}) = \int_{-T_{crit}}^{T_{crit}} f_T(t)dt = p \tag{7.37}$$

and it is a function of the number of degrees of freedom for the t distribution. Tables A.16, A.17, A.18, A.19, A.20, A.21, and A.22 report the value of p as function of the critical value T_{crit} for selected degrees of freedom, and Table A.23 compares the t distribution with the standard Gaussian.

Example 7.8   Assume now that the five measurements $(10, 12, 15, 11, 13)$ are distributed like a Gaussian of $\mu = 13$, but without reference to a parent variance. In this case we consider the t statistic and start by calculating the sample variance:

$$s^2 = \frac{1}{4} \sum (x_i - \bar{x})^2 = 3.7.$$

With this we can now calculate the t statistic,

$$t = \frac{\bar{x} - \mu}{s/\sqrt{5}} = \frac{12.2 - 13}{1.92/\sqrt{5}} = -0.93.$$

This value of t corresponds to a probability of approximately $\sim 40\%$ to exceed the absolute value of this measurement, using the t distribution with 4 degrees of freedom of Table A.23. It is clear that the estimation of the variance from the data has added a source of uncertainty in the comparison of the measurement with the parent distribution.   ◇

7.5.2   *Comparison of Two Sample Means and Hypothesis Testing*

The same distribution function is also applicable to the comparison between two sample means $\bar{x}_1$ and $\bar{x}_2$, derived from samples of size N_1 and N_2, respectively. In this case, we define the following statistic:

$$T = \frac{\overline{x_1} - \overline{x_2}}{s\sqrt{1/N_1 + 1/N_2}}. \tag{7.38}$$

where

$$\begin{cases} S^2 = S_1^2 + S_2^2 \\ s^2 = \dfrac{S^2}{N_1 + N_2 - 2} \\ S_1^2 = \displaystyle\sum_{i=1}^{N_1}(x_i - \overline{x_1}) \\ S_2^2 = \displaystyle\sum_{j=1}^{N_2}(x_j - \overline{x_2}). \end{cases}$$

We show that this statistic is distributed like a T distribution with $f = N_1 + N_2 - 2$ degrees of freedom, and therefore we can use the same distribution also for testing the agreement between two sample means.

Proof Under the hypothesis that all measurements are drawn from the same parent distribution, $X \sim N(\mu, \sigma)$, we know that

$$\begin{cases} \dfrac{\overline{x_1} - \mu}{\sigma/\sqrt{N1}} \sim N(0, 1) \\ \dfrac{\overline{x_2} - \mu}{\sigma/\sqrt{N2}} \sim N(0, 1) \end{cases}$$

and, from (7.20)

$$\begin{cases} \dfrac{S_1^2}{\sigma^2} \sim \chi^2(N_1 - 1) \\ \dfrac{S_2^2}{\sigma^2} \sim \chi^2(N_2 - 1) \end{cases}.$$

First, we find the distribution function for the variable $(\overline{x_1} - \mu)/\sigma - (\overline{x_2} - \mu)/\sigma$. Assuming that the measurements are independent, then the variable is a Gaussian with zero mean, with variances added in quadrature, therefore

$$X = \left(\frac{\overline{x_1} - \mu}{\sigma} - \frac{\overline{x_2} - \mu}{\sigma}\right) \Big/ \sqrt{\frac{1}{N_1} + \frac{1}{N_2}} \sim N(0, 1).$$

Next, since independent χ^2 variables are also distributed like a χ^2 distribution with a number of degrees of freedom equal to the sum of the individual degrees of freedom,

$$Z = \frac{S_1^2}{\sigma^2} + \frac{S_2^2}{\sigma^2} \sim \chi^2(N_1 + N_2 - 2).$$

We also know the distribution of $\sqrt{Z/f}$ from (7.35), with $f = N_1 + N_2 - 2$ the number of degrees of freedom for both datasets combined. As a result, the variable T can be written as

$$T = \frac{X}{\sqrt{Z/f}} \tag{7.39}$$

in a form that is identical to the T function for comparison of sample mean with the parent mean, and therefore we can conclude that the random variable defined in (7.38) is in fact a T variable with $f = N_1 + N_2 - 2$ degrees of freedom.   □

Example 7.9 Using the ten measurements $(10, 12, 15, 11, 13, 16, 12, 10, 18, 13)$, we have already calculated the sample mean of the first and second half of the measurements as $\overline{x}_1 = 12.2$ and $\overline{x}_2 = 13.8$, and the sample variances as $S_1^2 = 14.8$ and $S_2^2 = 40.8$. This results in a measurement of the t distribution for the comparison between two means of

$$t = \frac{\overline{x}_1 - \overline{x}_2}{\sqrt{s_1^2 + s_2^2}\sqrt{1/N_1 + 1/N_2}} = -0.97. \tag{7.40}$$

This number is to be compared with a t distribution with 8 degrees of freedom, and we conclude that the measurement is consistent, at any reasonable level of confidence, with the parent distribution. In this case, we are making a statement regarding the fact that the two sets of measurements may have the same mean, but without committing to a specific value.   ◇

Summary of Key Concepts for this Chapter

☐ *Hypothesis Testing*: A four-step process that consists of (1) defining a null hypothesis to test, (2) determine the relevant statistic (e.g., χ^2), (3) a confidence level (e.g., 90 %), and (4) whether the null hypothesis is discarded or not.

☐ χ^2 *distribution*: The theoretical distribution of the sum of the squares of independent z-scores,

$$\chi^2 = \sum \left(\frac{x_i - \mu_i}{\sigma_i}\right)^2.$$

(mean N and variance $2N$).

☐ *Sampling distribution of variance*: Distribution of sample variance $s^2 = S^2/(N-1)$,

$$S^2/\sigma^2 \sim \chi^2(N-1)$$

☐ *F Statistic*: Distribution of the ratio of independent χ^2 variables

$$F = \frac{\chi_1^2/f_1}{\chi_2^2/f_2}$$

(mean $f_2/(f_2 - 2)$ for $f_2 > 2$) also used to test for additional model components.

☐ *Student's t distribution*: Distribution for the variable

$$T = \frac{\bar{x} - \mu}{s/\sqrt{n}},$$

useful to compare the sample mean to the parent mean when the variance is estimated from the data.

Problems

7.1 Five students score 70, 75, 65, 70, and 65 on a test. Determine whether the scores are compatible with the following hypotheses:

(a) The mean is $\mu = 75$;
(b) the mean is $\mu = 75$ and the standard deviation is $\sigma = 5$.

Test both hypotheses at the 95 % or 68 % confidence levels, assuming that the scores are Gaussian distributed.

7.2 Prove that the mean and variance of the F distribution are given by the following relationships,

$$\begin{cases} \mu = \dfrac{f_2}{f_2 - 2} \\ \sigma^2 = \dfrac{2f_2^2(f_1 + f_2 - 2)}{f_1(f_2 - 2)^2(f_2 - 4)}, \end{cases}$$

where f_1 and f_2 are the degrees of freedom of the variables at the numerator and denominator, respectively.

7.3 Using the same data as Problem (7.1), test whether the sample variance is consistent with a parent variance of $\sigma^2 = 25$, at the 95 % level.

7.4 Using the J.J. Thomson experiment data of page 23, measure the ratio of the sample variances of the m/e measurements in Air for Tube 1 and Tube 2. Determine if the null hypothesis that the two measurements are drawn from the same distribution can be rejected at the 90 % confidence level. State all assumptions required to use the F distribution.

7.5 Consider a dataset $(10, 12, 15, 11, 13, 16, 12, 10, 18, 13)$, and calculate the ratio of the sample variance of the first two measurements with that of the last eight. In particular, determine at what confidence level for the null hypothesis both subsets are consistent with the same variance.

7.6 Six measurements of the length of a wooden block gave the following measurements: 20.3, 20.4, 19.8, 20.4, 19.9, and 20.7 cm.

(a) Estimate the mean and the standard error of the length of the block;
(b) Assume that the block is known to be of length $\mu = 20$ cm. Establish if the measurements are consistent with the known length of the block, at the 90 % probability level.

7.7 Consider Mendel's experimental data in Table 1.1 shown at page 9.

(a) Consider the data that pertain to the case of "Long vs. short stem." Write an expression for the probability of making that measurement, assuming Mendel's hypothesis of independent assortment. You do not need to evaluate the expression.
(b) Using the distribution function that pertains to that measurement, determine the mean and variance of the parent distribution. Using the Gaussian approximation for this distribution, determine if the null hypothesis that the measurement is drawn from the parent distribution is compatible with the data at the 68 % confidence level.

7.8 Consider Mendel's experimental data in Table 1.1 shown at page 9. Considering all seven measurements, calculate the probability that the mean fraction of dominant characters agrees with the expectation of 0.75. For this purpose, you may use the t statistic.

7.9 Starting with (7.36), complete the derivation of (7.34).

7.10 Show that the t distribution,

$$f_T(t) = \frac{1}{\sqrt{f\pi}} \frac{\Gamma((f+1)/2)}{\Gamma(f/2)} \times \left(1 + \frac{t^2}{f}\right)^{-\frac{1}{2}(f+1)}$$

becomes a standard Gaussian in the limit of large f. You can make use of the asymptotic expansion of the Gamma function (A.17).

Chapter 8
Maximum Likelihood Methods for Two-Variable Datasets

Abstract One of the most common tasks in the analysis of scientific data is to establish a relationship between two quantities. Many experiments feature the measurement of a quantity of interest as function of another control quantity that is varied as the experiment is performed. In this chapter we use the maximum likelihood method to determine whether a certain relationship between the two quantities is consistent with the available measurements and the best-fit parameters of the relationship. The method has a simple analytic solution for a linear function but can also be applied to more complex analytic functions.

8.1 Measurement of Pairs of Variables

A general problem in data analysis is to establish a relationship $y = y(x)$ between two random variables X and Y for which we have available a set of N measurements (x_i, y_i). The random variable X is considered to be the independent variable and it will be treated as having uncertainties that are much smaller than those in the dependent variable, i.e., $\sigma_x \ll \sigma_y$. This may not always be the case and there are some instances in which both errors need to be considered. The case of datasets with errors in both variables is presented in Chap. 12.

The starting point of the analysis of a two-dimensional dataset is an analytic form for $y(x)$, e.g., $y(x) = a + bx$. The function $f(x)$ has a given number of adjustable parameters $a_k, k = 1, \ldots, m$ that are to be constrained according to the measurements. When the independent variable X is assumed to be known exactly, then the two-variable data set can be described as a sequence of random variables $Y(X_i)$. For these variables we typically have a measurement of the standard error such that the two-variable data are of the form

$$(x_i, y_i \pm \sigma_i) \qquad i = 1, \ldots, N.$$

An example of this situation may be a dataset in which the size of an object is measured at different time intervals. In this example the time of measurement t_i is the independent variable, assumed to be known exactly, and $r_i \pm \sigma_i$ is the measurement of the size at that time interval. Although we call $y(x_i) = r_i \pm \sigma_i$ a

© Springer Science+Busines Media New York 2017
M. Bonamente, *Statistics and Analysis of Scientific Data*, Graduate Texts in Physics, DOI 10.1007/978-1-4939-6572-4_8

"measurement," it really may itself be obtained from a number of measurements from which one infers the mean and the variance of that random variable, as described in the earlier chapters. It is therefore reasonable to expect that the measurement provides also an estimate of the standard error.

Before describing the mathematical properties of the method used to estimate the best-fit parameters we need to understand the framework for the analysis. Consider as an example the case of a linear function between X and Y illustrated in Fig. 8.1. The main assumption of the method is that the function $y = y(x)$ is the correct description of the relationship between the two variables. This means that each random variable $y(x_i)$ is a Gaussian with the following parameters:

$$\begin{cases} \mu_i = y(x_i) & \text{the parent mean is determined by } y(x) \\ \quad \sigma_i^2 & \text{variance is estimated from the data.} \end{cases} \qquad (8.1)$$

Notice how this framework is somewhat of a hybrid: the parent mean is determined by the parent model $y(x)$ while the variance is estimated from the data. It should not be viewed as a surprise that the model $y = y(x)$ typically cannot determine by itself the variance of the variable. In fact, we know that the variance depends on the quality of the measurements made and therefore it is reasonable to expect that σ_i is estimated from the data themselves. In Sect. 8.2 we will use the assumption that Y has a Gaussian distribution, but this need not be the only possibility. In fact, in Sect. 8.8 we will show how data can be fit in alternative cases, such as when the variable has a Poisson distribution.

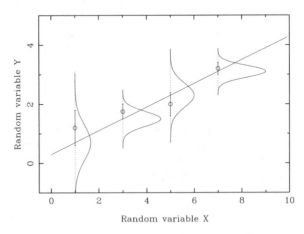

Fig. 8.1 In the fit of two-variable data to a linear function, measurements of the dependent variable Y are made for few selected points of the variable X (in this example $x_1 = 1, x_2 = 3, x_3 = 5$ and $x_4 = 7$). Each datapoint is marked by the *circle with error bars*. The independent variable X is assumed to be known exactly and the size of the *error bar* determines the value of the variance of $y(x_i)$

8.2   Maximum Likelihood Method for Gaussian Data

In many cases the variables $Y(X_i)$ have a Gaussian distribution, as illustrated in Fig. 8.1. The data are represented by points with an error bar and the model for each data point is a Gaussian centered at the value of the parent model $y(x_i)$. The model $y(x)$ can be any function and, as described in the previous section, the standard deviation σ_i is estimated from the data themselves.

The goal of fitting data to a model is twofold: to determine whether the model $y(x)$ is an accurate representation of the data and, at the same time, to determine what values of the adjustable parameters are compatible with the data. The two goals are necessarily addressed together. The starting point is the calculation of the likelihood $\mathcal{L}$ of the data with the model as

$$\mathcal{L} = P(\text{data/model}) = \prod_{i=1}^{N} \frac{1}{\sqrt{2\pi\sigma_i^2}} e^{-\frac{(y_i - y(x_i))^2}{2\sigma_i^2}} =$$

$$\left(\prod_{i=1}^{N} \frac{1}{\sqrt{2\pi\sigma_i^2}} \right) e^{-\sum_{i=1}^{N} \frac{(y_i - y(x_i))^2}{2\sigma_i^2}} \tag{8.2}$$

In the previous equation we have assumed that the measurements $y_i \pm \sigma_i$ are independent of one other, so that the Gaussian probabilities can be simply multiplied. Independence between measurements is a critical assumption in the use of the maximum likelihood method.

The core of the maximum likelihood method is the requirement that the unknown parameters a_k of the model $y = y(x)$ are those that maximize the likelihood of the data. This is the same logic used in the estimate of parameters for a single variable presented in Chap. 5. The method of maximum likelihood results in the condition that the following function has to be minimized:

$$\chi^2 = \sum_{i=1}^{N} \left(\frac{y_i - y(x_i)}{\sigma_i} \right)^2 \tag{8.3}$$

In fact, the factor in (8.2) containing the product of the sample variances is constant with respect to the adjustable parameters and maximization of the likelihood is obtained by minimization of the exponential term.

Equation (8.3) defines the goodness of fit statistic χ^2_{min}, which bears its name from the fact that it is distributed like a χ^2 variable. The number of degrees of freedom associated with this variable depends on the number of free parameters of the model $y(x)$, as will be explained in detail in Chap. 10. The simplest case is that of a model that has no free parameters. In that case, we know already that the minimum χ^2 has exactly N degrees of freedom. Given the form of (8.3), the maximum likelihood method, when applied to Gaussian distribution, is also known as the *least squares* method.

8.3   Least-Squares Fit to a Straight Line, or Linear Regression

When the fitting function is

$$y(x) = a + bx \tag{8.4}$$

the problem of minimizing the χ^2 defined in (8.3) can be solved analytically. The conditions of minimum χ^2 are written as partial derivatives with respect to the two unknown parameters:

$$
\begin{cases}
\dfrac{\partial}{\partial a}\chi^2 = -2\sum \dfrac{1}{\sigma_i^2}(y_i - a - bx_i) = 0 \\[2ex]
\dfrac{\partial}{\partial b}\chi^2 = -2\sum \dfrac{x_i}{\sigma_i^2}(y_i - a - bx_i) = 0
\end{cases}
\tag{8.5}
$$

$$
\Rightarrow
\begin{cases}
\sum \dfrac{y_i}{\sigma_i^2} = a\sum \dfrac{1}{\sigma_i^2} + b\sum \dfrac{x_i}{\sigma_i^2} \\[2ex]
\sum \dfrac{x_i y_i}{\sigma_i^2} = a\sum \dfrac{x_i}{\sigma_i^2} + b\sum \dfrac{x_i^2}{\sigma_i^2}
\end{cases}
\tag{8.6}
$$

which is a system of two equations in two unknowns. The solution is

$$
\begin{cases}
a = \dfrac{1}{\Delta}
\begin{vmatrix}
\sum \dfrac{y_i}{\sigma_i^2} & \sum \dfrac{x_i}{\sigma_i^2} \\[2ex]
\sum \dfrac{x_i y_i}{\sigma_i^2} & \sum \dfrac{x_i^2}{\sigma_i^2}
\end{vmatrix}; \\[5ex]
b = \dfrac{1}{\Delta}
\begin{vmatrix}
\sum \dfrac{1}{\sigma_i^2} & \sum \dfrac{y_i}{\sigma_i^2} \\[2ex]
\sum \dfrac{x_i}{\sigma_i^2} & \sum \dfrac{x_i y_i}{\sigma_i^2}
\end{vmatrix}.
\end{cases}
\tag{8.7}
$$

where

$$\Delta = \sum \frac{1}{\sigma_i^2}\sum \frac{x_i^2}{\sigma_i^2} - \sum \frac{x_i}{\sigma_i^2}\sum \frac{x_i}{\sigma_i^2}. \tag{8.8}$$

Equation (8.7) provides the solution for the best-fit parameters of the linear model. The determination of the parameters of the linear model is known as *linear regression*.

When all errors are identical, $\sigma_i = \sigma$, it is easy to show that the best-fit parameters estimated by the least-squares method are equivalent to

$$
\begin{cases}
b = \dfrac{Cov(X, Y)}{Var(X)} \\
a = E(Y) - bE(X)
\end{cases}
\tag{8.9}
$$

[see Problem (8.9)]. This means that, in the absence of correlation between the two variables, the best-fit slope will be zero and the value of a is simply the linear average of the measurements.

8.4 Multiple Linear Regression

The method outlined above in Sect. 8.3 can be generalized to a fitting function of the form

$$
y(x) = \sum_{k=1}^{m} a_k f_k(x).
\tag{8.10}
$$

Equation (8.10) describes a function that is linear in the m parameters. In this case one speaks of *multiple linear regression*, or simply multiple regression. The functions $f_k(x)$ can have any analytical form. The linear regression described in the previous section has only two such function, $f_1(x) = 1$ and $f_2(x) = x$. A common case is when the functions are polynomials,

$$
f_k(x) = x^k.
\tag{8.11}
$$

The important feature to notice is that the functions $f_k(x)$ do not depend on the parameters a_k.

We want to find an analytic solution to the minimization of the χ^2 with the fitting function in the form of (8.10). As we have seen, this includes the simple linear regression as a special case. In the process of χ^2 minimization we will also determine the variance and the covariances on the fitted parameters a_k, since no fitting is complete without an estimate of the errors and of the correlation between the coefficients. As a special case we will therefore also find the variances and covariance between the fit parameters a and b for the linear regression.

8.4.1   Best-Fit Parameters for Multiple Regression

Minimization of χ^2 with respect to the m parameters a_k is obtaining by taking partial derivatives over the m unknown parameters a_k.

This yields the following m equations:

$$\frac{\partial}{\partial a_l} \sum_{i=1}^{N} \left(\frac{(y_i - \sum_{k=1}^{m} a_k f_k(x_i))^2}{\sigma_i^2} \right) = 0$$

or

$$-2 \sum_{i=1}^{N} \left(\frac{y_i - \sum_{k=1}^{m} a_k f_k(x_i)}{\sigma_i^2} \right) f_l(x_i) = 0.$$

These equations can be written as

$$\sum_{i=1}^{N} \frac{f_l(x_i)}{\sigma_i^2} \left(y_i - \sum_{k=1}^{m} a_k f_k(x_i) \right) = 0 \qquad (8.12)$$

leading to

$$\sum_{i=1}^{N} \frac{f_l(x_i) y_i}{\sigma_i^2} = \sum_{k=1}^{m} a_k \sum_{i=1}^{N} \frac{f_k(x_i) f_l(x_i)}{\sigma_i^2} \qquad l = 1, \ldots, m. \qquad (8.13)$$

Equation (8.13) are m coupled equations in the parameters a_k, which can be solved using matrix algebra, as described below. Notice that the term $f_l(x_i)$ is the lth model component (thus the index l is not summed over), and the index i runs from 1 to N, where N is the number of data points.

The best-fit parameters are therefore obtained by defining the row vectors $\boldsymbol{\beta}$ and $\boldsymbol{a}$ and the $m \times m$ symmetric matrix $\boldsymbol{A}$ as

$$\begin{cases} \boldsymbol{\beta} &= (\beta_1, \ldots, \beta_m) \qquad \text{in which } \beta_k = \sum_{i=1}^{N} f_k(x_i) y_i / \sigma_i^2 \\ \boldsymbol{a} &= (a_1, \ldots, a_m) \qquad \text{(model parameters)} \\ A_{lk} &= \sum_{i=1}^{N} \frac{f_l(x_i) f_k(x_i)}{\sigma_i^2} \qquad \text{(l, k component of the m×m matrix } \boldsymbol{A}) \end{cases}$$

With these definitions, (8.13) can be rewritten in matrix form as

$$\beta = aA, \tag{8.14}$$

and therefore the task of estimating the best-fit parameters is that of inverting the matrix A, which can be done numerically. The m best-fit parameters a_k are placed in a row vector a (of dimensions $1 \times m$) and are given by

$$a = \beta A^{-1}. \tag{8.15}$$

The $1 \times m$ row vector β and the $m \times m$ matrix A can be calculated from the data and the fit functions $f_k(x)$.

8.4.2 *Parameter Errors and Covariances for Multiple Regression*

To calculate errors in the best-fit parameters, we treat parameters a_k as functions of the measurements, $a_k = a_k(y_i)$. Therefore we can use the error propagation method to calculate variances and covariances between parameters as:

$$\begin{cases} \sigma_{a_k}^2 = \sum_{i=1}^{N} \left(\dfrac{\partial a_k}{\partial y_i} \right)^2 \sigma_i^2 \\ \sigma_{a_l a_j}^2 = \sum_{i=1}^{N} \dfrac{\partial a_l}{\partial y_i} \dfrac{\partial a_j}{\partial y_i} \sigma_i^2. \end{cases} \tag{8.16}$$

We have used the fact that the error in each measurement y_i is given by σ_i and that the measurements are independent.

We show that the variance $\sigma_{a_l a_j}^2$ is given by the l, j term of the inverse of the matrix A, which we define as the *error matrix*

$$\varepsilon = A^{-1}. \tag{8.17}$$

The error matrix ε is a symmetric matrix, of which the diagonal terms contain the variances of the fitted parameters and the off-diagonal terms contain the covariances.

Proof Use the matrix equation $a = \beta \varepsilon$ to write

$$a_l = \sum_{k=1}^{m} \beta_k \varepsilon_{kl} = \sum_{k=1}^{m} \varepsilon_{kl} \sum_{i=1}^{N} \frac{y_i f_k(x_i)}{\sigma_i^2} \Rightarrow \frac{\partial a_l}{\partial y_i} = \sum_{k=1}^{m} \varepsilon_{kl} \frac{f_k(x_i)}{\sigma_i^2}.$$

The equation above can be used into (8.16) to show that

$$
\sigma_{a_l\,a_j}^2 = \sum_{i=1}^{N}\left[\sigma_i^2 \sum_{k=1}^{m}\left(\varepsilon_{jk}\frac{f_k(x_i)}{\sigma_i^2}\right) \times \sum_{p=1}^{m}\left(\varepsilon_{lp}\frac{f_p(x_i)}{\sigma_i^2}\right)\right]
$$

in which the indices k and p indicate the m model parameters, and the index i is used for the sum over the N measurements.

$$
\Rightarrow \sigma_{a_l\,a_j}^2 = \sum_{k=1}^{m}\varepsilon_{jk}\sum_{p=1}^{m}\varepsilon_{lp}\sum_{i=1}^{n}\frac{f_k(x_i)f_p(x_i)}{\sigma_i^2} = \sum_{k=1}^{m}\varepsilon_{jk}\sum_{p=1}^{m}\varepsilon_{lp}A_{pk}.
$$

Now recall that A is the inverse of ε, and therefore the expression above can be simplified to

$$
\sigma_{a_l\,a_j}^2 = \sum_{k}\varepsilon_{jk}1_{kl} = \varepsilon_{jl}. \tag{8.18}
$$

□

8.4.3   Errors and Covariance for Linear Regression

The results of Sect. 8.4.2 apply also to the case of linear regression as a special case. We therefore use these results to estimate the errors in the linear regression parameters a and b and their covariance. In this case, the functions $f_l(x_i)$ are given, respectively, by $f_1(x) = 1$ and $f_2(x) = x$ and therefore the matrix A is a 2×2 symmetric matrix with the following elements:

$$
\begin{cases}
A_{11} = \displaystyle\sum_{i=1}^{N} 1/\sigma_i^2 \\[2ex]
A_{12} = A_{21} = \displaystyle\sum_{i=1}^{N} x_i/\sigma_i^2 \\[2ex]
A_{22} = \displaystyle\sum_{i=1}^{N} x_i^2/\sigma_i^2.
\end{cases} \tag{8.19}
$$

The inverse matrix $A^{-1} = \varepsilon$ is given by

$$
\begin{cases}
\varepsilon_{11} = A_{22}/\Delta \\[1.5ex]
\varepsilon_{12} = \varepsilon_{21} = -A_{12}/\Delta \\[1.5ex]
\varepsilon_{22} = A_{11}/\Delta
\end{cases} \tag{8.20}
$$

in which Δ is the determinant of A. Using (8.14) we calculate β:

$$\begin{cases} \beta_1 = \sum y_i/\sigma_i^2 \\ \beta_2 = \sum y_i x_i/\sigma_i^2 \end{cases} \tag{8.21}$$

and thus proceed to calculating the best-fit parameters and their errors. The best-fit parameters, already found in Sect. 8.3, are given by

$$(a, b) = (\beta_1, \beta_2) \begin{bmatrix} \varepsilon_{11} & \varepsilon_{12} \\ \varepsilon_{21} & \varepsilon_{22} \end{bmatrix}$$

which give the same results as previously found in (8.7). We are now in a position to estimate the errors in the best-fit parameters:

$$\begin{cases} \sigma_a^2 = \varepsilon_{11} = \dfrac{1}{\Delta} \sum_{i=1}^{N} x_i^2/\sigma_i^2 \\[2ex] \sigma_b^2 = \varepsilon_{22} = \dfrac{1}{\Delta} \sum_{i=1}^{N} 1/\sigma_i^2 \\[2ex] \sigma_{ab}^2 = \varepsilon_{12} = -\dfrac{1}{\Delta} \sum_{i=1}^{N} x_i/\sigma_i^2. \end{cases} \tag{8.22}$$

The importance of (8.22) is that the errors in the parameters a and b and their covariance can be computed analytically from the N measurements. This simple solution make the linear regression very simple to implement.

8.5 Special Cases: Identical Errors or No Errors Available

It is common to have a dataset where all measurements have the same error. When all errors in the dependent variable are identical ($\sigma_i = \sigma$) (8.7) and (8.22) for the linear regression are simplified to

$$\begin{cases} a = \dfrac{1}{\Delta} \dfrac{1}{\sigma^4} \left(\sum_{i=1}^{N} y_i \sum_{i=1}^{N} x_i^2 - \sum_{i=1}^{N} x_i \sum_{i=1}^{N} x_i y_i \right) \\[2ex] b = \dfrac{1}{\Delta} \dfrac{1}{\sigma^4} \left(N \sum_{i=1}^{N} x_i y_i - \sum_{i=1}^{N} y_i \sum_{i=1}^{N} x_i \right) \\[2ex] \Delta = \dfrac{1}{\sigma^4} \left(N \sum_{i=1}^{N} x_i^2 - \left(\sum_{i=1}^{N} x_i \right)^2 \right) \end{cases} \tag{8.23}$$

and

$$
\begin{cases}
\sigma_a^2 = \dfrac{1}{\Delta}\dfrac{1}{\sigma^2}\displaystyle\sum_{i=1}^{N} x_i^2 \\[2ex]
\sigma_b^2 = \dfrac{1}{\Delta}\dfrac{N}{\sigma^2} \\[2ex]
\sigma_{ab}^2 = -\dfrac{1}{\Delta}\dfrac{1}{\sigma^2}\displaystyle\sum_{i=1}^{N} x_i.
\end{cases}
\tag{8.24}
$$

The important feature is that the best-fit parameters are *independent* of the value σ of the error.

For dataset that do not have errors available it is often reasonable to assume that all datapoints have the same error and calculate the best-fit parameters without the need to specify the value of σ. The variances, which depend on the error, cannot however be estimated. The absence of errors therefore limits the applicability of the linear regression method. It is in general not possible to reconstruct the errors σ_i a posteriori. In fact, the errors are the result of the experimental procedure that led to the measurement of the variables. A typical example is the case in which each of the variables $y(x_i)$ was measured via repeated experiments which led to the measurement of $y(x_i)$ as the mean of the measurements and its error as the square root of the sample variance. In the absence of the "raw" data that permit the calculation of the sample variance, it is simply not possible to determine the error in σ_i.

Another possibility to use a dataset that does not report the errors in the measurements is based on the assumption that the fitting function $y = f(x)$ *is* the correct description for the data. Under this assumption, one can estimate the errors, assumed to be identical for all variables in the dataset, via a *model sample variance* defined as

$$
\sigma^2 = \frac{1}{N-m}\sum_{i=1}^{N}(y_i - \hat{y}_i)^2
\tag{8.25}
$$

where $\hat{y}_i$ is the value of the fitting function $f(x_i)$ evaluated with the best-fit parameters, which must be first obtained by a fit assuming identical errors. The underlying assumption behind the use of (8.25) is to treat each measurement y_i as drawn from a parent distribution $f(x_i)$, $i = 1,\ldots N$, e.g., assuming that the model is the correct description for the data. In the case of a linear regression, $m = 2$, since two parameters (a and b) are estimated from the data. It will become clear in Sect. 10.1 that this procedure comes at the expenses of the ability to determine whether the dataset is in fact well fit by the function $y = f(x)$, since that is the working assumption.

In the case of no errors reported, it may not be clear which variable is to be treated as independent. We have shown in (8.9) that, when no errors are reported, the best-fit parameters can be written as

$$\begin{cases} b = \dfrac{Cov(X, Y)}{Var(X)} \\ a = E(Y) - bE(X). \end{cases}$$

This equation clearly shows that the best-fit linear regression model is *dependent* on the choice of which between x and y is considered the independent variable. In fact, if y is regarded as the independent variable, and the data fit to the model

$$x = a' + b'y \tag{8.26}$$

the least-squares method gives the best-fit slope of

$$b' = \frac{Cov(X, Y)}{Var(Y)}.$$

When the model is rewritten in the usual form

$$y = a_{X/Y} + b_{X/Y}x$$

in which the notation X/Y means "X given Y," the best-fit model parameters are

$$\begin{cases} b_{X/Y} = -\dfrac{1}{b'} = \dfrac{Var(Y)}{Cov(X, Y)} \\ a_{X/Y} = E(Y) - b_{X/Y}E(X) \end{cases}$$

and therefore the two linear models assuming x or y as independent variable will be different from one another. It is up to the data analyst to determine which of the two variables is to be considered as independent when there is a dataset of (x_i, y_i) measurements with no errors reported in either variable. Normally the issue is resolved by knowing how the experiment was performed, e.g., which variable had to be assumed or calculated first in order to calculate or measure the second. Additional considerations for the fit of two-variable datasets are presented in Chap. 12.

8.6 A Classic Experiment: Edwin Hubble's Discovery of the Expansion of the Universe

In the early twentieth century astronomers were debating whether "nebulae," now known to be external galaxies, were in fact part of our own Galaxy, and there was no notion of the Big Bang and the expansion of the universe. Edwin Hubble pioneered the revolution via a seemingly simple observation that a

(continued)

number of "nebulae" moved away from the Earth with a velocity v that is proportional to their distance d, known as *Hubble's law*

$$v = H_0 d. \tag{8.27}$$

The quantity H_0 is the *Hubble constant*, typically measured in the units of $km\,s^{-1}\,Mpc^{-1}$, where Mpc indicates a distance of 10^6 parsec. The data used by Hubble [21] is summarized in Table 8.1.

The quantity m is the apparent magnitude, related to the distance via the following relationship,

$$\log d = \frac{m - M + 5}{5} \tag{8.28}$$

where $M = -13.8$ is the absolute magnitude, also measured by Hubble as part of the same experiment, and considered as a constant for the purpose of this dataset, and d is measured in parsecs.

The first part of the experiment consisted in fitting the (v, m) dataset to a relationship that is linear in $\log v$,

$$\log v = a + b \cdot m \tag{8.29}$$

where a and b are the adjustable parameters of the linear regression. Instead of performing the linear regression described in Sects. 8.3 and 8.4.3, Hubble reported two different fit results, one in which he determined also the error in a,

$$\log v = (0.202 \pm 0.007) \cdot m + 0.472 \tag{8.30}$$

and one in which he fixed $a = 0.2$, and determined the error in b:

$$\log v = 0.2 \cdot m + 0.507 \pm 0.012. \tag{8.31}$$

Using (8.31) into (8.28), Hubble determined the following relationship between velocity and distance,

$$\log \frac{v}{d} = 0.2M - 0.493 = -3.253 \tag{8.32}$$

and this results in the measurement of his name-sake constant, $H_0 = v/d = 10^{-3.253} = 558 \times 10^{-6}\,km\,s^{-1}\,pc^{-1}$, or $558\,km\,s^{-1}\,Mpc^{-1}$.

(continued)

Table 8.1 Data from E. Hubble's measurements

Name of nebula	Mean velocity km s^{-1}	Number of velocities	Mean m
Virgo	890	7	12.5
Pegasus	3810	5	15.5
Pisces	4630	4	15.4
Cancer	4820	2	16.0
Perseus	5230	4	16.4
Coma	7500	3	17.0
Ursa Major	11,800	1	18.0
Leo	19,600	1	19.0
(No name)	2350	16	13.8
(No name)	630	21	11.6

Fig. 8.2 Best-fit linear regression model for the data in Table 8.1

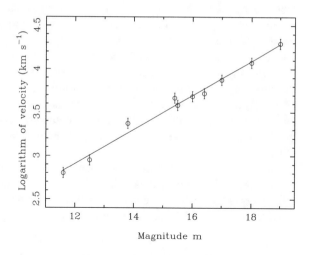

Example 8.1 The data from Hubble's experiment are a typical example of a dataset in which no errors were reported. A linear fit can be initially performed by assuming equal errors, and the best-fit line is reported in red in Fig. 8.2. Using (8.25), the common errors in the dependent variables $\log v(x_i)$ are found to be $\sigma = 0.06$, the best-fit parameters of the models are $a = 0.55 \pm 0.13$, $b = 0.197 \pm 0.0085$, and the covariance is $\sigma_{ab}^2 = -1.12 \times 10^{-3}$, for a correlation coefficient of -0.99. The uncertainties and the covariance are measured using the method of (8.23). The best-fit line is shown in Fig. 8.2 as a solid line.   ◇

8.7   Maximum Likelihood Method for Non-linear Functions

The method described in Sect. 8.4 assumes that the model is linear in the fitting parameters a_k. This requirement is, however, not necessary to apply the maximum likelihood criterion. We can assume that the relationship $y = f(x)$ has any analytic form and still apply the maximum likelihood criterion for the N measurements [see (8.3)]. The best-fit parameters are still those that minimize the χ^2 statistic. In fact, all considerations leading to (8.3) do not require a specific form for the fitting function $y = f(x)$. The assumption that must still be satisfied is that each variable y_i is Gaussian distributed, in order to obtain the likelihood in the form of (8.2).

The only complication for nonlinear functions is that an analytic solution for the best-fit values and the errors is in general no longer available. This is often not a real limitation, since numerical methods to minimize the χ^2 are available. The most straightforward way to achieve a minimization of the χ^2 as function of all parameters is to construct an m dimensional grid of all possible parameter values, evaluate the χ^2 at each point, and then find the global minimum. The parameter values corresponding to this minimum can be regarded as the best estimate of the model parameters. The direct grid-search method becomes rapidly unfeasible as the number of free parameters increases. In fact, the full grid consists of n^m points, where n is the number of discrete points into which each parameter is investigated. One typically wants a large number of n, so that parameter space is investigated with the necessary resolution, and the time to evaluate the entire space depends on how efficiently a calculation of the likelihood can be obtained. Among the methods that can be used to bypass the calculation of the entire grid, one of the most efficient and popular is the Markov chain Monte Carlo technique, which is discussed in detail in Chap. 16.

To find the uncertainties in the parameters using the grid search method requires a knowledge of the expected variation of the χ^2 around the minimum. This problem will be explained in the next chapter. The Markov chain Monte Carlo also technique provides estimates of the parameter errors and their covariance.

8.8   Linear Regression with Poisson Data

The two main assumptions made so far in the maximum likelihood method are that the random variables $y(x_i)$ are Gaussian and the variance of these variables are estimated from the data as the measured variance σ_i^2. In the following we discuss how the maximum likelihood method can be applied to data without making the assumption of a Gaussian distribution. One case of great practical interest is when variables have Poisson distribution, which is the case in many counting experiments. For simplicity we focus on the case of linear regression, although all considerations can be extended to any type of fitting function.

When $y(x_i)$ is assumed to be Poisson distributed, the dataset takes the form of (x_i, y_i), in which the values y_i are intended as integers resulting from a counting experiment. In this case, the value $y(x_i) = a + bx_i$ is considered as the parent mean for a given choice of parameters a and b,

$$\mu_i = y(x_i) = a + bx_i. \tag{8.33}$$

The likelihood is calculated using the Poisson distribution and, under the hypothesis of independent measurements, it is

$$\mathcal{L} = \prod_{i=1}^{N} \frac{y(x_i)^{y_i} e^{-y(x_i)}}{y_i!}. \tag{8.34}$$

Once we remove the Gaussian assumption, there is no χ^2 function to minimize, but the whole likelihood must be taken into account. It is convenient to minimize the logarithm of the likelihood,

$$\ln \mathcal{L} = \sum_{i=1}^{N} y_i \ln y(x_i) - \sum_{i=1}^{N} y(x_i) + A \tag{8.35}$$

where $A = -\sum \ln y_i!$ does not depend on the model parameters but only on the fixed values of the datapoints. Minimization of the logarithm of the likelihood is equivalent to a minimization of the likelihood, since the logarithm is a monotonic function of its argument. The principle of maximum likelihood requires that

$$\begin{cases} \dfrac{\partial}{\partial a} \ln \mathcal{L} = 0 \\ \dfrac{\partial}{\partial b} \ln \mathcal{L} = 0 \end{cases} \Rightarrow \begin{cases} N = \sum \dfrac{y_i}{a + bx_i} \\ \sum x_i = \sum \dfrac{x_i y_i}{a + bx_i}. \end{cases} \tag{8.36}$$

The fact that the minimization was done with respect to $\ln \mathcal{L}$ instead of χ^2 is a significant difference relative to the case of Gaussian data. For Poisson data we define the fit statistic C as

$$C = -2 \ln \mathcal{L} + B, \tag{8.37}$$

where B is a constant term. This is called the *Cash statistic*, after a paper by Cash in 1979 [9]. This statistic will be discussed in detail in Sect. 10.2 and it will be shown to have the property of being distributed like a χ^2 distribution with $N - m$ degrees of freedom in the limit of large N. This result is extremely important, as it allows to proceed with the Poisson fitting in exactly the same way as in the more common Gaussian case in order to determine the goodness of fit.

There are many cases in which a Poisson dataset can be approximated with a Gaussian dataset, and therefore use χ^2 as fit statistic. When the number of counts in each measurement y_i is approximately larger than 10 or so (see Sect. 3.4), the Poisson distribution is accurately described by a Gaussian of same mean and variance. When the number of counts is lower, one method to turn a Poisson dataset into a Gaussian one is to *bin* the data into fewer variables of larger count rates. There are, however, many situations in which such binning is not desirable, especially when the dependent variable y has particular behaviors for certain values of the independent variable x. In those cases, binning of the data smears those features, which we would like to retain in the datasets. In those cases, the best option is to use the Poisson fitting method described in this section, and use C as the fit statistic instead.

Example 8.2 Consider a set of $N = 4$ measurements (3,5,4,2) to be fit to a constant model, $y = a$. In this case, (8.36) become

$$a = \frac{1}{N} \sum_{i=1}^{N} y_i$$

which means that the maximum likelihood estimator of a constant model, for a Poisson dataset, is the average of the measurements. The maximum likelihood best-fit parameter is therefore $a = 3.5$.   ◇

Summary of Key Concepts for this Chapter

☐ *ML fit to two-dimensional data*: A method to find best-fit parameters of a model fit to x, y data assuming that one variable (typically x) is the independent variable.

☐ *Linear regression*: ML fit to a linear model, best-fit parameters when all errors are identical are

$$\begin{cases} b = \dfrac{Cov(X, Y)}{Var(X)} \\ a = E[Y] - bE[X] \end{cases}$$

(assuming x as independent variable).

☐ *Multiple linear regression*: An extension of the linear regression to models of the type

$$y = \sum a_k f_k(x).$$

☐ *Model sample variance*: When errors in the dependent variable (y) are not known, they can be estimated via the model sample variance

$$\sigma^2 = \frac{1}{N - m} \sum (y_i - \hat{y}_i)^2$$

where m is the number of model parameters.

Problems

8.1 Consider the data from Hubble's experiment in Table 8.1.

(a) Determine the best-fit values of the fit to a linear model for $(m, \log v)$ assuming that the dependent variables have a common value for the error.
(b) Using the best-fit model determined above, estimate the error from the data and the best-fit model, and then estimate the errors in the parameters a and b, and the correlation coefficient between a and b.
(c) Calculate the minimum χ^2 of the linear fit, using the common error as estimated in part (a).

8.2 Consider the following two-dimensional data, in which X is the independent variable, and Y is the dependent variable assumed to be derived from a photon-counting experiment:

x_i	y_i
0.0	25
1.0	36
2.0	47
3.0	64
4.0	81

(a) Determine the errors associated with the dependent variables Y_i.
(b) Find the best-fit parameters a, b of the linear regression curve

$$y(x) = a + bx;$$

also compute the errors in the best-fit parameters and the correlation coefficient between them;
(c) Calculate the minimum χ^2 of the fit, and the corresponding probability to exceed this value.

8.3 Consider the following Gaussian dataset in which the dependent variables are assumed to have the same unknown standard deviation σ,

x_i	y_i
0.0	0.0
1.0	1.5
2.0	1.5
3.0	2.5
4.0	4.5
5.0	5.0

The data are to be fit to a linear model.

(a) Using the maximum likelihood method, find the analytic relationships between $\sum x_i$, $\sum y_i$, $\sum x_i y_i$, $\sum x_i^2$, and the model parameters a and b.
(b) Show that the best-fit values of the model parameters are $a = 0$ and $b = 1$.

8.4 In the case of a maximum likelihood fit to a 2-dimensional dataset with equal errors in the dependent variable, show that the conditions for having best-fit parameters $a = 0$ and $b = 1$ are

$$
\begin{cases}
\sum_{i=1}^{N} y_i = \sum_{i=1}^{N} x_i \\
\sum_{i=1}^{N} x_i^2 = \sum_{i=1}^{N} x_i y_i.
\end{cases}
\tag{8.38}
$$

8.5 Show that the best-fit parameter b of a linear fit to a Gaussian dataset is insensitive to a change of all datapoints by the same amount Δx, or by the same amount Δy. You can show that this property applies in the case of equal errors in the dependent variable, although the same result applies also for the case of different errors.

8.6 The background rate in a measuring apparatus is assumed to be constant with time. N measurements of the background are taken, of which $N/2$ result in a value of $\bar{y}+\Delta$, and $N/2$ in a value $\bar{y}-\Delta$. Determine the sample variance of the background rate.

8.7 Find an analytic solution for the best-fit parameters of a linear model to the following Poisson dataset:

x	y
-2	-1
-1	0
0	1
1	0
2	2

8.8 Use the data provided in Table 6.1 to calculate the best-fit parameters a and b for the fit to the radius vs. pressure ratio data, and the minimum χ^2. For the fit, you can assume that the radius is known exactly, and that the standard deviation of the pressure ratio is obtained as a linear average of the positive and negative errors.

8.9 Show that, when all measurement errors are identical, the least squares estimators of the linear parameters a and b are given by $b = Cov(X, Y)/Var(X)$ and $a = E(Y) - bE(X)$.

Chapter 9
Multi-Variable Regression

Abstract In many situations a variable of interest depends on several other variables. Such multi-variable data is common across the sciences and in many other fields such as economics and business. Multi-variable analysis can be performed in a simple and effective way when the relationship that links the variable of interest to the other quantities is linear. In this chapter we study the method of multi-variable regression and show how it is related to the multiple regression described in Chap. 8 which applies to the traditional two-variable dataset. This chapter also presents methods for hypothesis testing on the multi-variable regression and its parameters.

9.1 Multi-Variable Datasets

Two-dimensional dataset studied so far include an independent variable (X) and a dependent variable (Y) and the data take the form of a collection of $(x_i, y_i \pm \sigma_i)$, where $i = 1, \ldots, N$ and N indicates the total number of measurements. In Chap. 8 we have developed a method to fit such two-dimensional data. In that case, the linear regression formula takes the form of $y(x) = a + bx$, where a and b are the parameters of the linear regression.

Datasets that have measurements for three or more variables are referred to as multi-variable datasets. An example of multi-variable dataset is presented in Sect. 9.2, which reports measurements of different characteristics of irises performed by Fisher and Anderson in 1936 [14]. Each of those measurement comprises four quantities: the sepal length, sepal width, petal length, and petal width of 50 irises. For several multi-variable datasets such as that of Fisher and Anderson it is often unclear which variable is the dependent one. It typically depends on what question we want to address with the data: if we want to determine the sepal length of an iris flower based on the sepal width, petal length, and petal width, then the sepal length becomes the dependent variable and the remaining three are the independent variables.

Using multi-variable datasets to predict or forecast the behavior of one quantity based on several other variables is a fundamental topic in data analysis. It is common throughout the sciences and especially used in such fields as economics or behavioral sciences, where a number of possible factors can be used to predict one quantity of interest. An example is to predict the score on a college-admission

© Springer Science+Busines Media New York 2017

M. Bonamente, *Statistics and Analysis of Scientific Data*, Graduate Texts in Physics, DOI 10.1007/978-1-4939-6572-4_9

test based on factors such as the grade-point average during the sophomore and the junior year, a measure of the motivation of the student and their economic status. Another example is to predict the price of a stock based, e.g., on the overall index of the stock exchange, a consumer's index for goods in the relevant class and the rate of treasury bonds. To address any such questions clearly requires a multi-variable dataset that has several measurements for all quantities of interest.

In this chapter we develop a method to determine the relationship between one of the quantities of a multi-dimensional datasets based on the others, assuming a linear relationship among the variables. This method will also let us study whether one or more of the quantities are in fact not useful in predicting the variable of interest. For example, we may find that the treasury bond rates are irrelevant in predicting the stock value of a given corporation and therefore we can focus only on those variables that are useful in predicting its stock price.

9.2   A Classic Experiment: The R.A. Fisher and E. Anderson Measurements of Iris Characteristics

R.A. Fisher is one of the fathers of modern statistics. In 1936 he published the paper *The Use of Multiple Measurements in Taxonomic Problems* reporting measurements of several characteristics of three species of the iris plant [14].

Figure 9.1 reproduces the original measurements, performed by E. Anderson, of the petal length and the sepal length of 150 iris plants of the species *Iris setosa*, *Iris versicolor*, and *Iris virginica*. The measurements are in millimeters (mm). Fisher's aim was to find a linear combination of the four characteristics that would be best suited to identify one species from the others. It is already clear from the data in Fig. 9.1 that one of the quantites (e.g., the sepal length) may be used as a discriminator among the three species. R.A. Fisher used this dataset to find a linear combination of the four quantities that would improve the classification of irises.

The dataset is a classic example of a multi-variate dataset, in which several variables are measured simultaneously and independently. In addition to Fisher's original purpose, these data can also be used to determine whether one of the characteristics, e.g., the sepal length, can be efficiently predicted based on any (or all) of the other characteristics. For example, one could expect that the length of the sepal (which is part of the calyx of the flower) is related linearly to its width, or to the length of the petal. Assuming a linear relationship among the variables, we set

$$SL = a + bSW + cPL + dPW \qquad (9.1)$$

where a, b, c, and d are coefficients that we can estimate from the data using the method described in Sect. 9.3.

(continued)

Throughout this chapter we use these data to study the linear regression of (9.1) for the species *Iris setosa*. We will find that the most important variable needed to predict the sepal length is the sepal width, while the measurements of characteristics of petals are not very important in predicting the sepal length.

Iris setosa				Iris versicolor				Iris virginica			
Sepal length	Sepal width	Petal length	Petal width	Sepal length	Sepal width	Petal length	Petal width	Sepal length	Sepal width	Petal length	Petal width
5.1	3.5	1.4	0.2	7.0	3.2	4.7	1.4	6.3	3.3	6.0	2.5
4.9	3.0	1.4	0.2	6.4	3.2	4.5	1.5	5.8	2.7	5.1	1.9
4.7	3.2	1.3	0.2	6.9	3.1	4.9	1.5	7.1	3.0	5.9	2.1
4.6	3.1	1.5	0.2	5.5	2.3	4.0	1.3	6.3	2.9	5.6	1.8
5.0	3.6	1.4	0.2	6.5	2.8	4.6	1.5	6.5	3.0	5.8	2.2
5.4	3.9	1.7	0.4	5.7	2.8	4.5	1.3	7.6	3.0	6.6	2.1
4.6	3.4	1.4	0.3	6.3	3.3	4.7	1.6	4.9	2.5	4.5	1.7
5.0	3.4	1.5	0.2	4.9	2.4	3.3	1.0	7.3	2.9	6.3	1.8
4.4	2.9	1.4	0.2	6.6	2.9	4.6	1.3	6.7	2.5	5.8	1.8
4.9	3.1	1.5	0.1	5.2	2.7	3.9	1.4	7.2	3.6	6.1	2.5
5.4	3.7	1.5	0.2	5.0	2.0	3.5	1.0	6.5	3.2	5.1	2.0
4.8	3.4	1.6	0.2	5.9	3.0	4.2	1.5	6.4	2.7	5.3	1.9
4.8	3.0	1.4	0.1	6.0	2.2	4.0	1.0	6.8	3.0	5.5	2.1
4.3	3.0	1.1	0.1	6.1	2.9	4.7	1.4	5.7	2.5	5.0	2.0
5.8	4.0	1.2	0.2	5.6	2.9	3.6	1.3	5.8	2.8	5.1	2.4
5.7	4.4	1.5	0.4	6.7	3.1	4.4	1.4	6.4	3.2	5.3	2.3
5.4	3.9	1.3	0.4	5.6	3.0	4.5	1.5	6.5	3.0	5.5	1.8
5.1	3.5	1.4	0.3	5.8	2.7	4.1	1.0	7.7	3.8	6.7	2.2
5.7	3.8	1.7	0.3	6.2	2.2	4.5	1.5	7.7	2.6	6.9	2.3
5.1	3.8	1.5	0.3	5.6	2.5	3.9	1.1	6.0	2.2	5.0	1.5
5.4	3.4	1.7	0.2	5.9	3.2	4.8	1.8	6.9	3.2	5.7	2.3
5.1	3.7	1.5	0.4	6.1	2.8	4.0	1.3	5.6	2.8	4.9	2.0
4.6	3.6	1.0	0.2	6.3	2.5	4.9	1.5	7.7	2.8	6.7	2.0
5.1	3.3	1.7	0.5	6.1	2.8	4.7	1.2	6.3	2.7	4.9	1.8
4.8	3.4	1.9	0.2	6.4	2.9	4.3	1.3	6.7	3.3	5.7	2.1
5.0	3.0	1.6	0.2	6.6	3.0	4.4	1.4	7.2	3.2	6.0	1.8
5.0	3.4	1.6	0.4	6.8	2.8	4.8	1.4	6.2	2.8	4.8	1.8
5.2	3.5	1.5	0.2	6.7	3.0	5.0	1.7	6.1	3.0	4.9	1.8
5.2	3.4	1.4	0.2	6.0	2.9	4.5	1.5	6.4	2.8	5.6	2.1
4.7	3.2	1.6	0.2	5.7	2.6	3.5	1.0	7.2	3.0	5.8	1.6
4.8	3.1	1.6	0.2	5.5	2.4	3.8	1.1	7.4	3.8	6.1	1.9
5.4	3.4	1.5	0.4	5.5	2.4	3.7	1.0	7.9	3.8	6.4	2.0
5.2	4.1	1.5	0.1	5.8	2.7	3.9	1.2	6.4	2.8	5.6	2.2
5.5	4.2	1.4	0.2	6.0	2.7	5.1	1.6	6.3	2.8	5.1	1.5
4.9	3.1	1.5	0.2	5.4	3.0	4.5	1.5	6.1	2.6	5.6	1.4
5.0	3.2	1.2	0.2	6.0	3.4	4.5	1.6	7.7	3.0	6.1	2.3
5.5	3.5	1.3	0.2	6.7	3.1	4.7	1.5	6.3	3.4	5.6	2.4
4.9	3.6	1.4	0.1	6.3	2.3	4.4	1.3	6.4	3.1	5.5	1.8
4.4	3.0	1.3	0.2	5.6	3.0	4.1	1.3	6.0	3.0	4.8	1.8
5.1	3.4	1.5	0.2	5.5	2.5	4.0	1.3	6.9	3.1	5.4	2.1
5.0	3.5	1.3	0.3	5.5	2.6	4.4	1.2	6.7	3.1	5.6	2.4
4.5	2.3	1.3	0.3	6.1	3.0	4.6	1.4	6.9	3.1	5.1	2.3
4.4	3.2	1.3	0.2	5.8	2.6	4.0	1.2	5.8	2.7	5.1	1.9
5.0	3.5	1.6	0.6	5.0	2.3	3.3	1.0	6.8	3.2	5.9	2.3
5.1	3.8	1.9	0.4	5.6	2.7	4.2	1.3	6.7	3.3	5.7	2.5
4.8	3.0	1.4	0.3	5.7	3.0	4.2	1.2	6.7	3.0	5.2	2.3
5.1	3.8	1.6	0.2	5.7	2.9	4.2	1.3	6.3	2.5	5.0	1.9
4.6	3.2	1.4	0.2	6.2	2.9	4.3	1.3	6.5	3.0	5.2	2.0
5.3	3.7	1.5	0.2	5.1	2.5	3.0	1.1	6.2	3.4	5.4	2.3
5.0	3.3	1.4	0.2	5.7	2.8	4.1	1.3	5.9	3.0	5.1	1.8

Fig. 9.1 Measurements of three iris species from the 1936 R.A. Fisher paper [14]. Measurements are in mm

9.3   The Multi-Variable Linear Regression

Consider a dataset of N measurements of $m + 1$ variables which we call $Y, X_1, \ldots,$ X_m. We can use the index i to indicate the measurement, $i = 1, \ldots, N$, and the index k for the variables X_k, $k = 1, \ldots, m$. Each set of measurements is therefore indicated as $(y_i \pm \sigma_i, x_{1i}, \ldots, x_{mi})$.

We write the variable Y as a linear function of the m variables X_i,

$$y(x) = a_0 + a_1 x_1 + \cdots + a_m x_m = a_0 + \sum_{k=1}^{m} a_k x_k. \tag{9.2}$$

The goal is to find the values for the $m + 1$ coefficients a_k, $k = 0, \ldots, m$ that minimize the χ^2 function

$$\chi^2 = \sum_{i=1}^{N} \left(\frac{y_i - y(x_i)}{\sigma_i} \right)^2. \tag{9.3}$$

The quantity $y(x_i) = a_0 + a_1 x_{1i} + \cdots + a_m x_{mi}$ is the value of $y(x)$ calculated for the i-th set of measurements of the X_k's. The coefficient a_0 is an overall offset, equivalent to the constant a for the two-dimensional linear regression function $y = a + bx$.

This form for the χ^2 function is the same as that used for the multiple linear regression of Sect. 8.4. The only change is that the measurements x_{ki} take the place of the functions $f_k(x_i)$. The quantity σ_i is interpreted as the error in the variable Y, which is the dependent quantity in this regression. As in the case of the two-variable dataset, we ignore the errors in the variables X_k (see Chap. 12 for an extension of the two-variable dataset regression with errors in both variables). When the multi-variable dataset has no errors, or if we choose to ignore the errors in the Y variable as well, we can omit the σ_i term in (9.3). This corresponds to assuming a uniform error for all measurements.

The similarity in form between the χ^2 functions to minimize for the present multi-variable linear regression and the multiple regression of Sect. 8.4 means that we have already at hand a solution for the coefficients of the regression and their errors. We need to make the following substitutions:

$$\begin{cases} f_1(x) = 1 \equiv x_0 \text{ (thus } x_{0i}\text{'s are not needed)} \\ f_{k+1}(x) = x_k, \ k = 1, \ldots, m. \end{cases} \tag{9.4}$$

and use the solution from Sect. 8.4 with $m + 1$ terms. The best-fit parameters a_k can be found via the matrix equation

$$a = \beta A^{-1}, \tag{9.5}$$

where the row vectors $\boldsymbol{\beta}$ and $\boldsymbol{a}$ and the $(m+1) \times (m+1)$ symmetric matrix $\boldsymbol{A}$ are given by

$$
\begin{cases}
\boldsymbol{\beta} & = (\beta_0, \beta_1, \ldots, \beta_m) \qquad \text{in which } \beta_k = \sum_{i=1}^{N} x_{ki} y_i / \sigma_i^2 \\[2mm]
\boldsymbol{a} & = (a_0, a_1, \ldots, a_m) \\[2mm]
A_{lk} & = \sum_{i=1}^{N} \dfrac{x_{li} x_{ki}}{\sigma_i^2} \qquad (l, k \text{ component of } \boldsymbol{A}).
\end{cases}
$$

The errors and covariances among parameters are likewise given by the error matrix $\epsilon = A^{-1}$. Assuming a constant value for the variance σ^2 (i.e., uniform measurement errors), the matrix A and the vector β can be written in extended form as

$$
A = \frac{1}{\sigma^2}
\begin{bmatrix}
N & \sum x_{1i} & \cdots & \sum x_{mi} \\
\sum x_{1i} & \sum x_{1i}^2 & \cdots & \sum x_{1i} x_{mi} \\
\cdots & & & \\
\sum x_{mi} & \sum x_{mi} x_{1i} & \cdots & \sum x_{mi}^2
\end{bmatrix}
\tag{9.6}
$$

$$
\beta = \frac{1}{\sigma^2} \left(\sum y_i, \sum x_{1i} y_i, \ldots, \sum x_{mi} y_i \right)
\tag{9.7}
$$

where all sums are over the N measurements. An estimate for the variance σ^2 is given by

$$
s^2 = \frac{1}{N - m - 1} \sum_{i=1}^{N} (y_i - \hat{y}_i)^2
\tag{9.8}
$$

where $\hat{y}_i = a_0 + a_1 x_{1i} + \cdots + a_m x_{mi}$ is calculated for the best-fit values of the coefficients a_k.

An alternative notation for finding the coefficients a_k makes use of the following definitions:

$$
y = \begin{bmatrix} y_1 \\ y_2 \\ \cdots \\ y_N \end{bmatrix} ; X = \begin{bmatrix} 1 & x_{11} & \cdots & x_{1m} \\ 1 & x_{21} & \cdots & x_{2m} \\ \cdots & & & \\ 1 & x_{N1} & \cdots & x_{Nm} \end{bmatrix} \text{ and } a = \begin{bmatrix} a_0 \\ a_1 \\ \cdots \\ a_m \end{bmatrix}
\tag{9.9}
$$

where X is called the *design matrix* and we have arranged the Y measurements and the vector of coefficients in column vectors. With this notation, the least-squares approach gives the following solution for the coefficients [41]:

$$
a = (X^T X)^{-1} X^T Y
\tag{9.10}
$$

It is easy to show that (9.5) and (9.10) are equivalent (see Problem 9.3). Using this notation, the error matrix is given by

$$\epsilon = s^2 (X^T X)^{-1}. \tag{9.11}$$

We therefore have two equivalent methods to calculate the coefficients of the multiple regression and their errors. The latter form (9.9) may be convenient if the data are already tabulated according to the form of matrix A and therefore a can be found using the matrix algebra of (9.10). The drawback is that the design matrix can be of very large size, $N \times (m + 1)$, where N is the number of measurements. The form of (9.5) is more compact, since the matrix A is $(m + 1) \times (m + 1)$, and the summation over the N measurements must be performed beforehand to obtain A.

9.4   Tests for Significance of the Multiple Regression Coefficients

The multi-variable linear regression model of (9.2) is specified by the $m + 1$ coefficients a_k. After determining their best-fit values and errors, it is necessary to establish whether the model is an accurate representation of the data and whether there are any independent variables X_i that do not provide significant contribution to the prediction of the Y variable. Both tasks can be performed using hypothesis testing on the relevant statistic. We discuss these tests of significance using the Fisher's data of Sect. 9.2

9.4.1   T-Test for the Significance of Model Components

It is necessary to test the significance of each of the $m + 1$ parameters of the multi-variable linear regression. The null hypothesis is that their true value is zero, i.e., the corresponding variable is not needed in the model. For this purpose, we show that the ratio of the parameter's best-fit value a_k and its standard deviation s_k,

$$t_k = \frac{a_k}{s_k} \tag{9.12}$$

is distributed like a Student's t distribution with $N - m - 1$ degrees of freedom.

Proof Following the derivation provided in Sect. 7.5.1 for the sample mean, we can write

$$t_k = \frac{(a_k - \mu_k)/\sigma_k}{s_k/\sigma_k} \tag{9.13}$$

where $\mu_k = 0$ is the null hypothesis and σ_k^2 is the unknown parent variance for the parameter. Recall that the sample variance of the parameter s_k^2 is obtained as a product of the diagonal term in the error matrix and the estimate of the data variance s^2. Accordingly we set

$$\frac{(N-m-1)s_k^2}{\sigma_k^2} \sim \frac{\sum(y_i-\hat{y}_i)^2}{\sigma_k^2} \sim \chi^2(N-m-1), \tag{9.14}$$

i.e., the denominator of t_k can be written as a function of a variable that is χ^2-distributed. It is also clear that, under the null hypothesis, $\mu_k = 0$ is the parent value of a_k, and therefore the numerator of t_k is distributed like a standard normal distribution.

It follows that t_k is distributed like a t distribution,

$$t_k \sim \frac{N(0,1)}{\sqrt{\chi^2(N-m-1)/(N-m-1)}} \sim t(N-m-1) \tag{9.15}$$

according to the definition of the t distribution of (7.33).   □

To test for the significance of coefficient a_k we therefore use the critical value for the t distribution for the appropriate number of degrees of freedom and the desired confidence level.

Example 9.1 (Multi-Variable Linear Regression on Iris setosa Data) The data of Fig. 9.1 for the *Iris setosa* species are fit to the linear model of (9.1), where the sepal length is used as the Y variable and the remaining three variables are the independent variables. Using (9.5) and the inverse of matrix A for the errors, we find the results shown in Table 9.1, including the t scores for the four parameters of the multiple regression.

For each parameter is reported the probability to exceed the absolute value of the measured t according to a t distribution with $f = 46$ degrees of freedom, where $f = N - m - 1$ with $N = 50$ measurements and $m = 3$ independent variable. It is clear that the parameters a_2 and a_3, corresponding to the petal length and width, are not significant because of the large probability p to exceed their value under the null hypothesis. Accordingly, it would be meaningful to repeat the linear regression using only the sepal width as an estimator for the sepal length.   ◇

Table 9.1 Multiple regression parameter for the *Iris setosa* data

Parameter	Best-fit value	Error	t score	p value
a_0	2.352	0.393	5.99	< 0.001
a_1	0.655	0.092	7.08	< 0.001
a_2	0.238	0.208	1.14	0.26
a_3	0.252	0.347	0.73	0.47

9.4.2 F-Test for Goodness of Fit

The purpose of the multi-variable linear model is to provide a fit to the data that is more accurate than a simple constant predictor, i.e., the average of the Y measurements. In other words, we want to establish whether any of the parameters $a_1, \ldots, a_m$ provides a significant improvement over the constant model with $a_1 = a_2 = \ldots = a_m = 0$.

For this purpose we write the total variance of the data as follows:

$$\sum_{i=1}^{N}(y_i - \bar{y})^2 = \sum_{i=1}^{N}(y_i - \hat{y}_i)^2 + \sum_{i=1}^{N}(\hat{y}_i - \bar{y})^2 \tag{9.16}$$

where $\hat{y}_i = y(x_i)$ is evaluated for the best-fit values of the parameters a_k. This equation can be shown to hold because the following property applies,

$$\sum_{i=1}^{N}(y_i - \hat{y}_i)(\hat{y}_i - \bar{y}) = 0 \tag{9.17}$$

(see Problem 9.7). The parent variance σ^2 of the data is unknown and it is not required for this test. We therefore ignore it for the considerations that follow by setting $\sigma^2 = 1$. The three terms in (9.16) are interpreted as follows. The left-hand side term is the total variance of the data and it is distributed like

$$S^2 = \sum_{i=1}^{N}(y_i - \bar{y})^2 \sim \chi^2(N - 1). \tag{9.18}$$

The total variance S^2 can be interpreted as the variance obtained using a model with $a_1 = \ldots = a_m = 0$, i.e., a constant model equal to the average of the Y measurements.

The first term on the right-hand side is the *residual variance* after the data are fit to the linear model and it follows the usual χ^2 distribution

$$S_r^2 = \sum_{i=1}^{N}(y_i - \hat{y}_i)^2 \sim \chi^2(N - m - 1) \tag{9.19}$$

because of the $m + 1$ parameters used in the fit. This is the usual variance obtained using the full model in which at least some of the a_k parameters are not equal to zero.

Finally, the second term on the right-hand side can be interpreted as the variance *explained* by the best-fit model and it is distributed like

$$S_e^2 = \sum_{i=1}^{N} (\hat{y}_i - \bar{y})^2 \sim \chi^2(m). \tag{9.20}$$

The distribution of the last term can be explained by the independence between the two variables on the right-hand side of the equation and the distribution of the left-hand side term, following a derivation similar to that of Sect. 7.3. Such derivation is not discussed in this book.

The variances described above can be used to define the variable

$$F = \frac{S_e^2/m}{S_r^2/(N - m - 1)} = \frac{\sum_{i=1}^{N} (\hat{y}_i - \bar{y})^2/m}{\sum_{i=1}^{N} (y_i - \hat{y}_i)^2/(N - m - 1)}, \tag{9.21}$$

which is distributed as an F variable with m, $N-m-1$ degrees of freedom under the null hypothesis that $a_1 = \ldots = a_m = 0$. The meaning of this variable is the ratio between the variance explained by the fit and the residual variance, each normalized by the respective degrees of freedom. A large value of this ratio is desirable, since it means that the model does a good job at explaining the variability of the data.

The measurement of F that results from the fit of a dataset to the multi-variable linear model can therefore be used to test the null hypothesis. If the measurement exceeds the critical value of the F distribution for the desired confidence level, the null hypothesis must be rejected and the linear model is considered acceptable.

Example 9.2 (F-Test of Iris setosa Data) The variances for the *Iris setosa* data are shown in Table 9.2. The variable F is

$$F = \frac{S_e^2/3}{S_r^2/46} = \frac{3.50/2}{2.59/46} = 20.76. \tag{9.22}$$

The 99 % ($p = 0.01$) critical value for an F distribution with 3, 46 degrees of freedom is 4.24. Therefore the null hypothesis that the linear model does *not* provide a significant improvement must be rejected. In practice, this means that the linear

Table 9.2 Variances and F-test results for the *Iris setosa* data

Variances	Value	d.o.f	F-test	Value	p value
S^2	6.09	$N - 1 = 49$			
S_r^2	2.59	$N - m - 1 = 46$			
S_e^2	3.50	$m = 3$			
			$\dfrac{S_e^2/m}{S_r^2/(N - m - 1)}$	20.76	1.2×10^{-8}

model is warranted. The probability to exceed the measured value of 20.7 for the test statistic is 1.2×10^{-8}, i.e., very small.   ◇

9.4.3 The Coefficient of Determination

The ratio of the explained variance S_e^2 to the total variance S^2, defined as

$$R^2 = \frac{S_e^2}{S^2} = \frac{\sum_{i=1}^{N}(\hat{y}_i - \bar{y})^2}{\sum_{i=1}^{N}(y_i - \bar{y})^2} \tag{9.23}$$

is a common measure of the ability of the linear model to describe the data. This ratio is called the *coefficient of (multiple) determination* and it is $0 \leq R^2 \leq 1$. A value close to 1 indicates that the model describes the data with little additional variance left unexplained.

It is possible to relate the coefficient R^2 to the F-test variable defined in (9.21) and obtain an equivalent test for the multi-variable regression based on R^2 instead of the F variable (see, e.g., [41] and [29]). Since the two quantities are related, it is sufficient to test the overall multiple regression using the F test provided in the previous section. The advantage of reporting explicitly a value for R^2 is that we can identify in a simple way the amount of variance that remains in the data after performing the multiple regression.

Example 9.3 (R^2 Value for the Iris setosa Data) We can use the data in Table 9.2 to calculate a coefficient of multiple determination $R^2 = 0.575$. This number means that 57.5 % of the total data variance is explained by the best-fit regression model.   ◇

In the case of the simple linear regression with just one independent variable, $y = a + bx$, the coefficient of determination is the same as the coefficient of linear correlation r defined earlier in (2.19) (see Problem 9.4). In this case it is possible to test the significance of the linear model using either the correlation coefficient r or the F test. The two tests will be equivalent.

Example 9.4 (Linear Fit to the Iris setosa Data Using a Single Independent Variable) In a previous example we have shown that the coefficients of multiple regression for the variables Petal Length and Petal Width were not statistically significant, according to the t test.

Excluding these two columns of data, a fit to the function $y = a + bx$, where Y is the Sepal Length and X the Sepal Width, can be shown to return the values $a = 2.64 \pm 0.31$ and $b = 0.69 \pm 0.09$ with a correlation coefficient of $r = 0.7425$ or a value of $F = 58.99$ for 1, 49 degrees of freedom (see Problem 9.5). The value of $r^2 = 0.551$ is very similar to that obtained from the full fit using the additional two variables. The fact that the reduction in r^2 is minimal between the $m = 3$ and the $m = 1$ case is an indication that the Sepal Length can be predicted with nearly the same precision using just the Sepal Width as an indicator. ◇

Summary of Key Concepts for this Chapter

☐ *Multi-variable dataset* Simultaneous measurements of several (> 2) variables, usually without reference to a specific independent variable.

☐ *Multi-variable linear regression*: Extension of the (multiple) linear regression to the case of multi-variable data. Best-fit coefficients are given by the matrix equation

$$a = (X^T X)^{-1} X^T Y.$$

☐ *Coefficient of determination*: The ratio between the explained variance and total variance $R^2 = S_e^2 / S^2 \leq 1$.

Problems

9.1 Calculate the best-fit parameters and uncertainties for the multi-variable regression of the *Iris setosa* data of Fig. 9.1.

9.2 Use an F test to determine whether the multi-variable regression of the *Iris setosa* data is justified or not.

9.3 Prove that (9.5) and (9.10) are equivalent. Take into consideration that in (9.5) the vectors a and β are row vectors. You may re-write (9.5) using column vectors.

9.4 Prove that the coefficient of determination R^2 for the simple linear regression $y = a + bx$ is equivalent to the sample correlation coefficient of (2.20).

9.5 Fit the *Iris setosa* data using the function $y = a + bx$, where Y is the Sepal Length and X the Sepal Width. For this fit, you will ignore the data associated with the petal. Determine the best-fit parameters of the linear model and their errors.

9.6 Using the results of Problem 9.5, determine whether there is sufficient evidence for the use of the simple $y = a + bx$ model for the data. Use a confidence level of 99 % to draw your conclusions.

9.7 Prove (9.17).

Chapter 10
Goodness of Fit and Parameter Uncertainty

Abstract After calculating the best-fit values of model parameters, it is necessary to determine whether the model is actually a correct description of the data, even when we use the best possible values for the free parameters. In fact, only when the model is acceptable are best-fit parameters meaningful. The acceptability of a model is typically addressed via the distribution of the χ^2 statistic or, in the case of Poisson data, of the Cash statistic. A related problem is the estimate of uncertainty in the best-fit parameters. This chapter describes how to derive confidence intervals for fit parameters, in the general case of Gaussian distributions that require χ^2 minimization, and for the case of Poisson data requiring the Cash statistic. We also study whether a linear relationship between two variables is warranted at all, providing a statistical test based on the linear correlation coefficient. This is a question that should be asked of a two-variable dataset prior to any attempt to fit with a linear or more sophisticated model.

10.1 Goodness of Fit for the χ^2_{min} Fit Statistic

For both linear and nonlinear Gaussian fits, one needs to establish if the set of best-fit parameters that minimize χ^2 are acceptable, i.e., if the fit was successful. For this purpose, we need to perform a hypothesis testing based on the minimum of the χ^2 statistic that was obtained for the given model. According to its definition,

$$\chi^2_{min} = \sum_{i=1}^{N} \frac{(y_i - \hat{y}_i)^2}{\sigma_i^2} \tag{10.1}$$

in which $\hat{y}_i = y(x_i)|_{\text{best-fit}}$ is the model calculated with the best-fit parameters. It is tempting to say that the χ^2_{min} statistic is distributed like a χ^2 random variable (Sect. 7.2), since it is the sum of N several random variables, each assumed to be distributed like a standard normal. If the function $y(x)$ has no free parameters, this is certainly the case, and it would be also clear that χ^2 will have N degrees of freedom.

The complication is that the fit function has m free parameters that were adjusted in such a way as to minimize the χ^2. This has two implications on the χ^2_{min} statistic: the free parameters will reduce the value of χ^2 with respect to the case in which no

© Springer Science+Busines Media New York 2017
M. Bonamente, *Statistics and Analysis of Scientific Data*, Graduate Texts in Physics, DOI 10.1007/978-1-4939-6572-4_10

free parameters were present, and, more importantly, the fit function $y(x)$ introduces a dependence among the N random variables in the sum. Given that the χ^2_{min} is no longer the sum of N independent terms, we cannot conclude that $\chi^2_{min} \sim \chi^2(N)$.

It can be shown that χ^2_{min} is in fact still distributed as a χ^2 variable, but with

$$f = N - m \tag{10.2}$$

degrees of freedom. This result applies to any type of function $f(x)$, under the assumptions that the m parameters are independent of one another, as is normally the case for "meaningful" fit functions. The general proof of this statement is rather elaborate, and can be found in the textbook by Cramer [11]. Here we limit ourselves to provide a proof for a specific case in which $f(x) = a$, meaning a one-parameter fit function that is a constant, to illustrate the reduction of degrees of freedom from N to $N - 1$ when there is just one free parameter that can be used to minimize χ^2.

Proof When performing a maximum likelihood fit to the function $y(x) = a$, we have shown that the best-fit parameter is estimated as

$$a = \bar{x} = \frac{1}{N} \sum_{i=1}^{N} x_i,$$

under the assumption that all measurements are drawn from the same distribution $N(\mu, \sigma)$ (see Sect. 5.1). Therefore, we can write

$$\chi^2 = \sum_{i=1}^{N} \frac{(x_i - \mu)^2}{\sigma^2} = \frac{(\bar{x} - \mu)^2}{\sigma^2/N} + \sum_{i=1}^{N} \frac{(x_i - \bar{x})^2}{\sigma^2} = \frac{(\bar{x} - \mu)^2}{\sigma^2/N} + \chi^2_{min}.$$

This equation is identical to the relationship used to derive the sampling distribution of the variance, (7.19), and therefore we can directly conclude that $\chi^2_{min} \sim \chi^2(N-1)$ and that χ^2_{min} and χ^2 are independent random variables. Both properties will be essential for the calculation of confidence intervals on fit parameters.   □

Now that the distribution function of the fit statistic χ^2_{min} is known, we can use the hypothesis testing methods of Sect. 7.2.3 to determine whether a value of the statistic is acceptable or not. The null hypothesis that the data are well fit by, or compatible with, the model, can be rejected at a confidence level p according to a one-tailed test defined by

$$1 - p = \int_{\chi^2_{crit}}^{\infty} f_{\chi^2}(f, x) dx = P(\chi^2(f) \geq \chi^2_{crit}). \tag{10.3}$$

The value χ^2_{crit} calculated from (10.3) for a specified value of p defines the rejection region $\chi^2_{min} \geq \chi^2_{crit}$. The data analyst must chose a value of p, say $p = 0.9$, and calculate the critical value χ^2_{crit} that satisfies (10.3), using Table A.7. If the χ^2 value measured from the data is higher than what calculated from (10.3), then the hypothesis should be rejected at the p, say 90 %, confidence level. On the other hand, if the χ^2 value measured from the data is lower than this critical value, the hypothesis should not be rejected, and the fit considered as consistent with the model or, more precisely, not rejected, at that confidence level.

Example 10.1 In Fig. 10.1 it is shown a linear fit using data from Table 6.1. The quantity Energy 1 is used as the independent variable, and its errors are neglected. The quantity Energy 2 is the dependent variable, and errors are calculated as the average of the positive and negative error bars. The best-fit linear model is represented as the dotted line, for a fit statistic of $\chi^2_{min} = 60.5$ for 23 degrees of freedom. The value of the fit statistic is too large, and the linear model must be discarded (see Appendix A.3 for critical values of the χ^2 distribution).

Despite failing the χ^2_{min} test, the best-fit model appears to be a reasonable match to the data. The large value of the test statistic are clearly caused by a few datapoints with small error bars, but there appears to be no systematic deviation from the linear model. One reason for the poor fit statistic could be that errors in the independent variables were neglected. In Chap. 12 we explain an alternative fitting method that takes into account errors in both variables. Another possibility for the poor fit is that there are other sources of error that are not accounted. This additional errors are often referred to as systematic errors. In Chap. 11 we address the presence of systematic errors and how one can handle the presence of such errors in the fit. ◇

Fig. 10.1 Linear fit to the data of Table 6.1. We assumed that the independent variable is Energy 1, errors for this variable were neglected in the fit. Note the logarithmic scale for both axes

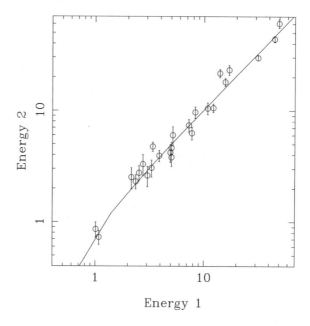

10.2   Goodness of Fit for the Cash C Statistic

In the case of a Poisson dataset (Sect. 8.8) the procedure to determine whether the best-fit model is acceptable is identical to that for Gaussian data, provided that the χ^2 fit statistic is replaced by the Cash statistic C, defined by

$$C = -2 \ln \mathscr{L} - B,$$

where

$$B = 2 \sum_{i=1}^{N} y_i - 2 \sum_{i=1}^{N} y_i \ln y_i + 2 \sum_{i=1}^{N} \ln y_i!.$$

We now prove that C is approximately distributed like a χ^2 distribution with $N - m$) degrees of freedom. This is an important result that lets us use the Cash statistic C in the same way as the χ^2 statistic.

Proof We start with

$$-2 \ln \mathscr{L} = -2 \left(\sum_{i=1}^{N} y_i \ln y(x_i) - \sum_{i=1}^{N} y(x_i) - \sum_{i=1}^{N} \ln y_i! \right).$$

and rewrite as

$$-2 \ln \mathscr{L} = 2 \sum_{i=1}^{N} \left(y(x_i) - y_i \ln \frac{y(x_i)}{y_i} - y_i \ln y_i + \ln y_i! \right).$$

In order to find an expression that asymptotically relates C to χ^2, define $d = y_i - y(x_i)$ as the "average" deviation of the measurement from the parent mean. It is reasonable to expect that

$$\frac{d}{y_i} \simeq \frac{1}{\sqrt{y_i}} \Rightarrow \frac{y(x_i)}{y_i} = \frac{y_i - d}{y_i} = 1 - \frac{d}{y_i}$$

where y_i is the number of counts in that specific bin. It follows that

$$-2 \ln \mathscr{L} = 2 \sum_{i=1}^{N} \left(y(x_i) - y_i \ln(1 - \frac{d}{y_i}) - y_i \ln y_i + \ln y_i! \right)$$

$$\simeq 2 \sum_{i=1}^{N} \left(y(x_i) - y_i \left(-\frac{d}{y_i} - \frac{1}{2} \left(\frac{d}{y_i} \right)^2 \right) - y_i \ln y_i + \ln y_i! \right)$$

$$= 2 \sum_{i=1}^{N} \left(y(x_i) + (y_i - y(x_i)) + \frac{1}{2} \frac{(y_i - y(x_i))^2}{y_i} - y_i \ln y_i + \ln y_i! \right)$$

The quadratic term can now be written in such a way that the denominator carries the term $y(x_i) = \sigma^2$:

$$\frac{(y_i - y(x_i))^2}{y_i} = \frac{(y_i - y(x_i))^2}{d + y(x_i)} = \frac{(y_i - y(x_i))^2}{y(x_i)} \left(\frac{d}{y(x_i)} + \frac{y(x_i)}{y(x_i)} \right)^{-1}$$

$$\simeq \frac{(y_i - y(x_i))^2}{y(x_i)} \left(1 - \frac{d}{y(x_i)} \right).$$

We therefore conclude that

$$-2 \ln \mathscr{L} = \sum_{i=1}^{N} \frac{(y_i - y(x_i))^2}{y(x_i)} \left(1 - \frac{d}{y(x_i)} \right)$$

$$+ \left(2 \sum_{i=1}^{N} y_i - 2 \sum_{i=1}^{N} y_i \ln y_i + 2 \sum_{i=1}^{N} \ln y_i! \right),$$

showing that, within the multiplicative terms $(1 - d/y(x_i))$, the variable $C = -2 \ln \mathscr{L} - B$ has a χ^2 distribution with $N - m$ degrees of freedom. □

For the purpose of finding the best-fit parameters via minimization of the fit statistic, the constant term B is irrelevant. However, in order to determine the goodness of fit and confidence intervals, it is important to work with a statistic that is distributed as a χ^2 variable. Therefore the Cash statistic is defined as

$$C = -2 \sum_{i=1}^{N} y_i \ln \frac{y(x_i)}{y_i} + 2 \sum_{i=1}^{N} (y(x_i) - y_i). \tag{10.4}$$

Example 10.2 Consider an ideal set of $N = 10$ identical measurements , $y_i = 1$. For a fit to a constant model, $y = a$, it is clear that the best-fit model parameter must be $a = 1$. Using the Cash statistic as redefined by (10.4), we find that $C = 0$, since the data and the model match exactly. A similar result would be obtained if we had assumed a Gaussian dataset of $y_i = 1$ and $\sigma_i = 1$, for which $\chi^2 = 0$. ◇

10.3 Confidence Intervals of Parameters for Gaussian Data

In this section we develop a method to calculate confidence intervals on model parameters assuming a Gaussian dataset. The results will also be applicable to Poisson data, provided that the χ^2 statistic is replaced with the Cash C statistic (see Sect. 10.4).

Under the assumption that a given model with m parameters is the correct description of the data, the fit statistic χ^2 calculated with these fixed true values,

$$\chi^2_{true} = \sum_{i=1}^{N} \left(\frac{y_i - y(x_i)|_{true}}{\sigma_i^2} \right)^2 \tag{10.5}$$

is distributed as $\chi^2(N)$, i.e., we expect random variations of the measurement of χ^2_{true} according to a χ^2 distribution with N degrees of freedom. This is so because the true parameters are fixed and no minimization of the χ^2 function can be performed. The quantity χ^2_{true} is clearly only a mathematical construct, since the true values of the parameters are unknown. One does not expect that $\chi^2_{true} = 0$, meaning a perfect match between the data and the model. In fact, even if the model was correct, statistical fluctuations will result in random deviations from the parent model.

On the other hand, when finding the best-fit parameters a_i, we calculate the statistic:

$$\chi^2_{min} = \sum_{i=1}^{N} \left(\frac{y_i - \hat{y}_i}{\sigma_i} \right)^2 \tag{10.6}$$

which minimizes χ^2 with respect to all possible free parameters. In this case, we know that $\chi^2_{min} \sim \chi^2(N-m)$ from the discussion in Sect. 10.1. It is also clear that the values of the best-fit parameters are not identical to the true parameters, again for the presence of random fluctuations of the datapoints.

After finding χ^2_{min}, any change in the parameters (say, from a_k to a'_k) will yield a larger value of the test statistic, $\chi^2 > \chi^2_{min}$. We want to test whether the new set of parameters a'_k can be the true (yet unknown) values of the parameters, e.g., whether the corresponding χ^2 can be considered χ^2_{true}. For this purpose we construct a new statistic:

$$\Delta \chi^2 \equiv \chi^2 - \chi^2_{min} \tag{10.7}$$

where χ^2 is obtained for a given set of model parameters and, by definition, $\Delta \chi^2$ is always positive. The hypothesis we want to test is that χ^2 is distributed like χ^2_{true}, i.e., the χ^2 calculated using a new set of parameters is consistent with χ^2_{true}. Since χ^2_{true} and χ^2_{min} are independent (see Sect. 10.1), we conclude that

$$\Delta \chi^2 \sim \chi^2(m) \tag{10.8}$$

when $\chi^2 \sim \chi^2_{true}$. Equation (10.8) provides a quantitative way to determine how much χ^2 can increase, relative to χ^2_{min}, and still the value of χ^2 remaining consistent with χ^2_{true}. The method to use (10.8) for confidence intervals on the model parameters is described below.

10.3.1 Confidence Interval on All Parameters

Equation (10.8) provides a quantitative method to estimate the confidence interval on the m best-fit parameters. The value of $\Delta\chi^2$ is expected to follow the $\chi^2(m)$ distribution. This means that one can tolerate deviations from the best-fit values of the parameters leading to an increase in χ^2, provided such increase is consistent with the critical value of the respective $\Delta\chi^2$ distribution. For example, in the case of a model with $m = 2$ free parameters, one can expect a change $\Delta\chi^2 \geq 4.6$ for $p = 0.9$ confidence, or for a model with $m = 1$ parameter a change $\Delta\chi^2 \geq 2.7$ (see Table A.7).

The method to determine the confidence interval on the parameters starts with the value of χ^2_{min}. From this, one constructs an m-dimensional volume bounded by the surface of $\Delta\chi^2 = \chi^2 - \chi^2_{min} \leq \chi^2_{crit}$, where χ^2_{crit} is the value that corresponds to a given confidence level p for m degrees of freedom, as tabulated in Table A.7. The surface of this m-dimensional volume marks the boundaries of the rejection region at the p level (say p=90%) for the m parameters, i.e., the parameters can vary within this volume and still remain an acceptable fit to the data at that confidence level. In practice, a surface at fixed $\Delta\chi^2 = \chi^2_{crit}$ can be calculated by a grid of points around the values that correspond to χ^2_{min}. This calculation can become computationally intensive as the number of parameters m increases. An alternative method to estimate confidence intervals on fit parameters that makes use of Monte Carlo Markov chains (see Chap. 16) will overcome this limitation.

Example 10.3 Consider the case of a linear fit to the data of Table 10.1. According to the data in Table 10.1, one can calculate the best-fit estimates of the parameters as $a = 23.54 \pm 4.25$ and $b = 13.48 \pm 2.16$, using (8.7) and (8.22). The best-fit line is shown in Fig. 10.2. There is no guarantee that these values are in fact the true values: they are only the best estimates based on the maximum likelihood method. For these best-fit values of the coefficients, the fit statistic is $\chi^2_{min} = 0.53$, for $f = 3$ degrees of freedom, corresponding to a probability $p = 0.09$ (i.e., a probability $P(\chi^2(3) \geq 0.53 = 0.91)$. The fit cannot be rejected at any reasonable confidence level, since the probability to exceed the measured χ^2_{min} is so high.

We now sample the parameter space, and determine variations in the fit statistic χ^2 around the minimum value. The result is shown in Fig. 10.3, in which the contours mark the $\chi^2_{min} + 1.0$, $\chi^2_{min} + 2.3$ and $\chi^2_{min} + 4.6$ boundaries. In this

Table 10.1 Data used to illustrate the linear regression, and the estimate of confidence intervals on fit parameters

x_i	y_i	σ_i
(Indep. variable)	(Dependent variable)	
0	25	5
1	36	6
2	47	6.85
3	64	8
4	81	9

Fig. 10.2 Best-fit linear model (*dashed line*) to the data of Table 10.1. The $\chi_{min} = 0.53$ indicates a very good fit which cannot be rejected at any reasonable confidence level

x_i	y_i	σ_i
(indep. variable)	(dependent variable)	
0	25	5
1	36	6
2	47	6.85
3	64	8
4	81	9

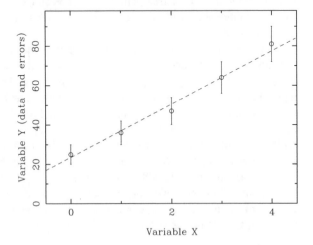

application, $m = 2$, a value of $\Delta\chi^2 = 4.6$ or larger is expected 10 % of the time. Accordingly, the $\Delta\chi^2 = 4.6$ contour marks the 90 % confidence surface: the true values of a and b are within this area 90 % of the time, if the null hypothesis that the model is an accurate description of the data is correct. This area is therefore the 90 % confidence area for the two fitted parameters.   ◇

10.3.2   Confidence Intervals on Reduced Number of Parameters

In the case of a large number m of free parameters, it is customary to report the uncertainty on each of the fitted parameters or, in general, on just a subset of $l < m$ parameters considered to be of interest. In this case, the l parameters $a_1, \ldots, a_l$ are said to be the *interesting* parameters, and the remaining $m - l$ parameters are said to be *uninteresting*. This can be thought of as reducing the number of parameters of the model from m to l, often in such a way that only one interesting parameter is investigated at a time ($l = 1$). This is a situation that is of practical importance for several reason. First, it is not convenient to display surfaces in more than two or three dimensions. Also, sometimes there are parameters that are truly uninteresting

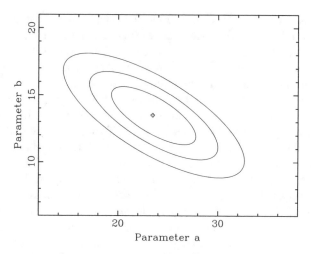

Fig. 10.3 Contours of $\Delta\chi^2 = 1.0$, 2.3 and 4.6 (from smaller to larger areas). For the example of $m = 2$ free parameters, the contours mark the area within which the true parameters a and b are expected to be with, respectively, 25, 67, and 90 % probability

to the interpretation of the data, although necessary for its analysis. One case of this is the presence of a measurement background, which must be taken into account for a proper analysis of the data, but it is of no interest in the interpretation of the results.

New considerations must be applied to χ^2_{true} and its parent distribution in this situation. We find χ^2_{min} in the usual way, that is, by fitting all parameters and adjusting them until the minimum χ^2 is found. Therefore χ^2_{min} continues to be distributed like $\chi^2(N - m)$. For χ^2_{true}, we want to ignore the presence of the uninteresting parameters. We do so by assuming that the l interesting parameters are fixed at the true values and *marginalize* over the $m - l$ uninteresting parameters. This process of marginalization means that we let the uninteresting parameters adjust themselves to the values that yield the lowest value of χ^2. This process ensures that $\chi^2_{true} \propto \chi^2(N - (m - l))$. Notice that the marginalization does *not* mean fixing the values of the uninteresting parameters to their best-fit values.

In summary, the change in χ^2 that can be tolerated will therefore be $\Delta\chi^2 = \chi^2_{true} - \chi^2_{min}$, in which $\chi^2_{true} \propto \chi^2(N - (m - l))$ and $\chi^2_{min} \propto \chi^2(N - m)$. Since the two χ^2 distributions are independent of one another, it follows that

$$\Delta\chi^2 \sim \chi^2(l) \tag{10.9}$$

where l is the number of interesting parameters. The process of finding confidence intervals for a reduced number of parameters is illustrated in the following example of a model with $m = 2$ free parameters, for which we also find confidence intervals for one interesting parameters at a time.

Example 10.4 Consider the case in Fig. 10.3, and assume that the interesting parameter is a. The χ^2_{min} for each value of a is done by searching the minimum χ^2 long a vertical line (i.e., for a fixed value of a). The best-fit value of a is already known, marked by a cross in Fig. 10.3. When seeking the 68 % confidence interval for the interesting parameter a, the limiting values of a are those on either side of the best-fit value that result in a minimum χ^2 value of $\chi^2_{min} + 1.0$ (where χ^2_{min} is the global minimum). Therefore, the 68 % confidence interval for a is found by *projecting* the $\chi^2_{min} + 1.0$ contour along the a axis. That is to say, we find the smallest and largest values of a along the $\chi^2_{min} + 1.0$ contour, which is the innermost contour in Fig. 10.3. Likewise the projection of the same contour along the b axis gives the 68 % confidence interval on b, when considered as the only interesting parameter.

On the other hand, the 2-dimensional 68 % confidence surface on a, b was given by the $\chi^2 + 2.3$ contour. It is important not to confuse those two confidence ranges, both at the same level of confidence of 68 %. The reason for the difference ($\chi^2_{min} + 1.0$ for one interesting parameter vs. $\chi^2 + 2.3$ for two interesting parameters) is the numbers of degrees of freedom of the respective $\Delta\chi^2$.   ◇

This procedure for estimation of intervals on a reduced number of parameters was not well understood until the work of Lampton and colleagues in 1976 [27]. It is now widely accepted as the correct method to estimate errors in a subset of the model parameters.

10.4   Confidence Intervals of Parameters for Poisson Data

The fit to Poisson data was described in Sect. 8.8. Since the Cash statistic C follows approximately the χ^2 distribution in the limit of a large number of datapoints, then the statistic

$$\Delta C = C_{true} - C_{min} \qquad (10.10)$$

has the same statistical properties as the $\Delta\chi^2$ distribution. Parameter estimation with Poisson statistic therefore follows the same rules and procedures as with the χ^2 statistic.

Example 10.5 Consider an ideal dataset of N identical measurement $y_i = 1$. We want to fit the data to a constant model $y = a$, and construct a 1-σ confidence interval on the fit parameter a using both the Poisson fit statistic C, and the Gaussian fit statistic χ^2. In the case of the Poisson statistic, we assume that the measurements are derived from a counting experiment, that is, a count of 1 was recorded in each case. In the case of Gaussian variables, we assume uncertainties of $\sigma_i = 1$.

In the case of Poisson data, we use the Cash statistic defined in (10.4). The best-fit value of the model is clearly $a = 1$, and we want to find the value δ corresponding to a change in C by a value of 1 with respect to the minimum value C_{min},

$$\Delta C = 1 \Rightarrow -2 \sum 1 \ln \frac{(1+\delta)}{1} + 2 \sum (1 + \delta - 1) = 1$$

Using the approximation $\ln(1 + \delta) \simeq (\delta - \delta^2/2)$, we find that

$$-20\delta + 10\delta^2 + 20\delta = 1 \Rightarrow \delta = \sqrt{\frac{1}{10}}.$$

This shows that the 68 % confidence range is between $1 - \sqrt{1/10}$ and $1 + \sqrt{1/10}$, or $1 \pm \sqrt{1/10}$.

Using Gaussian errors, we calculate $\Delta\chi^2 = 1$, leading to $10\delta^2 = 1$, and the same result as in the case of the Poisson dataset.   ◇

10.5   The Linear Correlation Coefficient

We want to define a quantity that describes whether there is a linear relationship between two random variables X and Y. This quantity is based on the slopes of two linear fits of X and Y, using each in turn as the independent variable. Call b the slope of the regression $y = a + bx$ (where X is the independent variable) and b' the slope of the regression $x = a' + b'y$ (where Y is the independent variable) and assume that there are N measurements of the two variables. The linear correlation coefficient r is defined as the product of the slopes of the two fits via

$$r^2 = bb' = \frac{\left(N \sum x_i y_i - \sum x_i \sum y_i\right)^2}{\left(N \sum x_i^2 - \left(\sum x_i\right)^2\right)\left(N \sum y_i^2 - \left(\sum y_i\right)^2\right)} \tag{10.11}$$

in which we have used the results of (8.23). It is easy to show that this expression can be rewritten as

$$r^2 = \frac{\left(\sum (x_i - \bar{x})(y_i - \bar{y})\right)^2}{\sum (x_i - \bar{x})^2 \sum (y_i - \bar{y})^2} \tag{10.12}$$

and therefore r is the sample correlation coefficient as defined in (2.20).

Consider as an example the data from Pearson's experiment at page 30. The measurement of mother's and father's height are likely to have the same uncertainty, since one expects that both women and men followed a similar procedure for the measurement. Therefore no precedence should be given to either when assigning the tag of "independent" variable. Instead, one can proceed with two separate fits: one in which the father's height (X) is considered as the independent variable, or the regression of Y on X (dashed line), and the other in which the mother's height (Y) is

Fig. 10.4 Linear regressions based on the data collected by Pearson, Table 2.3 at page 30. Larger circles indicate a higher number of occurrence for that bin

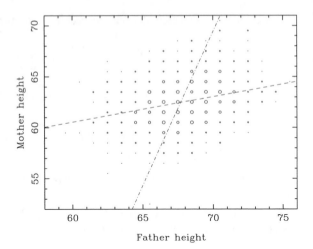

the independent variable, or linear regression of X on Y (dot-dash line). The two fits are reported in Fig. 10.4, obtained by maximum likelihood method assuming equal errors for the dependent variables.

If the two variables X and Y are uncorrelated, then the two best-fit slopes b and b' are expected to be zero. In fact, as one variable varies through its range, the other is not expected to either decrease (negative correlation) or increase (positive correlation), resulting in null best-fit slopes for the two fits. We therefore expect the sample distribution of r to have zero mean, under the null hypothesis of lack of correlation between X and Y. If there is a true linear correlation between the two variables, i.e., $y = a + bx$ is satisfied with $b \neq 0$, then it is also true that $x = a' + b'x = -a/b + 1/by$. In this case one therefore expects $bb' = r^2 = 1$.

10.5.1   The Probability Distribution Function

A quantitative test for the correlation between two random variables requires the distribution function $f_r(r)$. We show that the probability distribution of r, under the hypothesis that the two variables X and Y are uncorrelated, is given by

$$f_r(r) = \frac{1}{\sqrt{\pi}} \frac{\Gamma\left(\dfrac{f+1}{2}\right)}{\Gamma\left(\dfrac{f}{2}\right)} \left(\frac{1}{1-r^2}\right)^{-\frac{f-2}{2}} \tag{10.13}$$

where $f = N - 2$ is the effective number of degrees of freedom of a dataset with N measurements of the pairs of variables. The form of the distribution function is reminiscent of the t distribution, which in fact plays a role in the determination of this distribution.

Proof The proof starts with the determination of the probability distribution function of a suitable function of r, and then, by change of variables, the distribution of r is obtained.

The best-fit parameter b is given by

$$b^2 = \frac{(N \sum x_i y_i - \sum x_i \sum y_i)^2}{(N \sum x_i^2 - (\sum x_i)^2)^2} = \frac{(\sum (x_i - \bar{x})(y_i - \bar{y}))^2}{(\sum (x_i - \bar{x})^2)^2};$$

and accordingly we obtain

$$r^2 = \frac{(\sum (x_i - \bar{x})(y_i - \bar{y}))^2}{\sum (x_i - \bar{x})^2 \sum (y_i - \bar{y})^2} = b^2 \frac{\sum (x_i - \bar{x})^2}{\sum (y_i - \bar{y})^2}. \tag{10.14}$$

Also, using (8.5), the best-fit parameter a can be shown to be equal to $a = \bar{y} - b\bar{x}$, and therefore we obtain

$$S^2 \equiv \sum (y_i - a - bx_i)^2 = \sum (y_i - \bar{y})^2 - b^2 \sum (x_i - \bar{x})^2. \tag{10.15}$$

Notice that $S^2/\sigma^2 = \chi^2_{min}$, where σ^2 is the common variance of the Y measurements, and therefore using (10.14) and (10.15) it follows that

$$\frac{S^2}{\sum (y_i - \bar{y})^2} = 1 - r^2$$

or, alternatively,

$$\frac{r}{\sqrt{1 - r^2}} = \frac{b \sqrt{\sum (x_i - \bar{x})^2}}{S}. \tag{10.16}$$

Equation (10.16) provides the means to determine the distribution function of $r/\sqrt{1-r^2}$. First, notice that the variance of b is given by

$$\sigma_b^2 = \frac{N \sigma^2}{N \sum x_i^2 - (\sum x_i)^2} = \frac{\sigma^2}{\sum (x_i - \bar{x})^2}.$$

According to (8.35), $s^2 = S^2/(N-2)$ is the unbiased estimator of the variance σ^2, since two parameters have been fit to the data. Assuming that the true parameter for the slope of the distribution is β, then

$$\frac{b - \beta}{\sigma_b} = \frac{b - \beta}{\sigma / \sqrt{\sum (x_i - \bar{x})^2}} \sim N(0, 1)$$

is therefore distributed like a standard Gaussian. In the earlier equation, if we replace σ^2 with the sample variance $s^2 = S^2/(N-2)$, and enforce the null hypothesis that the variables X and Y are uncorrelated ($\beta = 0$), we obtain a new variable that is distributed like a t distribution with $N-2$ degrees of freedom,

$$\frac{b}{S}\sqrt{N-2}\sqrt{\sum(x_i - \bar{x})^2} \sim t(N-2).$$

Using (10.16), we find that the variable

$$\frac{r\sqrt{N-2}}{\sqrt{1-r^2}} \tag{10.17}$$

is distributed like a t distribution with $f = N-2$ degrees of freedom and, since it is a monotonic function of r, its distribution can be related to the distribution $f_r(r)$ via a simple change of variables, following the method described in Sect. 4.4.1.

Starting with $v = r\sqrt{N-2}/\sqrt{1-r^2}$, and

$$f_T(v) = \frac{1}{\sqrt{\pi f}}\frac{\Gamma((f+1)/2)}{\Gamma(f/2)}\left(1 - \frac{v^2}{f}\right)^{-\frac{f+1}{2}}$$

with

$$\frac{dv}{dr} = \frac{\sqrt{N-2}}{(1-r^2)^{3/2}},$$

the equation of change of variables $f_r(r) = f_T(v)dv/dr$ yields (10.13) after a few steps of algebra.   □

10.5.2 Hypothesis Testing

A test for the presence of linear relationship between two variables makes use of the distribution function of r derived in the previous section. In the absence of linear relationship, we expect a value of r close to 0, while values close to the extremes of ± 1 indicate a strong correlation between the two variables. Since the null hypothesis is that there is no correlation, we use a two-tailed test to define the critical value of r via

$$P(|r| > r_{crit}) = 1 - \int_{-r_{crit}}^{r_{crit}} f_r(r')dr' = 1 - p, \tag{10.18}$$

where p is intended, as usual, as a number close to 1 (e.g., p=0.9 or 90 % confidence). Critical values of r for various probability levels are listed in Table A.24.

If the measured value of r exceeds the critical value, the null hypothesis must be discarded. This is an indication that there is a linear relationship between the two quantities and further modelling of Y vs. X or X vs. Y is warranted. In practice, the linear correlation coefficient test should be performed prior to attempting any regression between the two variables.

Example 10.6 The two fits to the data from Pearson's experiment (page 30) are illustrated in Fig. 10.4. A linear regression provides a best-fit slope of $b = 0.25$ (dashed line) and of $b' = 0.33$ (dot-dash line), respectively, when using the father's stature (x axis) or the mother's stature as the independent variable. For these fits we use the data provided in Table 2.3. Each combination of father–mother heights is counted a number of times equal to its frequency of occurrence, for a total of $N = 1,079$ datapoints.

The linear correlation coefficient for these data is $r = 0.29$, which is also equal to $\sqrt{bb'}$. For $N = 1,079$ datapoints, Table A.24 indicates that the hypothesis of no correlation between the two quantities must be discarded at $> 99\%$ confidence, since the critical value at 99 % confidence is ~ 0.081, and our measurement exceeds it. As a result, we conclude that the two quantities are likely to be truly correlated. The origin of the correlation is probably with the fact that people have a preference to marry a person of similar height, or more precisely, a person of a height that is linearly proportional to their own. $\diamond$

Summary of Key Concepts for this Chapter

☐ *The χ^2_{min} statistic*: It applies to Gaussian data and it is distributed like a χ^2 distribution with $N - m$ degrees of freedom.

☐ *The Cash statistic*: It applies to Poisson data and it is defined as

$$C = -2 \sum y_i \ln(y(x_i)/y_i) + 2 \sum (y(x_i) - y_i).$$

It is approximately distributed like χ^2_{min}.

☐ *Confidence intervals for χ^2_{min} statistic*: They are obtained from the condition that $\Delta\chi^2 \sim \chi^2(m)$, where m is the number of parameters of interest.

☐ *Interesting parameters*: A subset of all model parameters for which we are interested in calculating confidence intervals.

☐ *Linear correlation coefficient*: The quantity $-1 \leq r \leq 1$ that determines whether there is a linear correlation between two variables.

Problems

10.1 Use the same data as in Problem 8.2 to answer the following questions.

(a) Plot the 2-dimensional confidence contours at 68 and 90 % significance, by sampling the (a,b) parameter space in a suitable interval around the best-fit values.

(b) Using a suitable 2-dimensional confidence contour, determine the 68 % confidence intervals on each parameter separately, and compare with the analytic results obtained from the linear regression method.

10.2 Find the minimum χ^2 of the linear fit to the radius vs. ratio data of Table 6.1 and the number of degrees of freedom of the fit. Determine if the null hypothesis can be rejected at the 99 % confidence level.

10.3 Consider a simple dataset with the following measurements, assumed to be derived from a counting process. Show that the best-fit value of the parameter a for

x	y
0	1
1	1
2	1

the model $y = e^{ax}$ is $a = 0$ and derive its 68 % confidence interval.

10.4 Consider the same dataset as in Problem 10.3 but assume that the y measurements are Gaussian, with variances equal to the measurements. Show that the confidence interval of the best-fit parameter $a = 0$ is given by $\sigma_a = \sqrt{1/5}$.

10.5 Consider the same dataset as in Problem 10.3 but assume a constant fit function, $y = a$. Show that the best-fit is given by $a = 1$ and that the 68 % confidence interval corresponds to a standard deviation of $\sqrt{1/3}$.

10.6 Consider the biometric data in Pearson's experiment (page 30). Calculate the *average* father height (X variable) for each value of the mother's height (Y variable), and the *average* mother height for each value of the father's height. Using these two averaged datasets, perform a linear regression of Y on X, where Y is the average value you have calculated, and, similarly, the linear regression of X on Y. Calculate the best-fit parameters a, b (regression of Y on X) and a', b' (regression of X on Y), assuming that each datapoint in your two sets has the same uncertainty. This problem is an alternative method to perform the linear regressions of Fig. 10.4, and it yields similar results to the case of a fit to the "raw" data, i.e., without averaging.

10.7 Calculate the linear correlation coefficient for the data of Hubble's experiment (logarithm of velocity, and magnitude m), page 157. Determine whether the hypothesis of uncorrelation between the two quantities can be rejected at the 99 % confidence level.

10.8 Use the data from Table 6.1 for the radius vs. ratio, assuming that the radius is the independent variable with no error. Draw the 68 and 90 % confidence contours on the two fit parameters a and b, and calculate the 68 % confidence interval on the b parameter.

Chapter 11
Systematic Errors and Intrinsic Scatter

Abstract Certain types of uncertainty are difficult to estimate and may not be accounted in the initial error budget. This sometimes leads to a poor goodness-of-fit statistic and the rejection of the model used to fit the data. These missing sources of uncertainty may either be associated with the data themselves or with the model used to describe the data. In both cases, we describe methods to account for these errors and ensure that hypothesis testing is not biased by them.

11.1 What to Do When the Goodness-of-Fit Test Fails

The first step to ensure that a dataset is accurately described by a model is to test that the goodness-of-fit statistic is acceptable. For example, when the data have Gaussian errors, χ^2_{min} can be used as the goodness-of-fit statistic. If the value of χ^2_{min} exceeds a critical value, it is recommended that one rejects the model. At that point, the standard option is to use an alternative model, and repeat the testing procedure.

There are cases when it is reasonable to try a bit harder and investigate further whether the model and the dataset may still be compatible, despite the poor goodness of fit. The general situation when additional effort is warranted is in the case of a model that generally follows the data without severe outliers, yet the best-fit statistic (such as χ^2_{min}) indicates that the model is not acceptable. An example of this situation is that of Fig. 10.1: the best-fit linear model follows the distribution of the data without systematic deviations, yet its high value of $\chi^2_{min} = 60.5$ for 23 degrees of freedom cannot be formally accepted at any level of confidence.

In this chapter we describe two types of analysis that can be performed when the fit of a dataset to a model is poor. The first method assumes that the model itself has a degree of uncertainty that results in an intrinsic scatter above and beyond the variance of the data (Sect. 11.2). The second investigates whether there are additional sources of error in the data that may not have been properly accounted (Sect. 11.3). The two methods are conceptually different but result in similar modifications to the analysis.

© Springer Science+Busines Media New York 2017
M. Bonamente, *Statistics and Analysis of Scientific Data*, Graduate Texts in Physics, DOI 10.1007/978-1-4939-6572-4_11

11.2   Intrinsic Scatter and Debiased Variance

When fitting a dataset to a model we assume that the data are drawn from a parent model that is described by a number of parameters. As such, we surmise that there are exact model parameters that describe the parent distribution of the data, although we don't know their precise values. We use the data to estimate them, typically through a maximum likelihood method that consists of finding model parameters that maximize the likelihood of the data being drawn from that model (Chap. 8). For Gaussian data, the maximum likelihood method consists of finding the minimum of the χ^2 statistic.

A possible reason for a poor value of the minimum χ^2 statistic is that the model itself, although generally accurate, may have an *intrinsic scatter* or variance that needs to be accounted in the determination of the fit statistic. In other words, the parent model may not be exact but it may feature an inherent degree of variability. The goal of this section is to provide a method to describe and measure such scatter.

11.2.1   Direct Calculation of the Intrinsic Scatter

Each measurement in a dataset can be described as the sum of two variables,

$$y_i = \eta_i + \epsilon_i, \tag{11.1}$$

where η_i represents the parent value from which the measurement y_i is drawn and ϵ_i is the variable representing the measurement error. Usually, we assume that $\eta_i = y(x_i)$ is a fixed number, estimated by the least-squares (or other) method. Since ϵ_i is a variable of zero mean, and its variance is simply the measurement variance σ_i^2, (11.1) implies that the variance of the measurement y_i is just σ_i^2.

The model η_i may, however, be considered a variable with non-zero variance. This is to describe the fact that the model is not known exactly, but has an intrinsic degree of variability measured by its variance $\sigma_{int}^2 = Var(\eta_i)$. For simplicity, we assume that this model variance is constant for all points along the model. Under the assumption that the measurement error and the model are independent, variances of the variables on the right-hand side of (11.1) add and this yields to

$$\sigma_{int}^2 = Var(y_i) - \sigma_i^2. \tag{11.2}$$

The equation means that the intrinsic variance is obtained as the difference of the data variance minus the variance due to measurement errors. In keeping up with the definitions of (11.1), $Var(y_i)$ refers to the total variance of the i-th variable at location x_i. It is meaningful to calculate the average variance for all the y_i's assuming that each measurement is drawn from a parent mean of $\hat{y}_i$, the best-fit value of the model $y(x_i)$. In so doing, we make use of the fact that the model is not constant but it varies at different positions. As a result, (11.2) can be used to calculate the intrinsic scatter or variance of the model σ_{int}^2 as

$$\sigma_{int}^2 = \frac{1}{N-m} \sum_{i=1}^{N} (y_i - \hat{y}_i)^2 - \frac{1}{N} \sum_{i=1}^{N} \sigma_i^2. \qquad (11.3)$$

where m is the number of model parameters. The intrinsic variance can also be referred to as the *debiased variance*, because of the subtraction of the expected scatter (due to measurement errors) from the total sample variance. Equation (11.3) can be considered a generalization of (2.11) in two ways. First, the presence of errors in the measurements of y_i leads to the addition of the last term on the right-hand side. Second, the total variance of the data are calculated *not* relative to the data mean $\bar{y}$ but to the parent mean of each measurement. It is possible that the second term in the right-hand side of (11.3) is larger than the first term, leading to a negative value for the intrinsic variance. This is an indication that, within the statistical errors σ_i, there is no evidence for an intrinsic scatter of the model. This method to estimate the intrinsic scatter is derived from [2] and [24].

It is important to remember that in calculating the intrinsic scatter we have made the assumption that the model *is* an accurate representation of the data. This means that we can no longer test for the null hypothesis that the model represents the parent distribution—we have already assumed this to be the case.

When the model is constant, with $\hat{y}_i = \bar{y}$ being the sample mean, the intrinsic scatter is calculated as

$$\sigma_{int}^2 = \frac{1}{N-1} \sum_{i=1}^{N} (y_i - \bar{y})^2 - \frac{1}{N} \sum_{i=1}^{N} \sigma_i^2. \qquad (11.4)$$

In this case, (11.4) is an unbiased estimate of the variance of Y.

11.2.2 Alternative Method to Estimate the Intrinsic Scatter

An alternative method to measure the amount of extra variance in a fit makes use of the fact that, for a Gaussian dataset, the expected value of the reduced χ_{min}^2 is one. A large value of the minimum χ^2 can be reduced by increasing the size of the errors until $\chi_{red}^2 \simeq 1$, or

$$\chi_{min}^2 = \sum_{i=1}^{N} \frac{(y_i - \hat{y}_i)^2}{\sigma_i^2 + \sigma_{int}^2} \simeq N - m \qquad (11.5)$$

where m is the number of free model parameters, and σ_{int} is the intrinsic scatter that makes the reduced χ^2 unity. In (11.5) we have made the following substitution relative to the standard use of the χ^2_{min} method:

$$\sigma_i^2 \rightarrow \sigma_i^2 + \sigma_{int}^2. \tag{11.6}$$

This method is only approximate, in that an acceptable model need not yield exactly a value of $\chi^2_{red} = 1$. This method to estimate the intrinsic scatter is nonetheless useful as an estimate of the level of scatter present in the data. Like in the earlier method, the analyst is making the assumption that the model fits the data and that the extra variance is attributed to an intrinsic variability of the model (σ_{int}^2).

Example 11.1 The example shown in Fig. 10.1 illustrates a case in which the data do not show systematic deviations from a best-fit model, and yet the χ^2 test would require a rejection of the model. The quantities Energy 1 (independent variable) and Energy 2 were fit to a linear model, the best-fit linear model yielded a fit statistic of $\chi^2_{min} = 60.5$ for 23 degrees of freedom and the model was therefore not acceptable.

Making use of the methods developed in this section, we can estimate the intrinsic scatter that makes the model consistent with the data. Using (11.3), the intrinsic scatter is estimated to be $\sigma_{int} = 2.5$. This means that the model has a typical uniform variability of 2.5 units (the units are those of the y axis, in this case used to measure energy). Using (11.5), a value of $\sigma_{int} = 1.6$ is needed to obtain a reduced χ^2_{min} of unity. The two methods were not expected to provide the same answer since they are based on different assumptions.   ◇

11.3 Systematic Errors

The errors described so far in this book are usually referred to as *random errors*, since they describe the uncertainties in the random variables of interest. There are many sources of random error. A common source of randome error is the Poisson or counting error which derives from measuring N counts in an experiment and results in an error of $\sqrt{N}$. Another source of error is due to the presence of a background that needs to be subtracted from the measured signal. In general, any instrument used to record data will have sources of error that causes the measurements to fluctuate randomly around its mean value.

One of the main tasks of a data analyst is to find all the important sources of error that contribute to the variance of the random variable of interest. A typical case is the measurement of a total signal T in the presence of a background B, where the random variable of interest is the background-subtracted signal S,

$$S = T - B. \tag{11.7}$$

If the background is measured independently from the signal T, then the variance of the source is

$$\sigma_S^2 = \sigma_T^2 + \sigma_B^2. \tag{11.8}$$

The lesson to learn is that the variance of the random variable of interest S *increases* when the background is subtracted. If one assumes that there is no background, or that the background is constant ($\sigma_B^2 = 0$), the random error associated with S may be erroneously underestimated.

The term *statistical error* is often used as a synonym of random error. Sometimes, however, it is used to designate the leading source of random error, such as the Poisson uncertainty in a counting experiment, not including other sources of random error that are equally statistical or random in nature. Such use is not accurate, but the reader should be aware that there is no universally accepted meaning for the term "statistical error."

The term *systematic error* designates sources of error that systematically shift the signal of interest either too high or too low. Sources of systematic errors need to be identified to correct the erroneous offset. A typical example is an instrument that is miscalibrated and systematically reports measurements that have an erroneous offset. Even after the correction for the offset, it is however quite likely that there still remains a source of error, for example associated with the fact that such correction may not be uniform for all datapoints. If the systematic error is additive in nature, i.e., it shifts the random variable X according to $X' = X \pm E$, then the variance of the data is to be modified according to

$$\sigma_i'^{\,2} = \sigma_i^2 + \sigma_E^2. \tag{11.9}$$

The term σ_E^2 denotes the variance of the systematic error E. If E is known exactly, then it would ideally have zero variance. But in all practical cases, there will be an additional source of variance from the correction of a systematic error that needs to be accounted. The modification of the error σ_i due to the presence of a source of systematic error is therefore identical in form to the presence of intrinsic error [compare (11.6) and (11.9)].

If the systematic error is multiplicative in nature, i.e., $X' = E \cdot X$, it may be convenient to use the logarithms, $\log X' = \log X + \log E$ and then proceed as in the case of a linear offset.

Example 11.2 Continuing with the example shown in Fig. 10.1, we can use the results provided in Example 11.1 to say that an additional error of $\sigma_E = 1.6$ would yield a fit statistic of $\chi_{min,red}^2 = 1$. This means that a possible interpretation for the large value of χ_{min}^2 is that we had neglected an additional source of error σ_E. This additional source of error would be in place of the intrinsic scatter, since either correction to the calculation of χ_{min}^2 is sufficient to bring the data in agreement with the model.

The errors of the data in Fig. 10.1 accounted for several sources of random error, including Poisson errors in the counting of photons from these sources, the background subtraction and for errors associated with the model used to describe the distribution of energy. The additional error of order $\sigma_E = 1.6$ for each datapoint may therefore be (a) an intrinsic error of the model (as described in Example 11.1), (b) an additional error from the correction of certain systematic errors that were performed in the process of the analysis or (c) an additional random error that were not already included in the original error budget. The magnitude of possible errors in cases (b) and (c) can be estimated based on the knowledge of the collection of the data and its analysis. If such errors cannot be as large as required to obtain an acceptable fit, the only remaining option is to attribute this error to an intrinsic variance of the model or to conclude that the model is not an accurate description of the data.   ◇

11.4   Estimate of Model Parameters with Systematic Errors or Intrinsic Scatter

In Sects. 11.2 and 11.3 we have assumed that intrinsic scatter or additional sources of systematic errors could be estimated using the best-fit values $\hat{y}_i$ obtained from the fit *without* these errors. Systematic errors or intrinsic scatter, however, do have an effect on the estimate of model parameters. The presence of systematic errors or intrinsic scatter, as discussed earlier in this chapter, is accounted with the addition of another source of variance to the data according to

$$\sigma_i'^{\,2} = \sigma_i^2 + \sigma^2. \tag{11.10}$$

The quantity σ is either the systematic error σ_E not accounted in the initial estimate of σ_i, or the intrinsic scatter σ_{int}. Both cases lead to the same effect on the overall error budget and the χ^2 fit statistic to minimize becomes

$$\chi^2 = \sum_{i=1}^{N} \frac{(y_i - y(x_i))^2}{\sigma_i^2 + \sigma^2}. \tag{11.11}$$

It is clear that repeating the fitting procedure with the larger σ_i' errors instead of the original error will lead to new best-fit values and new uncertainties for the model parameters. The effect of the larger errors is to de-weight datapoints that have small values of σ_i and in general to provide larger confidence intervals for the model parameters. An acceptable procedure to obtain truly *best-fit* values of model parameters and their confidence intervals is to first estimate the additional source of error σ (either an intrinsic scatter or additional statistical or systematic errors) and then repeat the fit.

Example 11.3 The linear fit to the data of Table 6.1 for Energy 1 (independent variable) and Energy 2 resulted in a $\chi^2_{min} = 60.5$ for 23 degrees of freedom. The fit was not acceptable at any level of confidence. In Example 11.1 we calculated that an additional variance of $\sigma^2 = 1.6$ yields a $\chi^2_{min} = 23$. We fit the data with the addition of this error to the dependent variable and find the best-fit values of $a = -0.085 \pm 0.48, b = 1.05 \pm 0.05$.

For comparison, the fit obtained with the original errors returned values of $a = -0.26 \pm 0.088$, $b = 1.04 \pm 0.27$. These values could not be properly called "best-fit," since the fit was not acceptable. Yet, comparison between these values and those for the $\chi^2_{red} = 1.0$ case shows that best-fit parameters are affected by the additional source of error and that the confidence intervals become larger with the increased errors, as expected. ◇

Summary of Key Concepts for this Chapter

☐ *Intrinsic scatter*: An uncertainty of the model that increases the measurement error according to $y_i = \eta_i + \epsilon_i$.

☐ *Debiased variance*: A correction to the measured variance that accounts for the presence of measurement errors,

$$\sigma^2_{int} = \frac{1}{N-m} \sum (y_i - \hat{y}_i)^2 - \frac{1}{N} \sum \sigma^2_i.$$

The square root provides a measure of the intrinsic scatter.

☐ *Systematic error*: A type of measurement error σ_E that systematically shifts the measurements (as opposed to the *statistical error* σ_i). The two errors typically are added in quadrature, $\sigma_i'^2 = \sigma_i^2 + \sigma^2$.

Problems

11.1 Fit the data from Table 6.1 for the radius vs. ratio using a linear model and calculate the intrinsic scatter using the best-fit linear model.

11.2 Using the same data as in Problem 11.1, provide an additional estimate of the intrinsic scatter using the $\chi^2_{red} \simeq 1$ method.

11.3 Justify the $1/(N-m)$ and $1/(N-1)$ coefficients in (11.3) and (11.4).

11.4 Using the data for the Hubble measurements of page 157, assume that each measurement of $\log v$ has an uncertainty of $\sigma = 0.01$. Estimate the intrinsic scatter in the linear regression of $\log v$ vs. m.

11.5 Using the data of Problem 8.2, estimate the intrinsic scatter in the linear fit of the X, Y data.

Chapter 12
Fitting Two-Variable Datasets with Bivariate Errors

Abstract The maximum likelihood method for the fit of a two-variable dataset described in Chap. 8 assumes that one of the variables (the independent variable X) has negligible errors. There are many applications where this assumption is not applicable and uncertainties in both variables must be taken into account. This chapter expands the treatment of Chap. 8 to the fit of a two-variable dataset with errors in both variables.

12.1 Two-Variable Datasets with Bivariate Errors

Throughout Chaps. 8 and 10 we have assumed a simple error model where the independent variable X is known without error, and all sources of uncertainty in the fit are due to the dependent variable Y. The two-variable dataset (X, Y) was effectively treated as a sequence of random variables of values $y_i \pm \sigma_i$ at a fixed location x_i with a parent model $y(x_i)$.

There are many applications, however, in which both variables have comparable uncertainties ($\sigma_x \simeq \sigma_y$) and there is no reason to treat one variable as independent. In general, a two-variable dataset is described by the datapoints

$$(x_i \pm \sigma_{xi}, y_i \pm \sigma_{yi})$$

and the covariance σ_{xyi}^2 between the two measurements. One example is the two measurements of energy in the data in Table 6.1, where it would be appropriate to account for errors in both measurements. There is in fact no particular reason why one measurement should be considered as the independent variable and the other the dependent variable.

There are several methods to deal with two-variable datasets with bivariate error. Given the complexity of the statistical model, there is not a uniquely accepted solution to the general problem of fitting data with bivariate errors. This chapter presents two methods for the linear fit to data with two-variable errors. The first method (Sect. 12.2) applies to a linear fit and it is an extension of the least-squares method of Sect. 8.3. The second method (Sect. 12.3) is based on an alternative definition of χ^2 and it applies to any type of fit function. Although this method

© Springer Science+Busines Media New York 2017
M. Bonamente, *Statistics and Analysis of Scientific Data*, Graduate Texts
in Physics, DOI 10.1007/978-1-4939-6572-4_12

does not have an analytic solution, it can be easily implemented using numerical methods such as Monte Carlo Markov chains described later in this book.

12.2   Generalized Least-Squares Linear Fit to Bivariate Data

In the case of identical measurement errors on the dependent variable Y and no error on the independent variable X, the least-squares method described in Sect. 8.3 estimated the parameters of the linear model as

$$
\begin{cases}
b = \dfrac{Cov(X, Y)}{Var(X)} = \dfrac{\sum_{i=1}^{N}(x_i - \bar{x})(y_i - \bar{y})}{\sum_{i=1}^{N}(x_i - \bar{x})^2} \\[3mm]
a = E(Y) - bE(X) = \dfrac{1}{N}\sum_{i=1}^{N}y_i - b\dfrac{1}{N}\sum_{i=1}^{N}x_i.
\end{cases}
\tag{12.1}
$$

A generalization of this least-squares method accounts for the presence of measurement errors in the estimate of the variances and the covariance in (12.1). The methods of analysis presented in this section were developed by Akritas and Bershady [2] and others [22, 24]. Those references can be used as source of additional information on these methods for bivariate data.

Measurements of the X and Y variables can be described by

$$
\begin{cases}
x_i = \eta_{xi} + \epsilon_{xi} \\
y_i = \eta_{yi} + \epsilon_{yi},
\end{cases}
\tag{12.2}
$$

each the sum of a *parent* quantity and a measurement error, as in (11.1). Accordingly, the variances of the parent variables are given by

$$
\begin{cases}
Var(\eta_{xi}) = Var(x_i) - \sigma_{xi}^2 \\
Var(\eta_{yi}) = Var(y_i) - \sigma_{yi}^2.
\end{cases}
\tag{12.3}
$$

This means that in (12.1) one must replace the sample covariance and variance by a *debiased* or *intrinsic* covariance and variance, i.e., quantities that take into account the presence of measurement errors.

The method of analysis that led to (12.1) assumes that the variable Y depends on X. In other words, we assumed that X is the independent variable. In this case, we talk of a fit of Y-given-X, or Y/X, and we write the linear model as

$$
y = a_{Y/X} + b_{Y/X}x.
\tag{12.4}
$$

Modification of (12.1) with (12.3) (and an equivalent formula for the covariance) leads to the following estimator for the slope and intercept of the linear Y/X model:

$$\begin{cases} b_{Y/X} = \dfrac{Cov(X,Y) - \overline{\sigma_{xy}^2}}{Var(X) - \overline{\sigma_x^2}} = \dfrac{\sum_{i=1}^{N}(x_i - \bar{x})(y_i - \bar{y}) - \sum_{i=1}^{N}\sigma_{xyi}^2}{\sum_{i=1}^{N}(x_i - \bar{x})^2 - \sum_{i=1}^{N}\sigma_{xi}^2} \\ a_{Y/X} = \bar{y} - b_{Y/X}\bar{x}. \end{cases} \tag{12.5}$$

In this equation the sample variance and covariance of (12.1) were replaced with the corresponding intrinsic quantities, and the subscript Y/X indicates that X was considered as the independent variable.

A different result is obtained if Y is considered as the independent variable. In that case, the X-given-Y (or X/Y) model is described as

$$x = a' + b'y. \tag{12.6}$$

The same equations above apply by exchanging the two variables X and Y:

$$\begin{cases} b' = \dfrac{\sum_{i=1}^{N}(x_i - \bar{x})(y_i - \bar{y}) - \sum_{i=1}^{N}\sigma_{xyi}^2}{\sum_{i=1}^{N}(y_i - \bar{y})^2 - \sum_{i=1}^{N}\sigma_{yi}^2} \\ a' = \bar{x} - b'\bar{y}. \end{cases}$$

It is convenient to compare the results of the Y/X and X/Y fits by rewriting the latter in the usual form with x as the independent variable:

$$y = a_{X/Y} + b_{X/Y}x = -\frac{a'}{b'} + \frac{x}{b'}$$

for which we find that the slope and intercept are given by

$$\begin{cases} b_{X/Y} = \dfrac{\sum_{i=1}^{N}(y_i - \bar{y})^2 - \sum_{i=1}^{N}\sigma_{yi}^2}{\sum_{i=1}^{N}(x_i - \bar{x})(y_i - \bar{y}) - \sum_{i=1}^{N}\sigma_{xyi}^2} \\ a_{X/Y} = \bar{y} - b_{X/Y}\bar{x}. \end{cases} \tag{12.7}$$

In general the two estimators Y/X and X/Y will give different results for the best-fit line. This difference highlights the importance of interpreting the data to determine which variable should be considered the independent quantity.

Uncertainties in the parameters a and b and the covariance between them have been calculated by Akritas and Bershady [2]. For the Y/X estimator they can be obtained via the following variables:

$$\xi_i = \frac{(x_i - \bar{x})(y_i - b_{Y/X}x_i - a_{Y/X}) + b_{Y/X}\sigma_{xi}^2 - \sigma_{xyi}^2}{\frac{1}{N}\sum(x_i - \bar{x})^2 - \frac{1}{N}\sum\sigma_{xi}^2}$$ (12.8)

$$\zeta_i = y_i - b_{Y/X}x_i - \bar{x}\xi_i.$$

With these, the variances of a and b and the covariance is given by

$$\begin{cases} \sigma_{bY/X}^2 = \frac{1}{N}\sum(\xi_i - \bar{\xi})^2 \\ \sigma_{aY/X}^2 = \frac{1}{N}\sum(\zeta_i - \bar{\zeta})^2 \\ \sigma_{ab}^2 = \frac{1}{N}\sum(\xi_i - \bar{\xi})(\zeta_i - \bar{\zeta}). \end{cases}$$ (12.9)

For the X/Y estimator there are equivalent formulas for the ξ and ζ variables that need to be used in place of (12.8):

$$\xi_i = \frac{(y_i - \bar{y})(y_i - b_{X/Y}x_i - a_{X/Y}) + b_{X/Y}\sigma_{xyi}^2 - \sigma_{yi}^2}{\frac{1}{N}\sum(x_i - \bar{x})(y_i - \bar{y}) - \frac{1}{N}\sum\sigma_{xyi}^2}$$ (12.10)

$$\zeta_i = y_i - b_{X/Y}x_i - \bar{x}\xi_i.$$

These values can then be used to calculate variances and the covariance of the parameters as in the Y/X fit.

Example 12.1 In Fig. 12.1 we illustrate the difference in the best-fit models when X is the independent variable (12.5) or Y is the independent variable (12.7), using the data of Table 6.1. The Y/X parameters are $a_{Y/X} = -0.367$ and $b_{Y/X} = 1.118$ and the X/Y parameters are $a_{X/Y} = -0.521$ and $b_{X/Y} = 1.132$. Unfortunately there is no definitive prescription to decide which variable should be regarded as independent. In this example each variable could be equally treated as the independent variable and the difference between the two best-fit models is relatively small. The difference between the two models for a value of the x axis of 1 is approximately 20 %. Note that the linear model and the data were plotted in a logarithmic scale to provide a more compact figure.

Also, the data of Table 6.1 do not report any covariance measurement and therefore the best-fit lines were calculated assuming independence between all measurements ($\sigma_{xyi}^2 = 0$). ◇

The example based on the data of Table 6.1 show that there is not just a single slope for the best-fit linear model, but that the results depend on which variable is assumed to be independent, as in the case of no measurement errors available

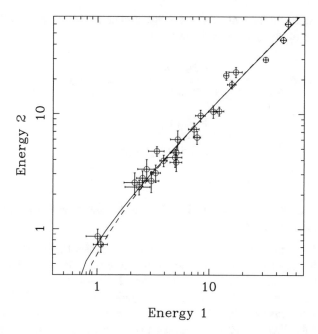

Fig. 12.1 Linear model fits to the data of Table 6.1 using the debiased variance method. The *solid line* is the model that uses Energy 1 as the independent variable X (12.4), the *dashed line* is the model that uses Energy 2 as the independent variable Y (12.6). Note the logarithmic scale for both axes

(Sect. 8.5). In certain cases it may be appropriate to use a model that is intermediate between the two Y/X and X/Y results. This is called the *bisector* model, which consists of the linear model that bisects the two lines obtained from the Y/X and X/Y fits described above. This method is also described by Akritas and Bershady [2] and Isobe and Feigelson [22] and the best-fit bisector line can be obtained from the following formulae:

$$
\begin{cases}
b_{bis} = \dfrac{b_{Y/X}b_{X/Y} - 1 + \sqrt{(1 + b_{Y/X}^2)(1 + b_{X/Y}^2)}}{b_{Y/X} + b_{X/Y}} \\
a_{bis} = \bar{y} - b_{bis}\bar{x}.
\end{cases}
\tag{12.11}
$$

The uncertainties in the slope and intercept parameters can also be obtained using this definition for the ξ and ζ variables:

$$\xi_i = \frac{(1 + b_{X/Y}^2)b_{bis}}{(b_{Y/X} + b_{X/Y})\sqrt{(1 + b_{Y/X}^2)(1 + b_{X/Y}^2)}}\xi_{Y/X}+$$

$$\frac{(1 + b_{Y/X}^2)b_{bis}}{(b_{Y/X} + b_{X/Y})\sqrt{(1 + b_{Y/X}^2)(1 + b_{X/Y}^2)}}\xi_{X/Y} \qquad (12.12)$$

$$\zeta_i = y_i - b_{bis}x_i - \bar{x}\xi_i,$$

where $\xi_{Y/X}$ is the ξ variable defined in (12.8) for the Y/X fit and $\xi_{X/Y}$ is the ξ variable defined in (12.10) for the X/Y fit.

Example 12.2 Figure 12.2 shows the fit to the variables Radius (X variable) and Ratio of thermal energies (Y variable) from Table 6.1. The solid line is the Y/X best-fit line with parameters $a = 1.1253$ and $b = -0.0005$, the dashed line is the X/Y best-fit line with parameters $a = 1.4260$ and $b = -0.0018$ and the dot-dash line is the bisector line with parameters $a = 1.2778$ and $b = -0.0011$. Notice how the Y/X and X/Y regressions give significantly different results. This is in part due to the presence of substantial scatter in the data, which results in several datapoints significantly distant from the best-fit regression lines. In the other

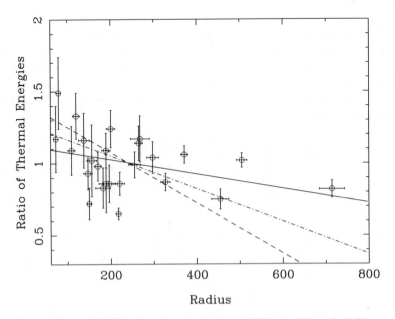

Fig. 12.2 Fit to the data of Table 6.1 using errors in both variables (see Example 12.2)

example of regression with errors in both variables (Fig. 12.1) the Y/X and X/Y best-fit lines were in better agreement.   ◇

12.3   Linear Fit Using Bivariate Errors in the χ^2 Statistic

An alternative method to fit a dataset with errors in both variables is to re-define the χ^2 statistic to account for the presence of errors in the X variable. In the case of a linear fit, the square of the deviation of each datapoint y_i from the model is given by

$$(y_i - a - bx_i)^2. \tag{12.13}$$

When there is no error in the X variable, the variance of the variable in (12.13) is simply the variance of Y, σ_{yi}^2. In the presence of a variance σ_{xi}^2 for X, the variance of the linear combination $y_i - a - bx_i$ is given by

$$Var(y_i - a - bx_i) = \sigma_{yi}^2 + b^2\sigma_{xi}^2,$$

where a and b are the parameters of the linear model and the variables X and Y are assumed to be independent. This suggests a new definition of the χ^2 function for this dataset [35, 40], namely

$$\chi^2 = \sum_{i=1}^N \frac{(y_i - a - bx_i)^2}{\sigma_{yi}^2 + b^2\sigma_{xi}^2}. \tag{12.14}$$

Since each term at the denominator is the variance of the term at the numerator, the new χ^2 variable defined in (12.14) is χ^2-distributed with $f = N - 2$ degrees of freedom.

The complication with the minimization of this function is that the unknown parameter b appears both at the numerator and the denominator of the function that needs to be minimized. As a result, an analytic solution to the maximum likelihood method cannot be given in general. Fortunately, the problem of finding the values of a and b that minimize (12.14) can be solved numerically. This method for the linear fit of two-variable data with errors in both coordinates is therefore of common use, and it is further described in [35].

Summary of Key Concepts for this Chapter

☐ *Data with bivariate errors*: A two-variable dataset that has errors in both variables. For these data there is no commonly accepted fit method.

☐ *Generalized least-squares fit to bivariate data*: An extension of the traditional ML fit to two-variable data. When x is the independent variable the best-fit parameters of the linear model are

$$\begin{cases} b_{Y/X} = \dfrac{Cov(X, Y) - \overline{\sigma_{xy}^2}}{Var(X) - \overline{\sigma_x^2}} \\ a_{Y/X} = \bar{y} - b_{Y/X}\bar{x}. \end{cases}$$

☐ *Bisector model*: A best-fit model for bivariate data that bisects the Y/X and X/Y models, intended to provide and intermediate model.

☐ *Use of bivariate errors in χ^2*: The χ^2 statistic can also be redefined to accommodate bivariate errors according to

$$\chi^2 = \sum_{i=1}^{N} \frac{(y_i - a - bx_i)^2}{\sigma_{yi}^2 + b^2\sigma_{xi}^2}.$$

Problems

12.1 Use the bivariate error data of Energy 1 and Energy 2 from Table 6.1. Calculate the best-fit parameters and errors of the linear model Y/X, where X is Energy 1 and Y is Energy 2.

12.2 Use the bivariate error data of Energy 1 and Energy 2 from Table 6.1. Calculate the best-fit parameters and errors of the linear model X/Y, where X is Energy 1 and Y is Energy 2.

12.3 For the Energy 1 and Energy 2 data of Table 6.1, use the results of Problems 12.1 and 12.2 to calculate the bisector model to the Energy 1 vs. Energy 2 data.

12.4 Repeat Problem 12.1 for the Ratio vs. Radius data of Table 6.1.

12.5 Repeat Problem 12.2 for the Ratio vs. Radius data of Table 6.1.

12.6 Repeat Problem 12.3 for the Ratio vs. Radius data of Table 6.1.

Chapter 13
Model Comparison

Abstract The availability of alternative models to fit a dataset requires a quantitative method for comparing the goodness of fit to different models. For Gaussian data, a lower reduced χ^2 of one model with respect to another is already indicative of a better fit, but the outstanding question is whether the value is *significantly* lower, or whether a lower value can be just the result of statistical fluctuations. For this purpose we develop the distribution function of the F statistic, useful to compare the goodness of fit between two models and the need for an additional "nested" model component, and the Kolmogorov–Smirnov statistics, useful in providing a quantitative measure of the goodness of fit, and in comparing two datasets regardless of their fit to a specific model.

13.1 The *F* Test

For Gaussian data, the χ^2 statistic is used for determining if the fit to a given parent function $y(x)$ is acceptable. It is possible that several different parent functions yield a goodness of fit that is acceptable. This may be the case when there are alternative models to explain the experimental data, and the data analyst is faced with the decision to determine what model best fits the experimental data. In this situation, the procedure to follow is to decide first a confidence level that is considered acceptable, say 90 or 99 %, and discard all models that do not satisfy this criterion. The remaining models are all acceptable, although a lower χ^2_{min} certainly indicates a better fit.

The first version of the F test applies to independent measurements of the χ^2 fit statistic, and its application is therefore limited to cases that compare different datasets. A more common application of the F test is to compare the fit of a given dataset between two models that have a *nested* component, i.e., one model is a simplified version of the other. For nested model components one can determine whether the additional component is really needed to fit the data.

© Springer Science+Busines Media New York 2017

M. Bonamente, *Statistics and Analysis of Scientific Data*, Graduate Texts
in Physics, DOI 10.1007/978-1-4939-6572-4_13

13.1.1 F-Test for Two Independent χ^2 Measurements

Consider the case of two χ^2_{min} values obtained by fitting data from a given experiment to two different functions, $y_1(x)$ and $y_2(x)$. If both models equally well approximate the parent model, then we would expect that the two values of χ^2 would be similar, after taking into consideration that they may have a different number of degrees of freedom. But if one is a better approximation to the parent model, then the value of χ^2 for such model would be significantly lower than for the other. We therefore want to proceed to determine whether both χ^2_{min} statistics are consistent with the null hypothesis that the data are drawn from the respective model. The statistic to use to compare the two values of χ^2 must certainly also take into account the numbers of degrees of freedom, which is related to the number of model parameters used in each determination of χ^2. In fact, a larger number of model parameters may result in fact result in a lower value of χ^2_{min}, simply because of the larger flexibility that the model has in following the data. For example, a dataset of N points will always be fitted perfectly by a polynomial having N terms, but this does not mean that a *simpler* model may not be just as good a model for the data, and the underlying experiment.

Following the theory described in Sect. 7.4, we define the F statistic as

$$F = \frac{\chi^2_{1,min}/f_1}{\chi^2_{2,min}/f_2} = \frac{\chi^2_{1,min,red}}{\chi^2_{2,min,red}}, \tag{13.1}$$

where f_1 and f_2 are the degrees of freedom of $\chi^2_{1,min}$ and $\chi^2_{2,min}$. Assuming that the two χ^2 statistics are *independent*, then F will be distributed like the F statistic with f_1, f_2 degrees of freedom, having a mean of approximately 1 [see (7.22) and (7.24)].

There is an ambiguity in the definition of which of the two models is labeled as 1 and which as 2, since two numbers can be constructed that are the reciprocal of each other, $F_{12} = 1/F_{21}$. The usual form of the F-test is that in which the value of the statistic is $F > 1$, and therefore we choose the largest of F_{12} and F_{21} to implement a one-tailed test of the null hypothesis with significance p,

$$1 - p = \int_{F_{crit}}^{\infty} f_F(f, x)dx = P(F \geq F_{crit}). \tag{13.2}$$

Critical values F_{crit} are reported in Tables A.8, A.9, A.10, A.11, A.12, A.13, A.14, and A.15 for various confidence levels p.

The null hypothesis is that the two values of χ^2_{min} are distributed following a χ^2 distributions; this, in turn, means that the respective fitting functions used to determine each χ^2_{min} are *both* good approximations of the parent distribution. Therefore the test based on this distribution can reject the hypothesis that both fitting functions are the parent distribution. If the test rejects the hypothesis at the desired confidence level, then only one of the models will still stand after the test—the one at the denominator with the lowest reduced χ^2—even if the value of χ^2_{min} alone was not able to discriminate between the two models.

Example 13.1 Consider the radius vs. ratio data of Table 6.1 (see also Problem 11.1). The linear fit to the entire dataset is not acceptable, and therefore a linear model for all measurements must be discarded. If we consider measurements 1 through 5, 6 through 10, and 11 through 15, a linear fit to these two subsets results in the values of best-fit parameters and χ^2 shown in the table, along with the probability to exceed the value of the fit statistic.

Measurements	a	b	χ^2_{min}	Probability
1–5	0.97 ± 0.09	-0.0002 ± 0.0002	5.05	0.17
6–10	1.27 ± 0.22	-0.0007 ± 0.0011	6.19	0.10
10–15	0.75 ± 0.09	-0.0002 ± 0.0003	18.59	0.0

The third sample provides an unacceptable fit to the linear model, and therefore this subset cannot be further considered. For the first two samples, the fits are acceptable at the 90 % confidence level, and we can construct the F statistic as

$$F = \frac{\chi^2_{min}(6-10)}{\chi^2_{min}(1-5)} = 1.23.$$

Both χ^2 have the same number of degrees of freedom (3), and Table A.13 shows that the value of 1.23 is certainly well within the 90 % confidence limit for the F statistics ($F_{crit} \simeq 5.4$). This test shows that both subsets are equally well described by a linear fit, and therefore the F-test cannot discriminate between them.

To illustrate the power of the F-test, assume that there is another set of five measurements that yield a $\chi^2_{min} = 1.0$ when fit to a linear model. This fit is clearly acceptable in terms of its χ^2 probability. Constructing an F statistic between this new set and set 6–10, we would obtain

$$F = \frac{\chi^2_{min}(6-10)}{\chi^2_{min}(new)} = 6.19.$$

In this case, the value of F is *not* consistent at the 90 % level with the F distribution with $f_1 = f_2 = 3$ degrees of freedom (the measured value exceeds the critical value). The F-test therefore results in the conclusion that, at the 90 % confidence level, the two sets are not equally likely to be drawn from a linear model, with the new set providing a better match.   ◇

It is important to note that the hypothesis of independence of the two χ^2 is not justified if the same data are used for both statistics. In practice, this means that the F statistic *cannot* be used to compare the fit of a given dataset to two different models. The test can still be used to test whether two different datasets, derived from the same experiment but with independent measurements, are equally well described by the same parametric model, as shown in the example above. In this case, the null hypothesis is that both datasets are drawn from the same parent model, and a rejection of the hypothesis means that both datasets cannot derive from the same distribution.

13.1.2   F-Test for an Additional Model Component

Consider a model $y(x)$ with m adjustable parameters, and another model $\bar{y}(x)$ obtained by fixing p of the m parameters to a reference (fixed) value. In this case, the $\bar{y}(x)$ model is said to be *nested* into the more general model, and the task is to determine whether the additional p parameters of the general model are required to fit the data.

Example 13.2 An example of nested models are polynomial models. The general model can be taken as a polynomial of second order,

$$y(x) = a + bx + cx^2$$

and the nested model as a linear model,

$$y(x) = a + bx.$$

The nested model is obtained from the general model with $c = 0$ and has one fewer degree of freedom than the general model.   ◇

Following the same discussion as in Chap. 10, we can say that

$$\begin{cases} \chi^2_{min} \sim \chi^2(N-m) & \text{(full model)} \\ \bar{\chi}^2_{min} \sim \chi^2(N-m+p) & \text{("nested" model).} \end{cases} \tag{13.3}$$

Clearly $\chi^2_{min} < \bar{\chi}^2_{min}$ because of the additional free parameters used in the determination of χ^2_{min}. A lower value of χ^2_{min} does not necessarily mean that the additional parameters of the general model are required. The nested model can in fact achieve an equal or even better fit relative to the parent distribution of the fit statistic, i.e., a lower χ^2_{red}, because of the larger number of degrees of freedom. In general, a model with fewer parameters is to be preferred to a model with larger number of parameters because of its more economical description of the data, provided that it gives an acceptable fit.

In Sect. 10.3 we discussed that, when comparing the true value of the fit statistic χ^2_{true} for the parent model to the minimum χ^2_{min} obtained by minimizing a set of p free parameters, $\Delta\chi^2 = \chi^2_{true} - \chi^2_{min}$ and χ^2_{min} are independent of one another, and that $\Delta\chi^2$ is distributed like χ^2 with p degrees of freedom. There are situations in which the same properties apply to the two χ^2 statistics described in (13.3), such that the statistic $\Delta\chi^2$ is distributed like

$$\Delta\chi^2 = \overline{\chi}^2_{min} - \chi^2_{min} \sim \chi^2(p), \tag{13.4}$$

and it is independent of χ^2_{min}. One such case of practical importance is precisely the one under consideration, i.e., when there is a nested model component described by parameters that are independent of the other model parameters. A typical example is an additional polynomial term in the fit function, as illustrated in the example above.

In this case, the null hypothesis we test is that $y(x)$ and $\overline{y}(x)$ are equivalent models, i.e., adding the p parameters does *not* constitute a significant change or improvement to the model. Under this hypothesis we can use the two independent statistics $\Delta\chi^2$ and χ^2_{min}, and construct a bona fide F statistic as

$$F = \frac{\Delta\chi^2/p}{\chi^2_{min}/(N-m)}. \tag{13.5}$$

This statistic tests the null hypothesis using an F distribution with $f_1 = p, f_2 = N-m$ degrees of freedom. A rejection of the hypothesis indicates that the two models $y(x)$ and $\overline{y}(x)$ are not equivalent. In practice, a rejection constitutes a positive result, indicating that the additional model parameters in the nested component *are* actually needed to fit the data. A common situation is when there is a single additional model parameter, $p = 1$, and the corresponding critical values of F are reported in Table A.8. A discussion of certain practical cases in which additional model components may obey (13.4) is provided in a research article by Protassov [36].

Example 13.3 The data of Table 10.1 and Fig. 10.2 are well fit by a linear model, while a constant model appears not to be a good fit to all measurements. Using only the middle three measurements, we want to compare the goodness of fit to a linear model, and that to a constant model, and determine whether the addition of the b parameter provides a significant improvement to the fit.

The best-fit linear model has a $\chi^2_{min} = 0.13$ which, for $f_2 = N-m = 1$ degree of freedom, with a probability to exceed this value of 72 %, i.e., it is an excellent fit. A constant model has a $\overline{\chi}^2_{min} = 7.9$, which, for 2 degrees of freedom, has a probability to exceed this value of ≥ 0.01, i.e., it is acceptable at the 99 % confidence level, but not at the 90 % level. If the analyst requires a level of confidence ≤ 90 %, then the constant model should be discarded, and no further analysis of the experiment is needed. If the analyst can accept a 99 % confidence level, we can determine whether the improvement in χ^2 between the constant and the linear model is significant. We construct the statistic

$$F = \frac{\overline{\chi}^2_{min} - \chi^2_{min}}{\chi^2_{min}} \frac{1}{1} = 59.4$$

which, according to Table A.8 for $f_1 = 1$ and $f_2 = 1$, is significant at the 99 %
(and therefore 95 %) confidence level, but not at 90 % or lower. In fact, the critical
value of the F distribution with $f_1 = 1, f_2 = 1$ at the 99 % confidence level is
$F_{crit} = 4,052$. Therefore a data analyst willing to accept a 99 % confidence level
should conclude that the additional model component b is *not* required, since there
is ≥ 1 % (actually, ≥ 5 %) probability that such an improvement in the χ^2 statistic is
due by chance, and not by the fact that the general model is truly a more accurate
description of the data.   ◇

The example above illustrates the principle of simplicity or parsimony in the
analysis of data. When choosing between two models, both with an acceptable fit
statistic at the same confidence level (in the previous example at the 99 % level),
one should prefer the model with fewer parameters, even if its fit statistic (e.g.,
the reduced χ^2_{min}) is inferior to that of the more complex model. This general
guiding principle is sometimes referred to as *Occam's razor*, after the Middle Ages
philosopher and Franciscan friar William of Occam.

13.2 Kolmogorov–Smirnov Tests

Kolmogorov–Smirnov tests are a different method for the comparison of a one-
dimensional dataset to a model, or for the comparison of two datasets to one another.
The tests make use of the cumulative distribution function, and are applicable to
measurements of a single variable X, for example to determine if it is distributed
like a Gaussian. For two-variable dataset, the χ^2 and F tests remain the most viable
option.

The greatest advantage the Kolmogorov–Smirnov test is that it does not require
the data to be binned, and, for the case of the comparison between two dataset, it
does not require any parameterization of the data. These advantages come at the
expense of a more complicated mathematical treatment to find the distribution function of the test statistic. Fortunately, numerical tables and analytical approximations
make these tests manageable.

13.2.1 Comparison of Data to a Model

Consider a random variable X with cumulative distribution function $F(x)$. The data
consist of N measurements, and for simplicity we assume that they are in increasing
order, $x_1 \leq x_2 \leq \ldots \leq x_N$. This condition can be achieved by re-labelling the
measurements, which preserves the statistical properties of the data. The goal is to
construct a statistic that describes the difference between the sample distribution of
the data and a specified distribution, to test whether the data are compatible with
this distribution.

Start with the sample cumulative distribution

$$F_N(x) = \frac{1}{N}[\text{\# of measurements} \leq x]. \tag{13.6}$$

By definition, $0 \leq F_N(x) \leq 1$. The test statistic we want to use is defined as

$$D_N = \max_x |F_N(x) - F(x)|, \tag{13.7}$$

where $F(x)$ is the parent distribution, and the maximum value of the difference between the parent distribution and the sample distribution is calculated for all values in the support of X.

One of the remarkable properties of the statistic D_N is that it has the same distribution for any underlying distribution of X, provided X is a continuous variable. The proof that D_N has the same distribution regardless of the distribution of X illustrates the properties of the cumulative distribution and of the quantile function presented in Sect. 4.8.

Proof We assume that $F(x)$ is continuous and strictly increasing. This is certainly the case for a Gaussian distribution, or any other distribution that does not have intervals where the distribution functions is $f(x) = 0$. We make the change of variables $y = F(x)$, so that the measurement x_k corresponds to $y_k = F(x_k)$. This change of variables is such that

$$F_N(x) = \frac{(\text{\# of } x_i < x)}{N} = \frac{(\text{\# of } y_k < y)}{N} = U_N(y)$$

where $U_N(y)$ is the sample cumulative distribution of Y and $0 \leq y \leq 1$. The cumulative distribution of Y is

$$U(y) = P(Y < y) = P(X < x) = F(x) = y.$$

The fact that the cumulative distribution is $U(y) = y$ shows that Y is a uniform distribution between 0 and 1. As a result, the statistic D_N is equivalent to

$$D_N = \max_{0 \leq y \leq 1} |U_N(y) - U(y)|$$

where Y is a uniform distribution. Since this is true no matter the original distribution X, D_N has the same distribution for any X. Note that this derivation relies on the continuity of X, and this assumption must be verified to apply the resulting Kolmogorov–Smirnov test.   □

The distribution function of the statistic D_N was determined by Kolmogorov in 1933 [25], and it is not easy to evaluate analytically. In the limit of large N, the cumulative distribution of D_N is given by

$$\lim_{N \to \infty} P(D_N < z/\sqrt{N}) = \sum_{r=-\infty}^{\infty} (-1)^r e^{-2r^2 z^2} \equiv \Phi(z). \tag{13.8}$$

Table 13.1 Critical points of
the Kolmogorov distribution
D_N for large values of N

Confidence level	
p	$\sqrt{N}D_N$
0.50	0.828
0.60	0.895
0.70	0.973
0.80	1.073
0.90	1.224
0.95	1.358
0.99	1.628

The function $\Phi(z)$ can also be used to approximate the probability distribution of D_N for small values of N, using

$$P(D_N < z/(\sqrt{N} + 0.12 + 0.11/\sqrt{N})) \simeq \Phi(z). \tag{13.9}$$

A useful numerical approximation for $P(D_N < z)$ is also provided in [30].

The probability distribution of D_N can be used to test whether a sample distribution is consistent with a model distribution. Critical values of the D_N distribution with probability p,

$$P(D_N \leq T_{crit}) = p \tag{13.10}$$

are shown in Table 13.1 in the limit of large N. For small N, critical values of the D_N statistic are provided in Table A.25. If the measured value for D_N is greater than the critical value, then the null hypothesis must be rejected, and the data are not consistent with the model. The test allows no free parameters, i.e., the distribution that represents the null hypothesis must be fully specified.

Example 13.4 Consider the data from Thomson's experiment to measure the ratio m/e of an electron (page 23). We can use the D_N statistic to test whether either of the two measurement of the variable m/e is consistent with a given hypothesis. It is necessary to realize that the Kolmogorov–Smirnov test applies to a fully specified hypothesis H_0, i.e., the parent distribution $F(x)$ cannot have free parameter that are to be determined by a fit to the data. We use a fiducial hypothesis that the ratio is described by a Gaussian distribution of $\mu = 5.7$ (the true value in units of 10^7 g Coulomb^{-1}, though the units are unnecessary for this test), and a variance of $\sigma^2 = 1$. Both measurements are inconsistent with this model, as can be seen from Fig. 13.1. See Problem 13.1 for a quantitative analysis of the results. $\diamond$

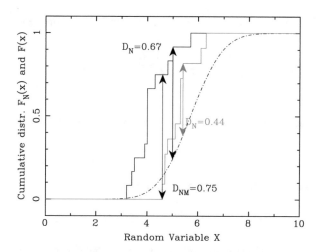

Fig. 13.1 Kolmogorov–Smirnov test applied to the measurements of the ratio m/e from Thomson's experiments described on page 23. The *black line* corresponds to the measurements for Tube 1, and the *red line* to those of Tube 2 (measurements have been multiplied by 10^7). The *dot-dashed line* is the cumulative distribution of a Gaussian with $\mu = 5.7$ (the correct value) and a fiducial variance of $\sigma^2 = 1$

13.2.2   Two-Sample Kolmogorov–Smirnov Test

A similar statistic can be defined to compare two datasets:

$$D_{NM} = \max_x |F_M(x) - G_N(x)| \tag{13.11}$$

where $F_M(x)$ is the sample cumulative distribution of a set of M observations, and $G_N(x)$ that of another independent set of N observations; in this case, there is no parent model used in the testing. The statistic D_{NM} measures the maximum deviation between the two cumulative distributions, and by nature it is a discrete distribution. In this case, we can show that the distribution of the statistic is the same as in (13.9), provided that the change

$$N \to \frac{MN}{M + N}$$

is made. This number can be considered as the effective number of datapoints of the two distributions. For the two-sample Kolmogorov–Smirnov D_{NM} test we can therefore use the same table as in the Kolmogorov–Smirnov one-sample test, provided N is substituted with $MN/(M + N)$ and that N and M are both large.

As N and M become large, the statistic approaches the following distribution:

$$\lim_{N,M\to\infty} P\left(D_{NM} < z/\sqrt{\frac{MN}{M+N}}\right) = \Phi(z). \qquad (13.12)$$

Proof We have already shown that for a sample distribution with M points,

$$F_M(x) - F(x) = U_M(y) - U(y),$$

where U is a uniform distribution in $(0,1)$. Since

$$F_M(x) - G_N(x) = F_M(x) - F - (G_N(x) - G),$$

where $F = G$ is the parent distribution, it follows that $F_M(x) - G_N(x) = U_N - V_N$, where U_M and V_N are the sample distribution of two uniform variables. Therefore the statistic

$$D_{NM} = \max_x |F_M(x) - G_N(x)|$$

is independent of the parent distribution, same as for the statistic D_N.

Next we show how the factor $\sqrt{1/N + 1/M}$ originates. It is clear that the expectation of $F_M(x) - G_N(x)$ is zero, at least in the limit of large N and M; the second moment can be calculated as

$$E[(F_M(x) - G_N(x))^2] = E[(F_M(x) - F(x))^2]$$
$$+E[(G_N(x) - G(x))^2] + 2E[(F_M(x) - F(x))(G_N(x) - G(x))]$$
$$= E[(F_M(x) - F(x))^2] + E[(G_N(x) - G(x))^2]$$

In fact, since $F_M(x) - F(x)$ is independent of $G_N(x) - G(x)$, their covariance is zero. Each of the two remaining terms can be evaluated using the following calculation:

$$E\left[(F_M(x) - F(x))^2\right] = E\left[\frac{1}{M}(\{\# \text{ of } x_i\text{'s} < x\} - MF(x))^2\right] =$$

$$\frac{1}{M^2}E\left[(\{\# \text{ of } x_i\text{'s} < x\} - E[\{\# \text{ of } x_i\text{'s} < x\}])^2\right].$$

For a fixed value of x, the variable $\{\# \text{ of } x_i\text{'s} < x\}$ is a binomial distribution in which "success" is represented by one measurement being $< x$, and the probability of success is $p = F(x)$. The expectation in the equation above is therefore equivalent to the variance of a binomial distribution with M tries, for which $\sigma^2 = Mp(1 - p)$, leading to

$$E\left[(F_M(x) - F(x))^2\right] = \frac{1}{M}F(x)(1 - F(x)).$$

It follows that

$$E[(F_M(x) - G_N(x))^2] = \left(\frac{1}{M} + \frac{1}{N}\right)F(x)(1 - F(x))$$

A simple way to make the mean square of $F_M(x) - G_N(x)$ independent of N and M is to divide it by $\sqrt{1/M + 1/N}$. This requirement is therefore a necessary condition for the variable $\sqrt{NM/(N + M)}D_{NM}$ to be independent of N and M.

Finally, we show that $\sqrt{NM/(N + M)}D_{NM}$ is distributed in the same way as $\sqrt{N}D_N$, at least in the asymptotic limit of large N and M. Using the results from the D_N distribution derived in the previous section, we start with

$$\max_x \left| \sqrt{\frac{MN}{M + N}}(F_M(x) - G_N(x)) \right| = \max_{0 \le y \le 1} \left| \sqrt{\frac{MN}{M + N}}(U_M - V_N)) \right|.$$

The variable can be rewritten as

$$\sqrt{\frac{MN}{M + N}}(U_M - U + (V - V_N)) = \sqrt{\frac{N}{M + N}}(\sqrt{M}(U_M - U))$$
$$+ \sqrt{\frac{M}{M + N}}(\sqrt{N}(V_N - V)).$$

Using the central limit theorem, it can be shown that the two variables $\alpha = \sqrt{M}(U_M - U)$ and $\beta = \sqrt{N}(V_N - V)$ have the same distribution, which tends to a Gaussian in the limit of large M. We then write

$$\sqrt{\frac{MN}{M + N}}(F_M(x) - G_N(x)) = \sqrt{\frac{N}{M + N}}\alpha + \sqrt{\frac{M}{M + N}}\beta$$

and use the property that, for two independent and identically distributed Gaussian variables α and β the variable $a \cdot \alpha + b \cdot \beta$ is distributed like α, provided that $a^2 + b^2 = 1$. We therefore conclude that, in the asymptotic limit,

$$D_{NM} = \max_x \left| \sqrt{\frac{MN}{M + N}}(F_M(x) - G_N(x)) \right| \sim \max_x \left| \sqrt{N}(V_N - V) \right| = D_N.$$

$\square$

Example 13.5 We can use the two-sample Kolmogorov–Smirnov statistic to compare the data from Tube #1 and Tube #2 of Thomson's experiment to measure the ratio m/e of an electron (page 23). The result, shown in Fig. 13.1, indicates that the two measurements are not in agreement with one another. See Problem 13.2 for a quantitative analysis of this test.   ◇

Summary of Key Concepts for this Chapter

□ *F Test*: A test to compare two independent χ^2 measurements,

$$F = \chi^2_{1,red}/\chi^2_{2,red}.$$

□ *F Test for additional component*: The significance of an additional model component with p parameters can be tested using

$$F = \frac{\Delta\chi^2/p}{\chi^2_{min}/(N-m)}$$

when the additional component is nested within the general model.

□ *Kolmogorov–Smirnov test*: A non-parametric test to compare a one-variable dataset to a model or two datasets with one another.

Problems

13.1 Using the data from Thomson's experiment at page 23, determine the values of the Kolmogorov–Smirnov statistic D_N for the measurement of Tube #1 and Tube #2, when compared with a Gaussian model for the measurement with $\mu = 5.7$ and $\sigma^2 = 1$. Determine at what confidence level you can reject the hypothesis that the two measurements are consistent with the model.

13.2 Using the data from Thomson's experiment at page 23, determine the values of the two-sample Kolmogorov–Smirnov statistic D_{NM} for comparison between the two measurements. Determine at what confidence level you can reject the hypothesis that the two measurements are consistent with one another.

13.3 Using the data of Table 10.1, determine whether the hypothesis that the last three measurements are described by a simple constant model can be rejected at the 99 % confidence level.

13.4 A given dataset with $N = 5$ points is fit to a linear model, for a fit statistic of $\overline{\chi}^2_{min}$. When adding an additional nested parameter to the fit, $p = 1$, determine by how much should the χ^2_{min} be reduced for the additional parameter to be significant at the 90 % confidence level.

13.5 A dataset is fit to model 1, with minimum χ^2 fit statistic of $\chi_1^2 = 10$ for 5 degrees of freedom; the same dataset is also fit to another model, with $\chi_2^2 = 5$ for 4 degrees of freedom. Determine which model is acceptable at the 90 % confidence, and whether the F test can be used to choose one of the two models.

13.6 A dataset of size N is successfully fit with a model, to give a fit statistic $\overline{\chi}_{min}^2$. A model with a nested component with 1 additional independent parameter for a total of m parameters is then fit to χ_{min}^2, providing a reduction in the fit statistic of $\Delta\chi^2$. Determine what is the minimum $\Delta\chi^2$ that, in the limit of a large number of degrees of freedom, provides 90 % confidence that the additional parameter is significant.

Chapter 14
Monte Carlo Methods

Abstract The term *Monte Carlo* refers to the use of random variables to evaluate quantities such as integrals or parameters of fit functions that are typically too complex to evaluate via other analytic methods. This chapter presents elementary Monte Carlo methods that are of common use in data analysis and statistics, in particular the bootstrap and jackknife methods to estimate parameters of fit functions.

14.1   What is a Monte Carlo Analysis?

The term *Monte Carlo* derives from the name of a locality in the Principality of Monaco known for its resorts and casinos. In statistics and data analysis Monte Carlo is an umbrella word that means the use of computer-aided numerical methods to solve a specific problem, typically with the aid of random numbers.

Traditional Monte Carlo methods include numerical integration of functions that can be graphed but that don't have a simple analytic solution and simulation of random variables using random samples from a uniform distribution. Another problem that benefits by the use of random numbers is the estimation of uncertainties in the best-fit parameters of analytical models used to fit data. There are cases when an analytical solution for the error in the parameters is not available. In many of those cases, the bootstrap or the jackknife methods can be used to obtain reliable estimates for those uncertainties.

Among many other applications, Monte Carlo Markov chains stand out as a class of Monte Carlo methods that is now commonplace across many fields of research. The theory of Markov chains (Chap. 15) dates to the early twentieth century, yet only over the past 20 years or so it has found widespread use as Monte Carlo Markov chains (Chap. 16) because of the computational power necessary to implement the method.

© Springer Science+Busines Media New York 2017   225
M. Bonamente, *Statistics and Analysis of Scientific Data*, Graduate Texts
in Physics, DOI 10.1007/978-1-4939-6572-4_14

14.2   Traditional Monte Carlo Integration

A common numerical task is the evaluation of the integral of a function $f(x)$ for which analytic solution is either unavailable or too complicated to calculate exactly,

$$I = \int_A f(x)dx. \tag{14.1}$$

We want to derive a method to approximate this integral by randomly drawing N samples from the support A. For simplicity, we assume that the domain of the variable $f(x)$ is a subset of real numbers between a and b. We start by drawing samples from a uniform distribution between these two values,

$$g(x) = \begin{cases} \dfrac{1}{b-a} & \text{if } a \le x \le b \\ 0 & \text{otherwise.} \end{cases} \tag{14.2}$$

Recall that for a random variable X with continuous distribution $f(x)$, the expectation (or mean value) is defined as

$$E[X] = \int_{-\infty}^{\infty} xg(x)dx \tag{14.3}$$

(2.6); we have also shown that the mean can be approximated as

$$E[X] \simeq \frac{1}{N} \sum_{i=1}^{N} x_i$$

where x_i are independent measurements of that variable. The expectation of the function $f(x)$ of a random variable is

$$E[f(x)] = \int_{-\infty}^{\infty} f(x)g(x)dx,$$

and it can be estimated using the Law of Large Numbers (Sect. 4.5):

$$E[f(x)] \simeq \frac{1}{N} \sum_{i=1}^{N} f(x_i). \tag{14.4}$$

These equations can be used to approximate the integral in (14.1) as a simple sum:

$$I = (b-a) \int_a^b f(x)g(x)dx = (b-a)E[f(x)] \simeq (b-a)\frac{1}{N} \sum_{i=1}^{N} f(x_i). \tag{14.5}$$

Equation (14.5) can be implemented by drawing N random uniform samples x_i from the support, then calculating $f(x_i)$, and evaluating the sum. This is the basic *Monte Carlo integration* method, and it can be easily implemented by using a random number generator available in most programming languages.

The method can be generalized to more than one dimension; if the support $A \subset \mathbb{R}^n$ has volume V, then the integration of an n-dimensional function $f(x)$ is given by the following sum:

$$I = \frac{V}{N} \sum_{i=1}^{N} f(x_i) \tag{14.6}$$

It is clear that the precision in the evaluation of the integral depends on the number of samples drawn. The error made by this method of integration can be estimated using the following interpretation of (14.6): the quantity $Vf(x)$ is the random variable of interest, and I is the expected value. Therefore, the variance of the random variable is given by the usual expression,

$$\sigma_I^2 = \frac{V^2}{N} \sum_{i=1}^{N} (f(x_i) - \bar{f})^2. \tag{14.7}$$

This means that the relative error in the calculation of the integral is

$$\frac{\sigma_I}{I} = \frac{1}{\sqrt{N}} \frac{\sqrt{\sum_{i=1}^{N}(f(x_i) - \bar{f})^2}}{\sum_{i=1}^{N} f(x_i)} \propto \frac{1}{\sqrt{N}}; \tag{14.8}$$

as expected, the relative error decreases like the square root of N, same as for a Poisson variable. Equation (14.8) can be used to determine how many samples are needed to estimate an integral with a given precision.

14.3   Dart Monte Carlo Integration and Function Evaluation

Another method to integrate a function, or to perform related mathematical operations, can be shown by way of an example. Assume that we want to measure the area of a circle of radius R. One can draw a random sample of N values in the (x, y) plane, as shown in Fig. 14.1, and count all the points that fall within the circle, $N(R)$. The area of the circle, or any other figure with known analytic function, is accordingly estimated as

$$A = \frac{N(R)}{N} \times V \tag{14.9}$$

in which V is the volume sampled by the two random variables. In the case of a circle of radius $R = 1$ we have $V = 4$, and since the known area is $A = \pi R^2$, this method provides an approximation to the number π.

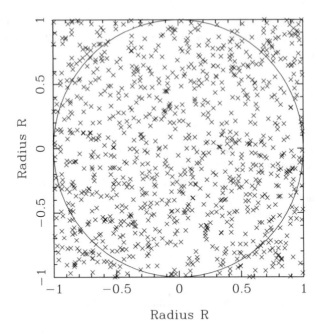

Fig. 14.1 Monte Carlo method to perform a calculation of the area of a *circle* (also a simulation of the number π), with $N = 1000$ iterations

Notice that (14.9) is equivalent to (14.6), in which the sum $\sum f(x_i)$ becomes $N(R)$, where $f(x_i) = 1$ indicates that a given random data point x_i falls within the boundaries of the figure of interest.

Example 14.1 (Simulation of the Number π)   Figure 14.1 shows a Monte Carlo simulation of the number π, using 1000 random numbers drawn in a box of linear size 2, encompassing a circle of radius $R = 1$. The simulation has a number $N(R) = 772$ of points within the unit circle, resulting in an estimate of the area of the circle as $\pi R^2 = 0.772 \times 4 = 3.088$. Compared with the exact result of $\pi = 3.14159$, the simulation has an error of 1.7 %. According to (14.8), a 1000 iteration simulation has an expected relative error of order 3.1 %, therefore the specific simulation reported in Fig. 14.1 is consisted with the expected error, and more numbers must be drawn to improve the precision.   ◇

14.4 Simulation of Random Variables

A method for the simulation of a random variable was discussed in Sect. 4.8. Since the generation of random samples from a uniform random variable was involved, this method also falls under the category of Monte Carlo simulations.

The method is based on (4.42):

$$X = F^{-1}(U),$$

in which F^{-1} represents the inverse of the cumulative distribution of the target variable X, and U represents a uniform random variable between 0 and 1. In Sect. 4.8 we provided the examples on how to use (4.42) to simulate an exponential distribution, which has a simple analytic function for its cumulative distribution.

The Gaussian distribution is perhaps the most common variable in many statistical applications, and its generation cannot be accomplished by (4.42), since the cumulative distribution is a special function and $F(x)$ does not have a close form. A method to overcome this limitation was discussed in Sect. 4.8.2, and it consists of using two uniform random variables U and V to simulate two standard Gaussians X and Y of zero mean and unit variance via (4.45),

$$\begin{cases} X = \sqrt{-2\ln(1-U)} \cdot \cos(2\pi V) \\ Y = \sqrt{-2\ln(1-U)} \cdot \sin(2\pi V). \end{cases} \quad (14.10)$$

A Gaussian X' of mean μ and variance σ^2 is related to the standard Gaussian X by the transformation

$$X = \frac{X' - \mu}{\sigma},$$

and therefore it can be simulated via

$$X' = \left(\sqrt{-2\ln(1-U)} \cdot \cos(2\pi V) \right) \sigma + \mu. \quad (14.11)$$

Figure 14.2 shows a simulation of a Gaussian distribution function using (14.11). Precision can be improved with increasing number of samples.

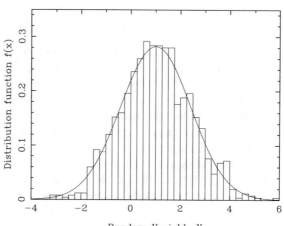

Fig. 14.2 Monte Carlo simulation of the probability distribution function of a Gaussian of $\mu = 1$ and $\sigma^2 = 2$ using 1000 samples according to (14.11)

14.5 Monte Carlo Estimates of Errors for Two-Variable Datasets

The two methods presented in this section, the bootstrap and the jackknife, are among the most common techniques to estimate best-fit parameters and their uncertainties in the fit to two-variable datasets. We have seen in previous chapters that the best-fit parameters and their uncertainties can be estimated analytically, for example, in the case of a linear regression with known errors in the dependent variable. In those cases, the exact analytical solution is typically the most straightforward to implement. When the analytic solution to a maximum likelihood fit is unavailable, then χ^2 minimization followed by the $\chi^2_{min} + \Delta\chi^2$ criterion can also be used to measure best-fit values and uncertainties in the parameters. Finally, Markov chain Monte Carlo methods to be presented in Chap. 16 can also be used in virtually any case for which the likelihood can be calculated.

The two methods presented in this section have a long history of use in statistical data analysis, and had been in use since well before the Markov chain Monte Carlo methods became of wide use. The bootstrap and jackknife methods are typically easier to implement than a Monte Carlo Markov chain. In particular, the bootstrap uses a large number of repetitions of the dataset, and therefore is computer intensive; the older jackknife method instead uses just a small number of additional random datasets, and requires less computing resources.

14.5.1 The Bootstrap Method

Consider a dataset Z composed of N measurements of either a random variable or, more generally, a pair of variables. The bootstrap method consists of generating as large a number of random, "synthetic" datasets based on the original set. Each set is then used to determine the distribution of the random variable (e.g., for the one-dimensional case) or of the best-fit parameters for the $y(x)$ model (for the two–dimensional case). The method has the following steps:

1. Draw at random N datapoints from the original set Z, with replacement, to form a synthetic dataset Z_i. The new dataset has therefore the same dimension as the original set, but a few of the original points may be repeated, and a few missing.
2. For each dataset Z_i, calculate the parameter(s) of interest a_i. For example, the parameters can be calculated using a χ^2 minimization technique.
3. Repeat this process as many times as possible, say N_{boot} times.
4. At the end of the process, the parameters a_n, $n = 1, \ldots, N_{boot}$, approximate the posterior distribution of the parameter of interest. These values can therefore be used to construct the sample distribution function for the parameters, and therefore obtain the best-fit value and confidence intervals.

Notice that one advantage of the bootstrap method is that it can be used even in cases in which the errors on the datapoints are not available, which is a very common occurrence. In this situation, the direct maximum likelihood method applied to the original set Z alone would not provide uncertainties in the best-fit parameters, as explained in Sect. 8.5. Since at each iteration the best-fit parameters alone must be evaluated, a dataset without errors in the dependent variable can still be fit to find the best-fit parameters, and the bootstrap method will provide an estimate of the uncertainties. This is one of the main reasons why the bootstrap method is so common.

Example 14.2 (Bootstrap Analysis of Hubble's Data)   We perform a bootstrap analysis on the data from Hubble's experiment of page 157. The dataset Z consists of the ten measurements of the magnitude m and logarithm of the velocity $\log v$, as shown in Fig. 8.2. We generate 10,000 random synthetic datasets of ten measurements each, for which typically a few of the original datapoints are repeated. Given that error bars on the dependent variable $\log v$ were not given, we assume that the uncertainties have a common value for all measurement (and therefore the value of the error is irrelevant for the determination of the best-fit parameters). For each dataset Z_i we perform a linear regression to obtain the best-fit values of the parameters a_i and b_i.

The sample distributions of the parameters are shown in Fig. 14.3; from them, we can take the median of the distribution as the "best-fit" value for the parameter, and the 68 % confidence interval as the central range of each parameter that contains 68 % of the parameter occurrences. It is clear that both distributions are somewhat asymmetric; the situation does not improve with a larger number of bootstrap samples, since there is only a finite number of synthetic datasets that

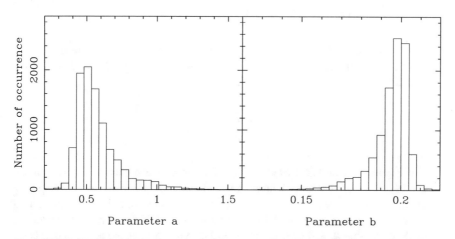

Fig. 14.3 Monte Carlo bootstrap method applied to the data from Hubble's experiment. (*Left*) Sample distribution of parameter a, with a median of $a = 0.54$ and a 68 % central range of 0.45–0.70. (*Right*) Distribution of b, with median $b = 0.197$ and a central range of 0.188–0.202. The best-fit values of the original dataset Z were found to be $a = 0.55$ and $b = 0.197$ (see page 159)

can be generated at random, with replacement, from the original dataset (see Problem 14.1). ◇

A key feature of the bootstrap method is that it is an unbiased estimator for the model parameters. We can easily prove this general property in the special case of a one-dimensional dataset, with the goal of estimating the sample mean and variance of the random variable X from N independent measurements. It is clear that we would normally not use the bootstrap method in this situation, since (2.8) and (5.4) provide the exact solution to the problem. The following proof is used to show that the bootstrap method provides unbiased estimates for the mean and variance of a random variable.

Proof The sample average calculated for a given bootstrap dataset Z_i is given by

$$\bar{x}_i = \frac{1}{N} \sum_{j=1}^{N} x_j n_{ji} \tag{14.12}$$

where n_{ji} is the number of occurrence of datapoint x_j in the synthetic set Z_i. If $n_{ji} = 0$ it means that x_j was not selected for the set, $n_{ji} = 1$ it means that there is just one occurrence of x_j (as in the original set), and so on. The number $n_{ji} \leq N$, and it is a random variable that is distributed like a binomial with $p = 1/N$, since the drawing for each bootstrap set is done at random, and with replacement. Therefore, we find that

$$\begin{cases} E[n_{ji}] = Np = 1 \\ Var(n_{ij}) \equiv \sigma_i^2 = Np(1-p) = \dfrac{N-1}{N} \end{cases} \tag{14.13}$$

where the expectation is calculated for a given dataset Z, drawing a large number of bootstrap sample based on that specific set. It follows that $\bar{x}_i$ is an unbiased estimator of the sample mean,

$$E[\bar{x}_i] = \frac{1}{N} \sum_{j=1}^{N} x_j E[n_{ji}] = \bar{x}. \tag{14.14}$$

The expectation operator used in the equation above relates to the way in which a specific synthetic dataset can be drawn, i.e., indicates an "average" over a specific dataset. The operation of expectation should also be repeated to average over all possible datasets Z consisting of N measurements of the random variable X, and that operation will also result in an expectation that is equal to the parent mean of X,

$$E[\bar{x}] = \mu. \tag{14.15}$$

Although we used the same symbol for the expectation of (14.14) and (14.15), the two operations are therefore different in nature.

The proof that the variance of the sample mean of dataset Z_i is an unbiased estimator of the parent variance σ^2/N is complicated by the fact that the random variables n_{ij} are not independent. In fact, they are related by

$$\sum_{i=1}^{N} n_{ij} = N, \tag{14.16}$$

and this enforces a negative correlation between the variables that vanishes only in the limit of very large N. It can be shown that the covariance of the n_{ij}'s (say, the covariance between n_{ij} and n_{ki}, were $i \neq k$, and i labels the dataset) is given by

$$\sigma_{jk}^2 = -\frac{1}{N}. \tag{14.17}$$

The proof of (14.17) is left as an exercise, and it is based on the use of (14.16), and (4.3) (see Problem 14.2).

The variance of $\bar{x}_i$ can be calculated using (4.3), since $\bar{x}_i$ is a linear combination of N random variables n_{ij}:

$$Var(\bar{x}_i) = Var\left(\frac{1}{N}\sum_{j=1}^{N} x_j n_{ji}\right) =$$

$$\frac{1}{N^2}\left(\sum_{j=1}^{N} x_j^2 \sigma_i^2 + 2\sum_{j=1}^{N}\sum_{k=j+1}^{N} x_j x_k \sigma_{jk}^2\right) =$$

$$\frac{1}{N^2}\left(\frac{N-1}{N}\sum_{j=1}^{N} x_j^2 - \frac{2}{N}\sum_{j=1}^{N}\sum_{k=j+1}^{N} x_j x_k\right)$$

in which we have used the results of (14.13) and (14.17). Next, we need to calculate the expectation of this variance, in the sense of varying the dataset Z itself:

$$E[Var(\bar{x}_i)] = \frac{N-1}{N^3}E\left[\sum_{j=1}^{N} x_j^2\right] - \frac{2}{N^3}\left(\frac{1}{2}\sum_{j \neq k} E[x_j x_k]\right) \tag{14.18}$$

The last sum in the equation above is over all pairs (j, k); the factor $1/2$ takes into account the double-counting of terms such as $x_j x_k$ and $x_k x_j$, and the sum contains a total of $N(N-1)$ identical terms. Since the measurements x_i, x_j are independent and identically distributed, $E[x_i x_k] = E[x_j]^2$, it follows that

$$E[Var(\bar{x}_i)] = \frac{N-1}{N^2}\left(E[x_i^2] - E[x_i]^2\right) = \frac{N-1}{N^2}\sigma^2 = \frac{N-1}{N}\sigma_\mu^2$$

where σ^2 is the variance of the random variable X, and $\sigma_\mu^2 = \sigma^2/N$ the variance of the sample mean. The equation states that $E[Var(\bar{x}_i)] = E[s^2]$, where s^2 is the sample variance of X. We showed in Sect. 5.1.2 that the sample variance is an unbiased estimator of the variance of the mean, provided it is multiplied by the known factor $N/(N-1)$. In practice, when calculating the variance from the N bootstrap samples, we should use the factor $1/(N-1)$ instead of $1/N$, as is normally done according to (5.6).   □

14.5.2   The Jackknife Method

The jackknife method is an older Monte Carlo method that makes use of just N resampled datasets to estimate best-fit parameters and their uncertainties. As in the bootstrap method, we consider a dataset Z of N independent measurements either of a random variable X or of a pair of random variables. The method consists of the following steps:

1. Generate a resampled dataset Z_j by deleting the jth element from the dataset. This resampled dataset has therefore dimension $N-1$.
2. Each dataset Z_j is used to estimate the parameters of interest. For example, apply the linear regression method to dataset Z_j and find the best-fit values of the linear model, a_j and b_j.
3. The parameters of interest are also calculated from the full-dimensional dataset Z, as one normally would. The best-fit parameters are called $\hat{a}$.
4. For each dataset Z_j, define the *pseudo-values* $a_j^\star$ as

$$a_j^\star = N\hat{a} - (N-1)a_j \qquad (14.19)$$

5. The jackknife estimate of each parameter of interest and its uncertainty are given by the following equations:

$$\begin{cases} a^\star = \dfrac{1}{N}\displaystyle\sum_{j=1}^{n} a_j^\star \\[2mm] \sigma_{a^\star}^2 = \dfrac{1}{N(N-1)}\displaystyle\sum_{j=1}^{N}(a_j^\star - a^\star)^2. \end{cases} \qquad (14.20)$$

To prove that (14.20) provide an accurate estimate for the parameters and their errors, we apply them to the simple case of the estimate of the mean from a sample of

N measurements. In this case we want to show that the expectation of the jackknife estimate of the mean $a^\star$ is equal to the parent mean μ, and that the expectation of its variance $\sigma_{a^\star}^2$ is equal to σ^2/N.

Proof For a sample of N measurements of a random variable x, the sample mean and its variance are given by

$$
\begin{cases}
\bar{x} = \dfrac{1}{N} \displaystyle\sum_{j=1}^{N} x_i \\
\dfrac{s^2}{N} = \dfrac{1}{N(N-1)} \displaystyle\sum_{j=1}^{N} (x_i - \bar{x})^2.
\end{cases}
\tag{14.21}
$$

The proof consists of showing that $a_j^\star = x_j$, so that $a^\star$ is the sample mean and $\sigma_{a^\star}^2$ is the sample variance. The result follows from:

$$
a_j = \frac{1}{N-1} \sum_{i \neq j} x_i, \; \hat{a} = \frac{1}{N} \sum_{i=1}^{N} x_i
$$

$$
\Rightarrow a_j^\star = N \frac{1}{N} \sum_{i=1}^{N} x_i - \frac{N-1}{N-1} \sum_{i \neq j} x_i = x_j.
$$

Notice that the factor of $1/(N-1)$ was used in the calculation of the sample variance, according to (5.6). □

Example 14.3 In the case of the Hubble experiment of page 157, we can use the jackknife method to estimate the best-fit parameters of the fit to a linear model of m versus $\log v$. According to (14.20), we find that $a^\star = 0.52$, $\sigma_{a^\star} = 0.13$, and $b^\star = 0.199$, $\sigma_{b^\star} = 0.008$. These estimates are in very good agreement with the results of the bootstrap method, and those of the direct fit to the original dataset for which, however, we could not provide uncertainties in the fit parameters. ◇

Summary of Key Concepts for this Chapter

☐ *Monte Carlo method*: Any numerical method that makes use of random variables to perform calculations that are too complex to be performed analytically, such as Monte Carlo integration and "dart" methods.

☐ *Bootstrap method*: A common method to estimate model parameters that uses a large number of synthetic datasets obtained by re-sampling of the original data.

☐ *Jackknife method*: A simple method to estimate model parameters that uses just N re-sampled datasets.

Problems

14.1 Calculate how many synthetic bootstrap datasets can be generated at random from a dataset Z with N unique datapoints. Notice that the order in which the datapoints appear in the dataset is irrelevant.

14.2 For a bootstrap dataset Z_j constructed from a set Z of N independent measurements of a variable X, show that the covariance between the number of occurrence n_{ji} and n_{jk} is given by (14.17),

$$\sigma_{ik}^2 = -\frac{1}{N}.$$

14.3 Perform a numerical simulation of the number π, and determine how many samples are sufficient to achieve a precision of 0.1 %. The first six significant digits of the number are $\pi = 3.14159$.

14.4 Perform a bootstrap simulation on the Hubble data presented in Fig. 14.3, and find the 68 % central confidence ranges on the parameters a and b.

14.5 Using the data of Problem 8.2, run a bootstrap simulation with $N = 1000$ iterations for the fit to a linear model. After completion of the simulation, plot the sample probability distribution function of the parameters a and b, and find the median and 68 % confidence intervals on the fit parameters. Describe the possible reason why the distribution of the fit parameters are not symmetric.

14.6 Use the data of Problem 8.2, but assuming that the errors in the dependent variable y are unknown. Run a bootstrap simulation with $N = 1000$ iterations, and determine the median and 68 % confidence intervals on the parameters a and b to the fit to a linear model.

14.7 Using the data of Problem 8.2, assuming that the errors in the dependent variable y are unknown, estimate the values of a and b to the fit to a linear model using a jackknife method.

14.8 Given two uniform random variables U_1 and U_2 between $-R$ and $+R$, as often available in common programming software, provide an analytic expression to simulate a Gaussian variable of mean μ and variance σ^2.

Chapter 15
Introduction to Markov Chains

Abstract The theory of Markov chains is rooted in the work of Russian mathematician Andrey Markov and has an extensive body of literature to establish its mathematical foundations. The availability of computing resources has recently made it possible to use Markov chains to analyze a variety of scientific data. Monte Carlo Markov chains are now one of the most popular methods of data analysis. This chapter presents the key mathematical properties of Markov chains, necessary to understand its implementation as Monte Carlo Markov chains.

15.1   Stochastic Processes and Markov Chains

This section presents key mathematical properties of Markov chains. The treatment is somewhat theoretical, but necessary to ensure that the applications we make to the analysis of data are consistent with the mathematics of Markov chains, which can be very complex. The goal is therefore that of defining and understanding a basic set of definitions and properties necessary to use Markov chains for the analysis of data, especially via the Monte Carlo simulations.

Markov chains are a specific type of stochastic processes, or sequence of random variables. A typical example of Markov chain is the so-called random walk, in which at each time step a person randomly takes a step either to the left, or to the right. As time progresses, the location of the person is the random variable of interest, and the collection of such random variables forms a Markov chain. The ultimate goal of a Markov chain is to determine the *stationary distribution* of the random variable. For the random walk, where we are interested in knowing the probability that at a given time the person is located n steps to the right or to the left of the starting point.

In the typical case of interest for the analysis of data, a dataset Z is fit to a parametric model. The goal is to create a Markov chain for each parameter of the model, in such a way that the stationary distribution for each parameter is the distribution function of the parameter. The chain will therefore result in the knowledge of the best-fit value of the parameter, and of confidence intervals, making use of the information provided by the dataset.

© Springer Science+Busines Media New York 2017   237
M. Bonamente, *Statistics and Analysis of Scientific Data*, Graduate Texts
in Physics, DOI 10.1007/978-1-4939-6572-4_15

15.2   Mathematical Properties of Markov Chains

A stochastic process is defined as a sequence of variables X_t,

$$\{X_t, \text{ for } t \epsilon T\} \tag{15.1}$$

where t labels the sequence. The domain for the index t is indicated as T to signify "time." The domain is usually a subset of the real numbers ($T \subset \mathbb{R}$) or of the natural numbers ($T \subset N$). As time progresses, the random variables X_t change value, and the stochastic process describes this evolution.

A Markov chain is a particular stochastic process that satisfies the following properties:

1. The time domain is the natural numbers ($T \subset \mathbb{N}$), and each random variable X_t can have values in a countable set, e.g., the natural numbers or even an n-dimensional space ($\mathbb{N}^n$), but not real numbers ($\mathbb{R}^n$). A typical example of a Markov chain is one in which $X_i = n$, where both i (the time index) and n (the value of the random variable) are natural numbers. Therefore a Markov chain takes the form of

$$X_1 \rightarrow X_2 \rightarrow X_3 \rightarrow \ldots \rightarrow X_n \rightarrow \ldots$$

   The random variable X_i describes the state of the system at time $t = i$. The fact that Markov chains must be defined by way of countable sets may appear an insurmountable restriction, since it would appear that the natural domain for an n-parameter space is $\mathbb{R}^n$. While a formal extension of Markov chains to $\mathbb{R}^n$ is also possible, this is not a complication for any practical application, since any parameter space can be somehow "binned" into a finite number of states. For example, the position of the person in a random walk was "binned" into a number of finite (or infinite but countable) positions, and a similar process can be applied to virtually any parameter of interest for a given model. This means that the variable under consideration can occupy one of a countable multitude of states $\varepsilon_1, \varepsilon_2, \ldots, \varepsilon_n, \ldots$, and the random variable X_i identifies the state of the system at time step i, $X_i = \varepsilon_n$.

2. A far more important property that makes a stochastic process a Markov chain is the fact that subsequent steps in the chain are only dependent on the current state of the chain, and not on any of its previous history. This "short memory" property is known as the *Markovian property*, and it is the key into the construction of Markov chains for the purpose of data analysis. In mathematical terms, given the present time $t = n$, the future state of the chain at $t = n + 1$ (X_{n+1}) depends only on the present time (X_n), but not on past history. Much of the efforts in the construction of a Monte Carlo Markov chain lies in the identification

of a *transition probability* from state ε_i to state ε_j between consecutive time steps,

$$p_{ij} = P(X_{n+1} = \varepsilon_j / X_n = \varepsilon_i). \tag{15.2}$$

A Markov chain requires that this probability be time-independent, and therefore a Markov chain has the property of time homogeneity. In Chap. 16 we will see how the transition probability takes into account the likelihood of the data Z with the model.

The two properties described above result in the fact that Markov chain is a sequence of states determined by transition probabilities p_{ij} (also referred to as *transition kernel*) that are fixed in time. The ultimate goal is to determine the probability to find the system in each of the allowed states. With an eye towards future applications for the analysis of data, each state may represent values of one or many parameters, and therefore a Markov chain makes it possible to reconstruct the probability distribution of the parameters.

Example 15.1 (Random Walk)   The *random walk* is a Markov chain that represents the location of a person who randomly takes a step of unit length forward with probability p, or a step backward with probability $q = 1 - p$ (typically $p = q = 1/2$). The state of the system is defined by the location i at which the person find itself at time $t = n$,

$$X_n = \{\text{Location } i \text{ along the } \mathbb{N}^+ \text{ axis}\}$$

where $\mathbb{N}^+$ indicates all positive and negative integers. For this chain, the time domain is the set of positive numbers ($T = \mathbb{N}$), and the position can be any negative or positive integer ($\mathbb{N}^+$). The transition probability describes the fact that the person can only take either a step forward or backward:

$$p_{ij} = \begin{cases} p & \text{if } j = i + 1, \text{ or move forward} \\ q & \text{if } j = i - 1, \text{ or move backward} \\ 0 & \text{otherwise.} \end{cases} \tag{15.3}$$

$$\diamond$$

The chain satisfies the Markovian property, since the transition probability depends only on its present position, and not on previous history.

Example 15.2   Another case of a Markov chain is a simple model of diffusion, known as the *Ehrenfest chain*. Consider two boxes with a total of m balls. At each time step, one selects a ball at random from either box, and replaces it in the other box. The state of the system can be defined via the random variable

$$X_n = \{\text{Number of balls in the first box}\}.$$

The random variable can have only a finite number of values $(0, 1, \ldots, m)$. At each time step, the transition probability is

$$
p_{ij} = \begin{cases} \dfrac{m-i}{m} & \text{if } j = i + 1 \text{ (box had } i \text{ balls, now has } i + 1) \\[2ex] \dfrac{i}{m} & \text{if } j = i - 1 \text{ (box had } i \text{ balls, now has } i - 1). \end{cases} \tag{15.4}
$$

For example, in the first case it means that we chose one of $m - i$ balls from the second box. The transition probabilities depend only on the number of balls in the first box at any given time, and are completely independent of how the box came to have that many balls. This chain therefore satisfies the Markovian property.   ◇

15.3   Recurrent and Transient States

We are interested in knowing how often a state is visited by the chain and, in particular, whether a given state can be visited infinitely often. Assume that the system is initially in state ε_i. We define u_k the probability that the system returns to the initial state in *exactly* k time steps, and v_n the probability that the system returns to the initial state at time n, with the possibility that it may have returned there other times prior to n. Clearly, it is true that $v_n \geq u_n$.

To determine whether a state is recurrent or transient, we define

$$
u \equiv \sum_{n=1}^{\infty} u_n \tag{15.5}
$$

as the probability of the system returning the initial state ε_i for the first time at some time n. The state can be classified as *recurrent* or *transient* according to the probability of returning to that state:

$$
\begin{cases} u = 1 & \text{state is recurrent;} \\ u < 1 & \text{state is transient.} \end{cases} \tag{15.6}
$$

Therefore a recurrent state is one that will certainly be visited again by the chain. Notice that no indication is given as to the time at which the system will return to the initial state.

We also state a few theorems that are relevant to the understanding of recurrent states. Proofs of these theorems can be found, for example, in the textbook by Ross [38] or other books on stochastic processes, and are not reported here.

Theorem 15.1 *With v_n the probability that the system returns to a state ε_i at time n,*

$$\text{state } \varepsilon_i \text{ is recurrent} \iff \sum_{n=1}^{\infty} v_n = \infty. \tag{15.7}$$

This theorem states that, if the system does return to a given state, then it will do so infinitely often. Also, since this is a necessary and sufficient condition, any transient state will not be visited by the chain an infinite number of times. This means that transient states will not be visited any more after a given time, i.e., they are only visited during an initial period. The fact that recurrent states are visited infinitely often means that it is possible to construct a sample distribution function for recurrent states with a precision that is function of the length of the chain. No information is, however, provided on the timing of the return to a recurrent state.

We also introduce the definition of *accessible* states: a state ε_j is said to be accessible from state ε_i if $p_{ij}(m) > 0$ for some natural number m, meaning that there is a non-zero probability of reaching this state from another state in m time steps. The following theorems establish properties of accessible states, and how the property of accessibility relates to that of recurrence.

Theorem 15.2 *If a state ε_j is accessible from a recurrent state ε_i, then ε_j is also recurrent, and ε_i is accessible from ε_j.*

This theorem states that once the system reaches a recurrent state, the states visited previously by the chain must also be recurrent, and therefore will be visited again infinitely often. This means that recurrent states form a network, or class, of states that share the property of recurrence, and these are the states that the chain will sample over and over again as function of time.

Theorem 15.3 *If a Markov chain has a finite number of states, then each state is accessible from any other state, and all states are recurrent.*

This theorem ensures that all states in a finite chain will be visited infinitely often, and therefore the chain will sample all states as function of time. This property is of special relevance for Monte Carlo Markov chain methods in which the states of the chain are possible values of the parameters. As the chain progresses, all values of the parameters are accessible, and will be visited in proportion of the posterior distribution of the parameters.

Example 15.3 (Recurrence of States of the Random Walk) Consider the random walk with transition probabilities given by (15.3). We want to determine whether the initial state of the chain is a recurrent or a transient state for the chain. The probability of returning to the initial state in k steps is clearly given by the binomial distribution,

$$p_{ii}(k) = \begin{cases} 0 & \text{if } k \text{ is odd} \\ C(n, k) p^n q^n & \text{if } k = 2n \text{ is even} \end{cases} \tag{15.8}$$

where

$$C(n,k) = \binom{k}{n} = \frac{k!}{(k-n)!n!} \tag{15.9}$$

is the number of combinations [of n successes out of $k = 2n$ tries, see (3.3)]. Using Stirling's approximation for the factorial function in the binomial coefficient,

$$n! \simeq \sqrt{2\pi n}n^n e^{-n},$$

the probability to return at time $k = 2n$ to the initial state becomes

$$v_k = p_{ii}(k) = \binom{2n}{n}p^n q^n = \frac{(2n)!}{(n!)^2}p^n q^n \simeq \frac{\sqrt{4\pi n}}{2\pi n}\frac{(2n)^{2n}e^{-2n}}{n^{2n}e^{-2n}}p^n q^n = \frac{(4pq)^n}{\sqrt{\pi n}}$$

which holds only for k even.

This equation can be used in conjunction with Theorem 15.1 to see if the initial state is transient or recurrent. Consider the series

$$\sum_{n=1}^{\infty} v_n = \sum_{n=1}^{\infty} \frac{1}{\sqrt{\pi n}}(4pq)^n.$$

According to Theorem 15.1, the divergence of this series is a necessary and sufficient condition to prove that the initial state is recurrent.

(a) $p \neq q$. In this case, $x = 4pq < 1$ and

$$\sum_{n=1}^{\infty} \frac{1}{\sqrt{\pi n}}(4pq)^n < \sum_{n=1}^{\infty} x^n = \frac{x}{1-x};$$

since $x < 1$, the series converges and therefore the state is transient. This means that the system may return to the initial state, but only for a finite number of times, even after an infinite time. Notice that as time progresses the state of the system will drift in the direction that has a probability $> 1/2$.

(b) $p = q = 1/2$, thus $4pq = 1$. The series becomes

$$\frac{1}{\sqrt{\pi}} \sum_{n=1}^{\infty} \frac{1}{n^{1/2}}. \tag{15.10}$$

It can be shown (see Problem 15.2) that this series diverges, and therefore a random walk with the same probability of taking a step to the left or to the right will return to the origin infinitely often.   ◇

15.4 Limiting Probabilities and Stationary Distribution

The ultimate goal of a Markov chain is to calculate the probability that a system occupies a given state ε_i after a large number n of steps. This probability is called the *limiting probability*. According to the frequentist approach defined in (1.2), it is given by

$$p_j^* = \lim_{n \to \infty} p_j(n), \qquad (15.11)$$

where $p_j(n)$ is the probability of the system to be found in state ε_j at time $t = n$. With the aid of the total probability theorem, the probability of the system to be in state ϵ_j at time $t = n$ is

$$p_j(n) = \sum_k p_k(n-1)p_{kj}. \qquad (15.12)$$

In fact $p_k(n-1)$ represents the probability of being in state ε_k at time $n-1$, and the set of probabilities $p_k(n-1)$ forms a set of mutually exclusive events which encompasses all possible outcomes, with the index k running over all possible states. This formula can be used to calculate recursively the probability $p_j(n)$ using the probability at the previous step and the transition probabilities p_{kj}, which do not vary with time.

Equation (15.12) can be written in a different form if the system is known to be in state ε_i at an initial time $t = 0$:

$$p_{ij}(n) = P(X_n = \varepsilon_j) = \sum_k p_{ik}(n-1)p_{kj} \qquad (15.13)$$

where $p_{ij}(n)$ is the probability of the system going from state ε_i to ε_j in n time steps.

The probabilities $p_j(n)$ and $p_{ij}(n)$ change as the chain progresses. The limiting probabilities p_j^*, on the other hand, are independent of time, and they form the *stationary* distribution of the chain. General properties for the stationary distribution can be given for Markov chains that have certain specific properties. In the following we introduce additional definitions that are useful to characterize Markov chains, and to determine the stationary distribution of the chain.

A number of states that are accessible from each other, meaning there is a non-zero probability to reach one state from the other ($p_{ij} > 0$), are said to *communicate*, and all states that communicate are part of the same *class*. The property of communication ($\leftrightarrow$) is an equivalence relation, meaning that it obeys the following three properties:

(a) The reflexive property: $i \leftrightarrow i$;
(b) The symmetric property: if $i \leftrightarrow j$, then $j \leftrightarrow i$; and
(c) The transitive property: if $i \leftrightarrow j$ and $j \leftrightarrow k$, then $i \leftrightarrow k$. Therefore, each
   class is separate from any other class of the same chain. A chain is said
   to be *irreducible* if it has only one class, and thus all states communicate
   with each other.

Another property of Markov chains is periodicity. A state is said to be
periodic with period T if $p_{ii}(n) = 0$ when n is not divisible by T, and T is the
largest such integer with this property. This means that the return to a given
state must occur in multiples of T time steps. A chain is said to be *aperiodic* if
$T = 1$, and return to a given state can occur at any time. It can be shown that
all states in a class share the same period.

The uniqueness of the stationary distribution and an equation that can be
used to determine it are established by the following theorems.

Theorem 15.4 *An irreducible aperiodic Markov chain belongs to either of
the following two classes:*

1. *All states are positive recurrent. In this case, $p_i^\star = \pi_i$ is the stationary
   distribution, and this distribution is unique.*
2. *All states are transient or null recurrent; in this case, there is no stationary
   distribution.*

This theorem establishes that a "well behaved" Markov chain, i.e., one with
positive recurrent states, does have a stationary distribution, and that this
distribution is unique. Positive recurrent states, defined in Sect. 15.3, are those
for which the expected time to return to the same state is finite, while the time
to return to a transient or null recurrent state is infinite. This theorem also
ensures that, regardless of the starting point of the chain, the same stationary
distribution will eventually be reached.

Theorem 15.5 *The limiting probabilities are the solution of the system of linear
equations*

$$p_j^\star = \sum_{i=1}^{N} p_i^\star p_{ij}. \tag{15.14}$$

Proof According to the recursion formula (15.12),

$$p_j(n) = \sum_{i=1}^{N} p_i(n-1)p_{ij}. \tag{15.15}$$

Therefore the result follows by taking the limit $n \to \infty$ of the above equation.
$\square$

If we consider a chain with a probability distribution at a time t_0 that satisfies

$$p_j(t_0) = \sum_{i=1}^{N} p_i(t_0)p_{ij},$$ (15.16)

then the probability distribution of the states satisfies (15.14), and the chain has reached the stationary distribution $p_j(n) = p_j^*$. Theorem 15.5 guarantees that, from that point on, the chain will maintain its stationary distribution. The importance of a stationary distribution is that, as time elapses, the chain *samples* this distribution. The sample distribution of the chain, e.g., a hystogram plot of the occurrence of each state, can therefore be used as an approximation of the posterior distribution.

Example 15.4 (Stationary Distribution of the Ehrenfest Chain)   We want to find a distribution function p_j^* that is the stationary distribution of the Ehrenfest chain. This case is of interest because the finite number of states makes the calculation of the stationary distribution easier to achieve analytically. The condition for a stationary distribution is

$$p_j^* = \sum_{i=1}^{N} p_i^* p_{ij}$$

where N is the number of states of the chain. The condition can also be written in matrix notation. Recall that the transition probabilities for the Ehrenfest chain are

$$p_{ij} = \begin{cases} \dfrac{m-i}{m} & \text{if } j = i+1 \\[2mm] \dfrac{i}{m} & \text{if } j = i-1, \end{cases}$$

and they can be written as a transition matrix $\boldsymbol{P}$

$$\boldsymbol{P} = [p_{ij}] = \begin{bmatrix} 0 & 1 & 0 & 0 & \dots & 0 & 0 \\ \dfrac{1}{m} & 0 & \dfrac{m-1}{m} & 0 & \dots & 0 & 0 \\ 0 & \dfrac{2}{m} & 0 & \dfrac{m-2}{m} & \dots & 0 & 0 \\ & & \dots\dots\dots\dots\dots \\ 0 & 0 & 0 & 0 & \dots & 1 & 0 \end{bmatrix}.$$ (15.17)

Notice that the sum of each line is one, since

$$\sum_{j} p_{ij} = 1$$

is the probability of going from state ε_i to *any* state ε_j. In (15.17) you can regard the vertical index to be $i = 0, \ldots, m$, and the horizontal index $j = 0, \ldots, m$.

The way in which we typically use (15.14) is simply to verify whether a distribution is the stationary distribution of the chain. In the case of the Ehrenfest chain, we try the binomial distribution as the stationary distribution,

$$p_i = \binom{m}{i} p^i q^{m-i} \qquad i = 0, \ldots, m,$$

in which p and q represent the probability of finding a ball in either box. At equilibrium one expects $p = q = 1/2$, since even an initially uneven distribution of balls between the two boxes should result in an even distribution at later times. It is therefore reasonable to expect that the probability of having i balls in the first box, out of a total of m, is equivalent to that of i positive outcomes in a binary experiment.

To prove this hypothesis, consider $p = [p_0, p_1, \ldots, p_m]$ as a row vector of dimension $m + 1$, and verify the equation

$$p = pP, \tag{15.18}$$

which is the matrix notation for the condition of a stationary distribution. For the Erhenfest chain, this condition is

$$[p_0, p_1, \ldots, p_m] = [p_0, p_1, \ldots, p_m] \begin{bmatrix} 0 & 1 & 0 & 0 & \ldots & 0 & 0 \\ \dfrac{1}{m} & 0 & \dfrac{m-1}{m} & 0 & \ldots & 0 & 0 \\ 0 & \dfrac{2}{m} & 0 & \dfrac{m-2}{m} & \ldots & 0 & 0 \\ & & \ldots\ldots\ldots\ldots & & & \\ 0 & 0 & 0 & 0 & \ldots & 1 & 0 \end{bmatrix}.$$

For a given state i, only two terms (at most) contribute to the sum,

$$p_i = p_{i-1} p_{i-1,i} + p_{i+1} p_{i+1,i}. \tag{15.19}$$

From this we can prove that the $p = q = 1/2$ binomial is the stationary distribution of the Ehrenfest chain (see Problem 15.1).   ◇

Summary of Key Concepts for this Chapter

☐ *Markov chain*: A stochastic process or sequence of random variables as function of an integer time variable.

☐ *Markovian property*: It is the key property of Markov chains, stating that the state of the system at a given time depends only on the state at the previous time step.

☐ *Recurrent and transient state*: A recurrent state occurs infinitely often while a transient state only occurs a finite number of times in the Markov chain.

☐ *Stationary distribution*: It is the asymptotic distribution of each variable, obtained after a large number of time steps of the Markov chain. When the variable represents a model parameter, the stationary distribution is the posterior distribution of the parameter.

Problems

15.1 Consider the Markov chain for the Ehrenfest chain described in Example 15.4. Show that the stationary distribution is the binomial with $p = q = 1/2$.

15.2 Show that the random walk with $p = q = 1/2$ (15.10) returns to the origin infinitely often, and therefore the origin is a recurrent state of the chain.

15.3 For the random walk with $p \neq p$, show that the origin is a transient state.

15.4 Assume that the diffusion model of Example 15.2 is modified in such a way that at each time step one has the option to choose one box at random from which to replace a ball to the other box.

(a) Determine the transition probabilities p_{ij} for this process.
(b) Determine whether this process is a Markov chain.

15.5 Using the model of diffusion of Problem 15.4, determine if the binomial distribution with $p = q = 1/2$ is the stationary distribution.

Chapter 16
Monte Carlo Markov Chains

Abstract Monte Carlo Markov Chains (MCMC) are a powerful method to analyze scientific data that has become popular with the availability of modern-day computing resources. The basic idea behind an MCMC is to determine the probability distribution function of quantities of interest, such as model parameters, by repeatedly querying datasets used for their measurement. The resulting sequence of values form a Markov chain that can be analyzed to find best-fit values and confidence intervals. The modern-day data analyst will find that MCMCs are an essential tool that permits tasks that are simply not possible with other methods, such as the simultaneous estimate of parameters for multi-parametric models of virtually any level of complexity, even in the presence of correlation among the parameters.

16.1 Introduction to Monte Carlo Markov chains

A typical data analysis problem is the fit of data to a model with adjustable parameters. Chapter 8 presented the maximum likelihood method to determine the best-fit values and confidence intervals for the parameters. For the linear regression to a two-variable dataset, in which the independent variable is assumed to be known and the dependent variable has errors associated with its measurements, we found an analytic solution for the best-fit parameters and its uncertainties (Sect. 8.3). Even the case of a multiple linear regressions is considerably more complex to solve analytically (Chap. 9) and most fits to non-linear functions do not have analytic solutions at all.

When an analytic solution is not available, the χ^2_{min} method to search for best-fit parameters and their confidence intervals is still applicable, as described in Sect. 10.3. The main complication is the computational cost of sampling the parameter space in search of χ^2_{min} and surfaces of constant $\chi^2_{min} + \Delta\chi^2$. Consider, for example, a model with 10 free parameters: even a very coarse sampling of 10 values for each parameter will result in 10^{10} evaluations of the likelihood, or χ^2, to cover the entire parameter space. Moreover, it is not always possible to improve the situation by searching for just a few interesting parameters at a time, e.g., fixing the value of the background while searching for the flux of the source. In fact, there may

© Springer Science+Busines Media New York 2017   249
M. Bonamente, *Statistics and Analysis of Scientific Data*, Graduate Texts
in Physics, DOI 10.1007/978-1-4939-6572-4_16

be correlation among parameters and this requires that the parameters be estimated simultaneously.

The Monte Carlo Markov chain (MCMC) methods presented in this chapter provide a way to bypass altogether the need for a uniform sampling of parameter space. This is achieved by constructing a Markov chain that only samples the interesting region of parameters space, i.e., the region near the maximum of the likelihood. The method is so versatile and computationally efficient that MCMC techniques have become the leading analysis method in many fields of data analysis.

16.2   Requirements and Goals of a Monte Carlo Markov Chain

A Monte Carlo Markov chain makes use of a dataset Z and a model with m adjustable parameters, $\theta = (\theta_1, \ldots, \theta_m)$, for which it is possible to calculate the likelihood

$$\mathscr{L} = P(Z/\theta). \tag{16.1}$$

Usually, the calculation of the likelihood is the most intensive task for an MCMC. It necessary to be able to evaluate the likelihood for all possible parameter values.

According to Bayesian statistic, one is allowed to have a *prior* knowledge on the parameters, even before they are measured (see Sect. 1.7). The prior knowledge may come from experiments that were conducted beforehand, or from any other type a priori belief on the parameters. The prior probability distribution will be referred to as $p(\theta)$.

The information we seek is the probability distribution of the model parameters *after* the measurements are made, i.e., the posterior distribution $P(\theta/Z)$. According to Bayes' theorem, the posterior distribution is given by

$$P(\theta/Z) = \frac{P(\theta)P(Z/\theta)}{P(Z)} = \frac{P(\theta) \cdot \mathscr{L}}{P(Z)}, \tag{16.2}$$

where the quantity $P(Z) = \int P(Z/\theta)P(\theta)d\theta$ is a normalization constant.

Taken at face value, (16.2) appears to be very complicated, as it requires a multi-dimensional integration of the term $P(Z)$. The alternative provided by a Monte Carlo Markov chain is the construction of a sequence of *dependent* samples for the parameters θ in the form of a Markov chain. Such Markov chain is constructed in such a way that each parameter value appears in the chain in proportion to this posterior distribution. With this method, it will be shown that the value of the normalization constant $P(Z)$ becomes unimportant, thus alleviating significantly the computational burden. The goal of a Monte Carlo Markov chain is therefore that of creating a sequence of parameter values that has as its stationary distribution the posterior distribution of the parameters. After the chain is run for a large number of iterations, the posterior distribution is obtained via the sample distribution of the parameters in the chain.

There are several algorithms to sample the parameter space that satisfy the requirement of having the posterior distribution of the parameters $P(\theta/Z)$ as the stationary distribution of the chain. A very common algorithm that can be used in most applications is that of Metropolis and Hastings [19, 32]. It is surprisingly easy to implement, and therefore constitutes a reference for any MCMC implementation. Another algorithm is that of Gibbs, but its use is limited by certain specific requirements on the distribution function of the parameters.Both algorithms presented in this chapter provide a way to sample values of the parameters and describe a way to accept them into the Markov chain.

16.3   The Metropolis–Hastings Algorithm

The Metroplis–Hastings algorithm [19, 32] was devised well before personal computers became of widespread use. In this section we first describe the algorithm and then prove that the resulting Markov chain has the desired stationary distribution. The method has the following steps.

1. The Metropolis–Hastings algorithm starts with an arbitrary choice of the initial values of the model parameters, $\theta_0 = (\theta_1^0, \ldots, \theta_m^0)$. This initial set of parameters is automatically accepted into the chain. As will be explained later, some of the initial links in the MCMC will later be discarded to offset the arbitrary choice of the starting point.
2. A *candidate* for the next link of the chain, θ', is then drawn from a *proposal (or auxiliary) distribution* $q(\theta'/\theta_n)$, where θ_n is the current link in the chain. The distribution $q(\theta'/\theta_n)$ is the probability of drawing a given candidate θ', given that the chain is in state θ_n. There is a large amount of freedom in the choice of the auxiliary distribution, which can depend on the current state of the chain θ_n, according to the Markovian property, but not on its prior history. One of the simplest choices for a proposal distribution is an m-dimensional uniform distribution of fixed width in the neighborhood of the current parameter. A uniform prior is very simple to implement, and it is the default choice in many applications. More complex candidate distributions can be implemented using, e.g., the method of simulation of variables described in Sect. 4.8.
3. A *prior distribution* $p(\theta)$ has to be assumed before a decision can be made whether the candidate is accepted into the chain or rejected. The Metropolis–Hastings algorithm gives freedom on the choice of the prior distribution as well. A typical choice of prior is another uniform distribution between two hard limits, enforcing a prior knowledge that a given parameter may not exceed certain boundaries. Sometimes the boundaries are set by nature of the parameter itself, e.g., certain parameter may only be positive numbers, or in a fixed interval range.

Other priors may be more restrictive. Consider the case of the measurement of the slope of the curve in the Hubble experiment presented on page 157. It is clear that, after a preliminary examination of the data, the slope parameter b will not be a negative number, and will not be larger than, say, $b = 2$. Therefore one can safely assume a prior on this parameter equal to $p(b) = 1/2$, for $0 \leq b \leq 2$. Much work on priors has been done by Jeffreys [23], in search of mathematical functions that express the lack of prior knowledge, known as *Jeffreys priors*. For many applications, though, simple uniform prior distributions are typically sufficient.

4. After drawing a random candidate θ', we must decide whether to accept it into the chain or reject it. This choice is made according to the following *acceptance probability*, which is the heart of the Metropolis–Hastings algorithm:

$$\alpha(\theta'/\theta_n) = \min \left\{ \frac{\pi(\theta')q(\theta_n/\theta')}{\pi(\theta_n)q(\theta'/\theta_n)}, 1 \right\}, \tag{16.3}$$

The acceptance probability $\alpha(\theta'/\theta_n)$ determines the probability of going from θ_n to the new candidate state θ', where $q(\theta'/\theta_n)$ is the proposal distribution, and $\pi(\theta') = P(\theta/Z)$ is the intended stationary distribution of the chain. Equation (16.3) means that the probability of going to a new value in the chain, θ', is proportional to the ratio of the posterior distribution of the candidate to that of the previous link. The acceptance probability can also be re-written by making use of Bayes' theorem (16.2), as

$$\alpha(\theta'/\theta_n) = \min \left\{ \frac{p(\theta')P(Z/\theta')q(\theta_n/\theta')}{p(\theta_n)P(Z/\theta_n)q(\theta'/\theta_n)}, 1 \right\} \tag{16.4}$$

In this form, the acceptance probability can be calculated based on known quantites. The term $p(\theta_n)q(\theta'/\theta_n)$ at the denominator represents the probability of occurrence of a given candidate θ'; in fact, the first term is the prior probability of the n-th link in the chain, and the second term is the probability of generating the candidate, once the chain is at that state. The other term, $\mathcal{L} = P(Z/\theta_n)$, is the likelihood of the current link in the chain. At the numerator, all terms have reverse order of conditioning between the current link and the candidate. Therefore all quantities in (16.4) are known, since $p(\theta_n)$ and $q(\theta'/\theta_n)$ (and their conjugates) are chosen by the analyst and the likelihood can be calculated for all model parameters.

Acceptance probability means that the candidate is accepted in the chain in proportion to the value of $\alpha(\theta'/\theta_n)$. Two cases are possible:

- $\alpha = 1$: This means that the candidate will be accepted in the chain, since the probability of acceptance is 100%. The candidate becomes the next link in the chain, $\theta_{n+1} = \theta'$. The min operator guarantees that the probability is never greater than 1, which would not be meaningful.

- $\alpha < 1$: This means that the candidate can only be accepted in the chain with a probability α. To enforce this probability of acceptance, it is sufficient to draw a random number $0 \leq u \leq 1$ and then accept or reject the candidate according to the following criterion:

$$
\begin{cases}
\text{if } \alpha \geq u \Rightarrow \text{candidate is accepted, } \theta_{n+1} = \theta' \\
\text{if } \alpha < u \Rightarrow \text{candidate is rejected, } \theta_{n+1} = \theta_n .
\end{cases}
\tag{16.5}
$$

It is important to notice that if the candidate is rejected, then the chain doesn't move from its current location and a new link equal to the previous one is added to the chain. This means that at each time step in the chain a new link is added, either by repeating the last link (if the candidate is rejected) or by adding a different link (if the candidate is accepted).

The logic of the Metropolis–Hastings algorithm can be easily understood in the case of uniform priors and auxiliary distributions. In that case, the candidate is accepted in proportion to just the ratio of the likelihoods, since all other terms in (16.3) cancel out:

$$
\alpha(\theta'/\theta_n) = \min\left\{ \frac{\mathcal{L}(\theta')}{\mathcal{L}(\theta_n)}, 1 \right\}.
\tag{16.6}
$$

If the candidate has a higher likelihood than the current link, it is automatically accepted. If the likelihood of the candidate is lower than the likelihood of the current link, then it is accepted in proportion to the ratio of the likelihoods of the candidate and of the current link. The possibility of accepting a parameter of *lower* likelihood permits a sampling of the parameter space, instead of a simple search for the point of maximum likelihood which would only result in a point estimate.

We now show that use of the Metropolis–Hastings algorithm creates a Markov chain that has $\pi(\theta_n) = P(\theta_n/Z)$ as its stationary distribution. For this purpose, we will show that the posterior distribution of the parameters satisfies the relationship

$$
\pi(\theta_n) = \sum_j \pi(\theta_j) p_{jn},
\tag{16.7}
$$

where p_{jn} are the transition probabilities of the Markov chain and the index j runs over all possible states.

Proof (Justification of the Metropolis–Hastings Algorithm) To prove that the Metropolis–Hastings algorithm leads to a Markov chain with the desired stationary distribution, consider the time-reversed chain:

original chain: $X_0 \rightarrow X_1 \rightarrow X_2 \rightarrow \ldots \rightarrow X_n \rightarrow \ldots$

time-reversed chain: $X_0 \leftarrow X_1 \leftarrow \ldots X_n \leftarrow X_{n+1} \leftarrow \ldots.$

The time-reversed chain is defined by the transition probability $p_{ij}^\star$:

$$p_{ij}^\star = P(X_n = \varepsilon_j / X_{n+1} = \varepsilon_i) = \frac{P(X_n = \varepsilon_j, X_{n+1} = \varepsilon_i)}{P(X_{n+1} = \varepsilon_i)} =$$

$$\frac{P(X_{n+1} = \varepsilon_i / X_n = \varepsilon_j) P(X_n = \varepsilon_j)}{P(X_{n+1} = \varepsilon_i)},$$

leading to the following relationship with the transition probability of the original chain:

$$\Rightarrow p_{ij}^\star = \frac{p_{ji} \pi(\theta_j)}{\pi(\theta_i)} \tag{16.8}$$

If the original chain is *time-reversible*, then $p_{ij}^\star = p_{ij}$, and the time-reversed process is also a Markov chain. In this case, the stationary distribution will follow the relationship

$$\pi(\theta_i) \cdot p_{ij} = p_{ji} \cdot \pi(\theta_j) \tag{16.9}$$

known as the equation of *detailed balance*. The detailed balance is the hallmark of a time-reversible Markov chain, stating that the probability to move forward and backwards is the same, once the stationary distribution is reached. Therefore, if the transition probability of the Metropolis–Hastings algorithm satisfies this equation, with $\pi(\theta) = P(\theta/Z)$, then the chain is time reversible, and with the desired stationary distribution. Moreover, Theorem 15.4 can be used to prove that this distribution is unique.

The Metropolis–Hastings algorithm enforces a specific transition probability between states θ_i and θ_j,

$$p_{ij} = q(\theta_j/\theta_i)\alpha(\theta_j/\theta_i) \qquad \text{if } \theta_i \neq \theta_j \tag{16.10}$$

where q is the probability of generating the candidate (or proposal distribution), and α the probability of accepting it. One can also show that the probability of remaining at the same state θ_i is

$$p_{ii} = 1 - \sum_{j \neq i} q(\theta_j/\theta_i)\alpha(\theta_j/\theta_i).$$

where the sum is over all possible states.

According to the transition probability described by (16.3),

$$\alpha(\theta_j/\theta_i) = \min\left\{\frac{p(\theta_j)P(Z/\theta_j)q(\theta_i/\theta_j)}{p(\theta_i)P(Z/\theta_i)q(\theta_j/\theta_i)}, 1\right\} = \min\left\{\frac{\pi(\theta_j)q(\theta_i/\theta_j)}{\pi(\theta_i)q(\theta_j/\theta_i)}, 1\right\}$$

in which we have substituted $\pi(\theta_i) \equiv p(\theta_i/Z) = P(Z/\theta_i)p(\theta_i)/p(Z)$ as the posterior distribution. Notice that the probability $p(Z)$ cancels out, therefore its value does not play a role in the construction of the chain.

It is clear that, if $\alpha(\theta_j/\theta_i) < 1$, then $\alpha(\theta_i/\theta_j) = 1$, thanks to the min operation. Assume, without loss of generality, that $\alpha(\theta_i, \theta_j) < 1$:

$$\alpha(\theta_j/\theta_i) = \frac{\pi(\theta_j)q(\theta_i/\theta_j)}{\pi(\theta_i)q(\theta_j/\theta_i)}$$

$$\Rightarrow \alpha(\theta_j/\theta_i)\pi(\theta_i)q(\theta_j/\theta_i) = \pi(\theta_j)q(\theta_i/\theta_j) \cdot \alpha(\theta_i/\theta_j)$$

Now, since we assumed $\alpha(\theta_j/\theta_i) < 1$, the operation of min becomes redundant. Using (16.10) the previous equation simplifies to

$$p_{ij} \cdot \pi(\theta_i) = p_{ji} \cdot \pi(\theta_j)$$

which shows that the Metropolis–Hastings algorithm satisfies the detailed balance equation; it thus generates a time-reversible Markov chain, with stationary distribution equal to the posterior distribution.   □

Example 16.1 The data from Hubble's experiment (page 157) can be used to run a Monte Carlo Markov chain to obtain the posterior distribution of the parameters a and b. This fit was also performed using a maximum likelihood method (see page 159) in which the common uncertainty in the dependent variable, $\log v$, was estimated according to the method described in Sect. 8.5.

Using these data, a chain is constructed using uniform priors on the two fit parameters a and b:

$$\begin{cases} p(a) = \dfrac{10}{7} & \text{for } 0.2 \leq b \leq 0.9 \\ p(b) = 10 & \text{for } 0.15 \leq a \leq 0.25. \end{cases}$$

The proposal distributions are also uniform distributions, respectively, of fixed width 0.1 and 0.02 for a and b, and centered at the current value of the parameters:

$$\begin{cases} p(\theta_{n+1}/a_n) = 5 & \text{for } a_n - 0.1 \leq \theta_{n+1} \leq a_n + 0.1 \\ p(\theta_{n+1}/b_n) = 25 & \text{for } b_n - 0.02 \leq \theta_{n+1} \leq b_n + 0.02 \end{cases}$$

in which a_n and b_n are, respectively, the n-th links of the chain, and θ_{n+1} represent the candidate for the $(n + 1)$-th link of the chain, for each parameter.

In practice, once the choice of a uniform distribution with fixed width is made, the actual value of the prior and proposals distributions are not used explicitly. In fact, the acceptance probability becomes simply a function of the ratio of the likelihoods, or of the χ^2's:

$$\alpha(\theta'/\theta_n) = \min\left\{\frac{\mathscr{L}(\theta')}{\mathscr{L}(\theta_n)}, 1\right\} = \min\left\{e^{\frac{\chi^2(\theta_n)-\chi^2(\theta')}{2}}, 1\right\}$$

$\diamond$

where $\chi^2(\theta_n)$ and $\chi^2(\theta')$ are the minimum χ^2's calculated, respectively, using the n-th link of the chain and the candidate parameters (Fig. 16.1).

A few steps of the chain are reported in Table 16.1. Where two consecutive links in the chain are identical, it is an indication that the candidate parameter drawn at that iteration was rejected, and the previous link was therefore repeated. Figure 16.2 shows the sample distributions of the two fit parameters from a chain with 100,000 links. A wider prior on parameter a would make it possible to explore further the tails of the distribution.

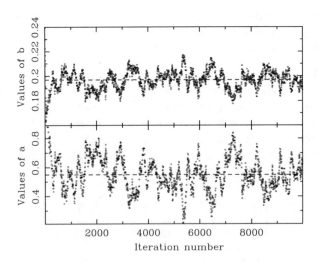

Fig. 16.1 MCMC for parameters a, b of linear model fit to the data in Table 8.1. The chain was run for 10,000 iterations, using uniform priors on both parameters (between 0.15 and 0.25 for a, and 0.2 and 0.9 for b). The chain started at $a = 0.90$ and $b = 0.25$. The proposal distributions were also uniform, with width of, respectively, 0.2 for a and 0.04 for b, centered at the current value of the chain

Table 16.1 Sample of MCMC chain for the Hubble data

n	a	b	$\chi^2(\theta_n)$				
1	0.90000	0.25000	3909.55420	136	0.80627	0.18064	11.47313
2	0.94116	0.24395	3563.63110	137	0.77326	0.18284	10.63887
3	0.96799	0.23951	3299.28149	138	0.77326	0.18284	10.63887
4	0.96799	0.23951	3299.28149	139	0.77326	0.18284	10.63887
5	0.96799	0.23951	3299.28149	140	0.77326	0.18284	10.63887
6	0.96799	0.23951	3299.28149		....		
7	0.97868	0.22983	2503.21655	1141	0.42730	0.20502	8.90305
8	0.97868	0.22983	2503.21655	1142	0.42730	0.20502	8.90305
9	0.96878	0.22243	1885.28088	1143	0.42174	0.20494	8.68957
10	1.01867	0.21679	1714.54456	1144	0.42174	0.20494	8.68957
	....			1145	0.42174	0.20494	8.68957
21	1.08576	0.19086	563.56506	1146	0.42174	0.20494	8.68957
22	1.06243	0.19165	536.47919	1147	0.42174	0.20494	8.68957
23	1.06243	0.19165	536.47919	1148	0.42174	0.20494	8.68957
24	1.06559	0.18244	254.36528	1149	0.42174	0.20494	8.68957
25	1.06559	0.18244	254.36528	1150	0.43579	0.20323	8.65683
26	1.06559	0.18244	254.36528		....		
27	1.06559	0.18244	254.36528	9991	0.66217	0.19189	12.43171
28	1.06559	0.18244	254.36528	9992	0.62210	0.19118	8.52254
29	1.04862	0.17702	118.84048	9993	0.62210	0.19118	8.52254
30	1.04862	0.17702	118.84048	9994	0.62210	0.19118	8.52254
	....			9995	0.62210	0.19118	8.52254
131	0.84436	0.17885	13.11242	9996	0.62210	0.19118	8.52254
132	0.84436	0.17885	13.11242	9997	0.62210	0.19118	8.52254
133	0.84436	0.17885	13.11242	9998	0.62210	0.19118	8.52254
134	0.80627	0.18064	11.47313	9999	0.64059	0.18879	11.11325
135	0.80627	0.18064	11.47313	10,000	0.64059	0.18879	11.11325

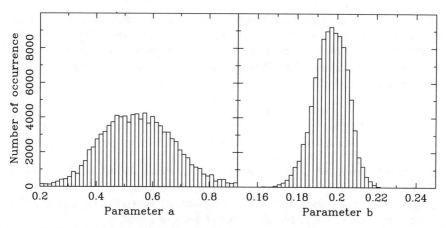

Fig. 16.2 Sample distribution function for parameters a and b, constructed using a histogram plot of 100,000 samples of a MCMC ran with the same parameters as Fig. 16.1

16.4   The Gibbs Sampler

The Gibbs sampler is another algorithm that creates a Markov chain having as stationary distribution the posterior distribution of the parameters. This algorithm is based on the availability of the *full conditional distribution*, defined as

$$\pi_i(\theta_i) = \pi(\theta_i | \theta_j, j \neq i) \tag{16.11}$$

The full conditional distribution is the (posterior) distribution of a given parameter, given that the values of all other parameters are known. If the full conditional distributions are known and can be sampled from, then a simple algorithm can be implemented:

1. Start the chain at a given value of the parameters, $\theta_0 = (\theta_0^1, \ldots, \theta_0^m)$.
2. Obtain a new value in the chain through successive generations:

$$\theta_1^1 \text{ drawn from } \pi(\theta_1 | \theta_0^2, \theta_0^3, \ldots)$$

$$\theta_1^2 \text{ drawn from } \pi(\theta_2 | \theta_1^1, \theta_0^3, \ldots)$$

$$\ldots$$

$$\theta_1^m \text{ drawn from } \pi(\theta_m | \theta_1^1, \theta_1^2, \ldots, \theta_1^{m-1})$$

3. Iterate until convergence to stationary distribution is reached.

The justification of this method can be found in [15]. In the case of data fitting with a dataset Z and a model with m adjustable parameters, usually it is not possible to know the full conditional distributions, thus this method is not as common as the Metropolis–Hastings algorithm. The great advantage of the Gibbs sampler is the fact that the acceptance is 100 %, since there is no rejection of candidates for the Markov chain, unlike the case of the Metropolis–Hastings algorithm.

Example 16.2 This example reproduces an application presented by Carlin et al. [8], and illustrates a possible application in which the knowledge of the full conditional distribution results in the possibility of implementing a Gibbs sampler.

Consider the case in which a Poisson dataset of n numbers, $y_i, i = 1, \ldots, n$, is fit to a step-function model:

$$\begin{cases} y = \lambda \text{ if } i \leq m \\ y = \mu \text{ if } i > m \end{cases} \tag{16.12}$$

The model therefore has three parameters, the values λ, μ, and the point of discontinuity, m. This situation could be a set of measurements of a quantity that may suddenly change its value at an unknown time, say the voltage in a given portion of an electric circuit after a switch has been opened or closed.

Assume that the priors on the parameters are, respectively, a gamma distributions for λ and μ, $p(\lambda) = G(\alpha, \beta)$ and $p(\mu) = G(\gamma, \delta)$, and a uniform distribution for m, $p(m) = 1/n$ (see Sect. 7.2 for definition of the gamma distribution). According to Bayes' theorem, the posterior distribution is proportional to the product of the likelihood and the priors:

$$\pi(\lambda, \mu, m) \propto P(y_1, \ldots, y_n/\lambda, \mu, m) \cdot p(\lambda)p(\mu)p(m). \tag{16.13}$$

The posterior is therefore given by

$$\pi(\lambda, \mu, m) \propto \prod_{i=1}^{m} e^{-\lambda} \lambda^{y_i} \prod_{i=m+1}^{n} e^{-\mu} \mu^{y_i} \cdot \lambda^{\alpha-1} e^{-\beta\lambda} \cdot \mu^{\gamma-1} e^{-\delta\mu} \cdot \frac{1}{n}$$

$$\Rightarrow \pi(\lambda, \mu, m) \propto \lambda^{\left(\alpha + \sum_{i=1}^{m} y_i - 1\right)} e^{-(\beta+m)\lambda} \cdot \mu^{\left(\gamma + \sum_{i=m+1}^{n} y_i - 1\right)} e^{-(\delta+n-m)\mu}.$$

The equation above indicates that the conditional posteriors, obtained by fixing all parameters except one, are given by

$$\begin{cases} \pi_\lambda(\lambda) = G\left(\alpha + \sum_{i=1}^{m} y_i, \beta + m\right) \\ \\ \pi_\mu(\mu) = G\left(\gamma + \sum_{i=m+1}^{n} y_i, \delta + n - m\right) \\ \\ \pi_m(m) = \dfrac{\lambda^{\left(\alpha + \sum_{i=1}^{m} y_i - 1\right)} e^{-(\beta+m)\lambda} \cdot \mu^{\left(\gamma + \sum_{i=m+1}^{n} y_i - 1\right)} e^{-(\delta+n-m)\mu}}{\sum_{i=1}^{n} \left(\lambda^{\left(\alpha + \sum_{i=1}^{m} y_i - 1\right)} e^{-(\beta+m)\lambda} \cdot \mu^{\left(\gamma + \sum_{i=m+1}^{n} y_i - 1\right)} e^{-(\delta+n-m)\mu}\right)}. \end{cases} \tag{16.14}$$

This is therefore an example of a case where the conditional posterior distributions are known, and therefore the Gibbs algorithm is applicable. The only complication is the simulation of the three conditional distributions, which can be achieved using the methods described in Sect. 4.8. ◇

16.5 Tests of Convergence

It is necessary to test that the MCMC has reached convergence to the stationary distribution before inference on the posterior distribution can be made. Convergence indicates that the chain has started to sample the posterior distribution, so that the MCMC samples are representative of the distribution of interest, and are not biased by such choices as the starting point of the chain. The period of time required for the chain to reach convergence goes under the name of *burn-in* period, and varies from chain to chain according to a variety of factors, such as the choice of prior and proposal distributions. We therefore must identify and remove such initial period from the chain prior to further analysis. The *Geweke z-score test* and the *Gelman-Rubin test* are two of the most common tests used to identify the burn-in period.

Another important consideration is that the chain must be run for a sufficient number of iterations, so that the sample distribution becomes a good approximation of the true posterior distribution. It is clear that the larger the number of iterations after the burn-in period, the more accurate will be the estimates of the parameters of the posterior distribution. In practice it is convenient to know the minimum *stopping time* that enables to estimate the posterior distribution with the required precision. The *Raftery-Lewis test* is designed to give an approximate estimate of both the burn-in time and the minimum required stopping time.

Typical considerations concerning the burn-in period and the stopping time of a chain can be illustrated with the example of three chains based on the data from Table 10.1. The chains were run, respectively, with a uniform proposal distribution of 1, 10, and 100 for both parameters of the linear model, starting at the same point (Figs. 16.3, 16.4 and 16.5). The chain with a narrower proposal distribution requires a longer time to reach the stationary value of the parameters, in part because at each time interval the candidate can be chosen in just a limited neighborhood of the

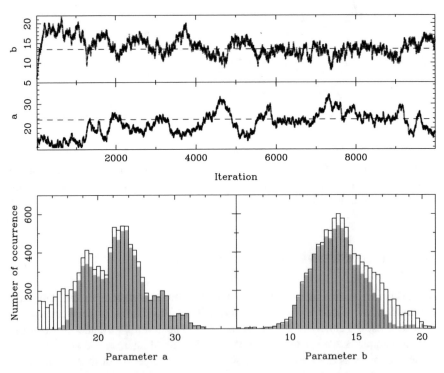

Fig. 16.3 MCMC for parameters *a*, *b* of linear model fit to the data in Table 10.1, using a uniform proposal distribution with width of 1 for both parameters. The chain started at $a = 12$ and $b = 6$. In *grey* is the sample distribution obtained by removing the initial 2000 iterations, the ones that are most affected by the arbitrary choice of starting point

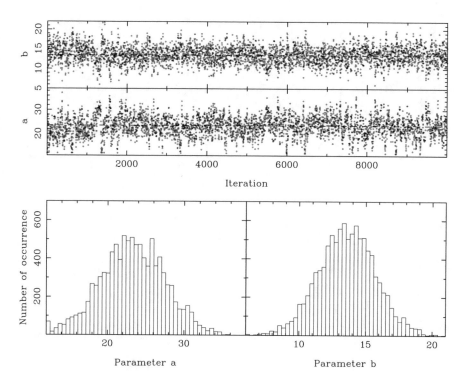

Fig. 16.4 MCMC for parameters a, b of linear model fit to the data in Table 10.1, using a uniform proposal distribution with width of 10 for both parameters. The chain started at same values as in Fig. 16.3

previous link. Moreover, the sampling of parameter space is less uniform because the chain requires longer time to span the entire parameter range. The intermediate value for the proposal distribution results in an almost immediate convergence, and the sampling of parameter space is clearly more uniform. An increase in the size of the proposal distribution, however, may eventually lead to slow convergence and poor sampling, as indicated by the chain with the largest value of the proposal width. In this case, candidates are drawn from regions of parameter space that have very low likelihood, or large χ^2, and therefore the chain has a tendency to remain at the same location for extended periods of time, with low acceptance rate. The result is a chain with poor coverage of parameter space and poorly determined sample distribution for their parameters. A smoother distribution is preferable, because it leads to a more accurate determination of the median, and of confidence ranges on the parameters.

Another consideration is that elements in the chain are more or less correlated to one another, according to the choice of the proposal distribution, and other choices in the construction of the chain. Links in the chains are always correlated by construction, since the next link in the chain typically depends on the current state of the chain. In principle a Markov chain can be constructed that does not

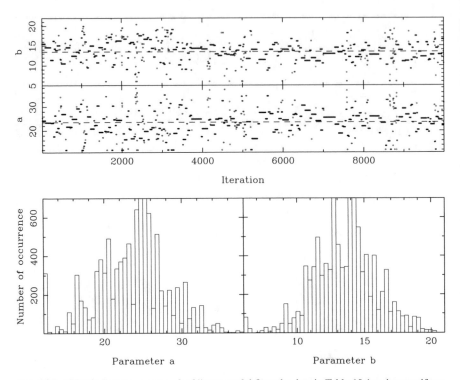

Fig. 16.5 MCMC for parameters a, b of linear model fit to the data in Table 10.1, using a uniform proposal distribution with width of 50 for both parameters. The chain started at same values as in Fig. 16.3

depend on the current state of the chain, but in most cases it is convenient to make full use of the Markovian property that allows to make use of the current state of the chain. The chains in Figs. 16.3, 16.4 and 16.5 illustrate the fact that the degree of correlation varies with the proposal distribution choice. For example, the chain with the narrowest proposal distribution appears more correlated than that with the intermediate choice for the width; also, the chain with the largest width appears to have periods with the highest degree of correlation, when the chain does not move for hundreds of iterations. This shows that the degree of correlation is a nonlinear function of the proposal distribution width, and that fine-tuning is always required to obtain a chain with good *mixing* properties. The degree of correlation among elements of the chain will become important when we desire to estimate the variance of the mean from a specific segment of the chain, since the formulas derived earlier in Chap. 4 apply only to independent samples.

Testing for convergence and stopping time of the chain are critical tasks for a Monte Carlo Markov chain. The tests discussed below are some of the more common analytic tools and can be implemented with relative ease.

16.5.1 *The Geweke Z Score Test*

A simple test of convergence is provided by the difference of the mean of two segments of the chain. Under the null hypothesis that the chain is sampling the same distribution during both segments, the sample means are expected to be drawn from the same parent mean. Consider segment A at the beginning of the chain, and segment B at the end of the chain; for simplicity, consider one parameter ψ at a time. If the chain is of length N, the prescription is to use an initial segment of $N_A = 0.1N$ elements, and a final segment with $N_B = 0.5N$ links, although those choices are somewhat arbitrary, and segments of different length can also be used.

The mean of each parameter in the two segments A and B is calculated as

$$\begin{cases} \overline{\psi_A} = \dfrac{1}{N_A} \sum_{j=1}^{N_A} \psi_j \\ \overline{\psi_B} = \dfrac{1}{N_B} \sum_{j=N-N_B+1}^{N} \psi_j. \end{cases} \tag{16.15}$$

To compare the two sample means, it is also necessary to estimate their sample variances $\sigma^2_{\psi_A}$ and $\sigma^2_{\psi_B}$. This task is complicated significantly by the fact that one *cannot* just use (2.11), because of the correlation between links of the chain. One possibility to overcome this difficulty is to *thin* the chain by using only every n-th iteration, so that the thinned chain better approximates independent samples.

The test statistic is the Z *score* of the difference between the means of the two segments:

$$Z_G = \frac{\overline{\psi_B} - \overline{\psi_A}}{\sqrt{\sigma^2_{\psi_A} + \sigma^2_{\psi_B}}}. \tag{16.16}$$

Under the assumption that the two means follow the same distribution and that they are uncorrelated, the Z-score is distributed as a standard Gaussian, $Z_G \sim N(0, 1)$. For this reason the two segments of the chain are typically separated by a large number of iterations. An application of the Geweke Z score is to step the start of segment A forward in time, until the Z_G scores don't exceed approximately ± 3, which correspond to a $\pm 3\sigma$ deviation in the means of the two segments. The burn-in period that needs to be excised is that before the Z scores stabilize around the expected values. An example of the use of this test statistic is provided in Fig. 16.6, in which Z_G was calculated from the chain with proposal width 10. An initial segment of length 20 % of the total chain length is compared to the final 40 % of the chain, by stepping the beginning of the initial segment until it overlaps with the final segment. By using all links in the chain to estimate the variance of the

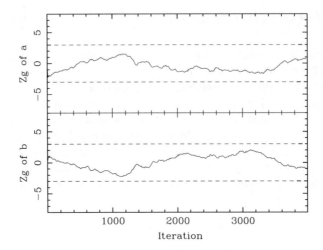

Fig. 16.6 Geweke Z scores with segment A and segment B, respectively, 20 and 40 % of the total chain length. The results correspond to the chain run with a proposal width of 10. The Z scores are calculated by using only every other 10-th iteration

mean, the variance would be underestimated because of the correlation among links, leading to erroneously large values of Z_G. If the chain is thinned by a factor of 10, then the estimate of the variance using (2.11) is more accurate, and the resulting Z scores show that the chains converge nearly immediately, as is also clear by a visual inspection of the chain from Fig. 16.4.

The effect of the starting point in the evaluation of the burn-in period is shown in Fig. 16.3, in which it is apparent that it takes about 2000 iterations for the chain to forget the initial position, and to start sampling the posterior distribution, centered at the dashed lines. A larger proposal distribution, as in Fig. 16.4, makes it easier to reach the posterior distribution more rapidly, to the point that no burn-in period is visible in this case. In the presence of a burn-in period, the sample distribution must be constructed by excising the initial portion of the chain, as shown in the grey histogram plot of Fig. 16.3.

16.5.2   The Gelman–Rubin Test

The Gelman–Rubin test investigates the effect of initial conditions on the convergence properties of the MCMC and makes use of m parallel chains starting from different initial points. Initially, the m chain will be far apart because of the different starting points. As the chains start sampling the stationary distribution, they will have the same statistical properties.

The test is based on two estimates of the variance, or variability, of the chains: the *within-chain* variance for each of the m chains W, and the *between-chain* variance B. At the beginning of the chain, W will underestimate the true variance of the model parameters, because the chains have not had time to sample all possible values. On the other hand, B will initially overestimate the variance, because of the different starting points. The test devised by Gelman and Rubin [17] defines the ratio of the within-to-between variance as a test to measure convergence of the chains, to identify an initial burn-in period that should be removed because of the lingering effect of initial conditions.

For each parameter, consider m chains of N iterations each, where $\bar{\psi}_i$ is the mean of each chain $i = 1, \ldots, m$ and $\bar{\psi}$ the mean of the means:

$$\begin{cases} \bar{\psi}_i = \dfrac{1}{N} \sum_{j=1}^{N} \psi_j \\ \bar{\psi} = \dfrac{1}{m} \sum_{i=1}^{m} \bar{\psi}_i. \end{cases} \qquad (16.17)$$

The between-chain variance B is defined as the average of the variances of the m chains,

$$B = \frac{N}{m-1} \sum_{i=1}^{m} (\bar{\psi}_i - \bar{\psi})^2 \qquad (16.18)$$

Notice that, in (16.18), B/N is the variance of the means $\bar{\psi}_i$. The within-chain variance W is defined by

$$s_i^2 = \frac{1}{N-1} \sum_{j=1}^{N} (\psi_j - \bar{\psi}_i)^2$$

$$W = \frac{1}{m} \sum_{i=1}^{m} s_i^2. \qquad (16.19)$$

The quantity $\hat{\sigma}_\psi^2$, defined as

$$\hat{\sigma}_\psi^2 = \left(\frac{N-1}{N} \right) W + \frac{1}{N} B \qquad (16.20)$$

is intended to be an unbiased estimator of the variance of the parameter ψ under the hypothesis that the stationary distribution is being sampled. At the beginning of a chain—before the stationary distribution is reached—$\hat{\sigma}_\psi^2$ overestimates the variance, because of the different initial starting points. It was suggested by Brooks and Gelman [6] to add an additional term to this estimate of the variance, to account for the variability in the estimate of the means, so that the estimate of the within-chain variance to use becomes

$$\hat{V} = \hat{\sigma}_\psi^2 + \frac{B}{mN}. \tag{16.21}$$

Convergence can be monitored by use of the following statistic:

$$\sqrt{\hat{R}} \equiv \sqrt{\frac{\hat{V}}{W}}, \tag{16.22}$$

which should converge to 1 when the stationary distribution in all chains has been reached. A common use of this statistic is to repeat the calculation of the Gelman–Rubin statistic after excising an increasingly longer initial portion of the chain, until approximately

$$\sqrt{\hat{R}} \leq 1.2. \tag{16.23}$$

A procedure to test for convergence of the chain using the Gelman–Rubin statistic is to divide the chain into segments of length b, such that the N iterations are divided into N/b batches. For each segment starting at iteration $i = k \times b, k = 0, \ldots, N/b - 1$, we can calculate the value $\hat{R}$ and claim convergence of the chains when (16.23) is satisfied. Figure 16.7 shows results of this test run on $m = 2$ chains

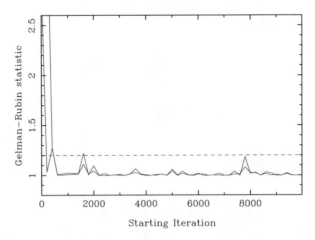

Fig. 16.7 Gelman–Rubin statistic $\hat{R}$ calculated from $m = 2$ chains with the same distributions as in Fig. 16.3, one starting at $a = 12$, $b = 6$, and the other at $a = 300$ and $b = 300$. The chain rapidly converges to its stationary distribution, and appears to forget about the starting point after approximately 500 iterations. The values of $\hat{R}$ were calculated in segments of length $b = 200$, starting at iteration $i = 0$

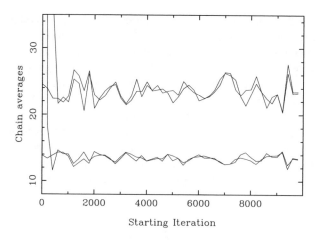

Fig. 16.8 Plot of the average of the parameters *a* and *b* for the two chains used in Fig. 16.7 (*top lines* are for parameter *a*, *bottom lines* for *b*). For both parameters, the two chains sample the same posterior distribution after approximately 500 iterations

based on the data of Table 10.1, starting at different values: one chain starting at a value that is close to the posterior mean of the parameters, and one starting at values that were intentionally chosen to be much larger than the parent values. After approximately 500 or so iterations, the within-chain and between-chain estimates of the variance become comparable, and the value of $\hat{R}$ approaches the value of one.

Another related tool that aids in assessing convergence is the plot of the mean of the parameters, shown in Fig. 16.8: it takes approximately 500 iterations for the chain starting with high initial values, to begin sampling the stationary distribution. It is clear that, from that point on, both chains hover around similar values of the parameters. One should also check that, individually, both $\hat{V}$ and W also stabilize to a common value as function of starting point of the batch, and not just their ratio $\hat{R}$. In fact, under the hypothesis of convergence, both within-chain and between-chain variances should converge to a common value.

Similar procedures to monitor convergence using the Gelman–Rubin may instead use batches of increasing length, starting from one of length *b*, and increasing to 2*b*, etc., optionally discarding the first half of each batch. Moreover, thinning can be implemented when calculating means, to reduce the effect of correlation among the samples. In all cases, the goal is to show that eventually the value of $\hat{R}$ stabilizes around unity.

16.5.3   The Raftery–Lewis Test

An ideal test for the convergence of MCMCs is one that determines the length of the burn-in period, and how long should the chain be run to achieve a given precision in the estimate of the model parameters. The Raftery–Lewis test provides estimates of both quantities, based on just a short test run of the chain. The test was developed by Raftery and Lewis [37], and it uses the comparison of the short sample chain with an uncorrelated chain to make inferences on the convergence properties of the chain. In this section we describe the application of the test and refer the reader interested in its justification to [37].

The starting point to use the Raftery–Lewis test is to determine what inferences we want to make from the Markov chain. Typically we want to estimate confidence intervals at a given significance for each parameter, which means we need to estimate two values θ_{min} and θ_{max} for each parameter θ such that their interval contains a probability $1 - q$ (e.g., respectively, $q = 0.32, 0.10$ or 0.01 for confidence level 68, 90 or 99 %),

$$1 - q = \int_{\theta_{min}}^{\theta_{max}} \pi(\theta)d\theta.$$

Consider, for example, the case of a 90 % confidence interval: the two parameter values θ_{min} and θ_{max} are respectively the $q = 0.95$ and the $q = 0.05$ quantiles, so that the interval $(\theta_{min}, \theta_{max})$ will contain 90 % of the posterior probability for that parameter.

One can think of each quantile as a statistic, meaning that we can only approximately estimate their values $\hat{\theta}_{min}$ and $\hat{\theta}_{max}$. The Raftery–Lewis test lets us estimate any quantile $\hat{\theta}_q$ such that it satisfies $P(\theta \le \hat{\theta}_q) = 1 - q$ to within $\pm r$, with probability s (say 95 % probability, $s = 0.95$). We have therefore introduced two additional probabilities, r and s, which should not be confused with the quantile q. Consider, for example, that the requirement is to estimate the $q = 0.05$ quantile, with a precision of $r = 0.01$ and a probability of achieving this precision of $s = 0.95$. This corresponds to accepting that the 90 % confidence interval resulting from such estimate of the $q = 0.05$ quantile (and of the $q = 0.95$ as well) may in reality be a 88 % or a 92 % confidence interval, 95 % of the time.

The Raftery–Lewis test uses the information provided by the sample chain, together with the desired quantile q and the tolerances r and s, and returns the number of burn-in iterations, and the required number of iterations N. A justification for this test can be found in [37], and the test can be simply run using widely available software such as the *gibbsit* code or the the *CODA* software [28, 34]. Note that the required number of iterations are a function of the quantile to be estimated, with estimation of smaller quantiles typically requiring longer iterations.

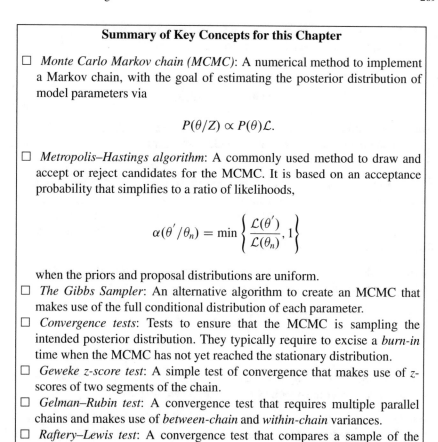

Problems

16.1 Prove that, in the presence of positive correlation among MCMC samples, the variance of the sample mean is larger than that of an independent chain.

16.2 Using the data of $\log m$ and velocity from Table 8.1 of Hubble's experiment, construct a Monte Carlo Markov chain for the fit to a linear model with 10,000 iterations. Use uniform distributions for the prior and proposal distributions of the two model parameters a and b, the latter with widths of 0.1 and 0.02, respectively, for a and b in the neighborhood of the current value. You can start your chain at values of $a = 0.2$ and $b = 0.9$. After completion of the chain, plot the sample distribution of the two model parameters.

16.3 A one-parameter chain is constructed such that in two intervals A and B the following values are accepted into the chain:

$$A : 10, 11, 13, 11, 10$$
$$B : 7, 8, 1, 11, 10, 8;$$

where A is an initial interval, and B an interval at the end of the chain. Not knowing how the chain was constructed, use the Geweke z score to determine whether the chain *might* have converged.

16.4 Using the data of Table 10.1, construct a Monte Carlo Markov chain for the parameters of the linear model, with 10,000 iterations. Use uniform distributions for the prior and proposal distributions, the latter with a width of 10 for both parameters. Start the chain at $a = 12$ and $b = 6$. After completion of the chain, plot the sample distribution of the two model parameters.

16.5 Consider the following portions of two one-parameter chains, run in parallel and starting from different initial positions:

$$7, 8, 1, 11, 10, 8$$
$$11, 11, 8, 10, 9, 12.$$

Using two segments of length $b = 3$, calculate the Gelman–Rubin statistic $\sqrt{\hat{R}}$ for both segments under the hypothesis of uncorrelated samples.

16.6 Consider the step-function model described in Example 16.2, and a dataset consisting of n measurements. Assuming that the priors on the parameters λ, μ and m are uniform, show that the full conditional distributions are given by

$$
\begin{cases}
\pi_\lambda(\lambda) = G\left(\sum_{i=1}^{m} y_i + 1, m \right) \\
\pi_\mu(\mu) = G\left(\sum_{i=m+1}^{n} y_i + 1, n - m \right) \\
\pi_m(m) = \dfrac{e^{-m\lambda} \lambda^{\sum_{i=1}^{m} y_i} e^{-(n-m)\mu} \mu^{\sum_{i=m+1}^{n} y_i}}{\sum_{l=1}^{n} e^{-l\lambda} \lambda^{\sum_{i=1}^{l} y_i} e^{-(n-l)\mu} \mu^{\sum_{i=l+1}^{n} y_i}},
\end{cases}
\qquad (16.24)
$$

where G represents the gamma distribution.

16.7 Consider the step-function model described in Example 16.2, and a dataset consisting of the following five measurements:

$$0, 1, 3, 4, 2.$$

Start a Metropolis–Hastings MCMC at $\lambda = 0$, $\mu = 2$ and $m = 1$, and use uniform priors on all three parameters. Assume for simplicity that all parameters can only be integer, and use uniform proposal distributions that span the ranges $\Delta\lambda = \pm 2$, $\Delta\mu = \pm 2$ and $\Delta m = \pm 2$, and that the following numbers are drawn in the first three iterations:

Iteration	$\Delta\lambda$	$\Delta\mu$	Δm	α
1	+1	−1	+1	0.5
2	+1	+2	+1	0.7
3	−1	−2	+1	0.1

With this information, calculate the first four links of the Metropolis–Hastings MCMC.

16.8 Consider a Monte Carlo Markov chain constructed with a Metropolis–Hastings algorithm, using uniform prior and proposal distribution. At a given iteration, the chain is at the point of maximum likelihood or, equivalently, minimum χ^2. Calculate the probability of acceptance of a candidate that has, respectively, $\Delta\chi^2 = 1, 2$, and 10.

Appendix: Numerical Tables

A.1 The Gaussian Distribution and the Error Function

The Gaussian distribution (3.11) is defined as

$$f(x) = \frac{1}{\sqrt{2\pi\sigma^2}} e^{-\frac{(x-\mu)^2}{2\sigma^2}}. \tag{A.1}$$

The maximum value is obtained at $x = \mu$, and the value of x where the Gaussian is a times the peak value is given by

$$z \equiv \frac{x - \mu}{\sigma} = \sqrt{-2\ln a}. \tag{A.2}$$

Figure A.1 shows a standard Gaussian normalized to its peak value, and values of a times the peak value are tabulated in Table A.1. The Half Width at Half Maximum (HWHM) has a value of approximately 1.18σ.

The error function is defined in (3.13) as

$$\text{erf } z = \frac{1}{\sqrt{\pi}} \int_{-z}^{z} e^{-x^2} dx \tag{A.3}$$

and it is related to the integral of the Gaussian distribution defined in (3.12),

$$A(z) = \int_{\mu-z\sigma}^{\mu+z\sigma} f(x)dx = \frac{1}{\sqrt{2\pi}} \int_{-z}^{z} e^{-\frac{x^2}{2}} dx. \tag{A.4}$$

The relationship between the two integrals is given by

$$\text{erf}\left(z/\sqrt{2}\right) = A(z). \tag{A.5}$$

© Springer Science+Busines Media New York 2017
M. Bonamente, *Statistics and Analysis of Scientific Data*, Graduate Texts
in Physics, DOI 10.1007/978-1-4939-6572-4

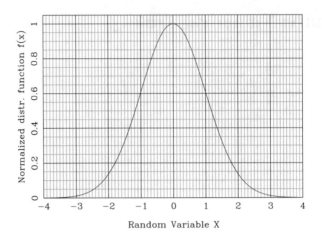

Fig. A.1 Normalized values of the probability distribution function of a standard Gaussian ($\mu = 0$ and $\sigma = 1$)

Table A.1 Values of a times the peak value for a Gaussian distribution

a	z	a	z	a	z	a	z	a	z
0.980	0.201	0.960	0.286	0.940	0.352	0.920	0.408	0.900	0.459
0.880	0.506	0.860	0.549	0.840	0.591	0.820	0.630	0.800	0.668
0.780	0.705	0.760	0.741	0.740	0.776	0.720	0.811	0.700	0.845
0.680	0.878	0.660	0.912	0.640	0.945	0.620	0.978	0.600	1.011
0.580	1.044	0.560	1.077	0.540	1.110	0.520	1.144	0.500	1.177
0.480	1.212	0.460	1.246	0.440	1.281	0.420	1.317	0.400	1.354
0.380	1.391	0.360	1.429	0.340	1.469	0.320	1.510	0.300	1.552
0.280	1.596	0.260	1.641	0.240	1.689	0.220	1.740	0.200	1.794
0.180	1.852	0.160	1.914	0.140	1.983	0.120	2.059	0.100	2.146
0.080	2.248	0.060	2.372	0.040	2.537	0.020	2.797	0.010	3.035

The function $A(z)$ describes the integrated probability of a Gaussian distribution to have values between $\mu - z\sigma$ and $\mu + z\sigma$. The number z therefore represents the number of σ by which the interval extends in each direction. The function $A(z)$ is tabulated in Table A.2, where each number in the table corresponds to a number z given by the number in the left column (e.g., 0.0, 0.1, etc.), and for which the second decimal digit is given by the number in the top column (e.g., the value of 0.007979 corresponds to $z = 0.01$).

The cumulative distribution of a standard Gaussian function was defined in (3.14) as

$$B(z) = \int_{-\infty}^{z} \frac{1}{\sqrt{2\pi}} e^{-\frac{t^2}{2}} \, dt; \tag{A.6}$$

Table A.2 Values of the integral $A(z)$ as a function of z, the number of standard errors σ

	0	1	2	3	4	5	6	7	8	9
0.0	0.000000	0.007979	0.015957	0.023933	0.031907	0.039878	0.047844	0.055806	0.063763	0.071713
0.1	0.079656	0.087591	0.095517	0.103434	0.111340	0.119235	0.127119	0.134990	0.142847	0.150691
0.2	0.158519	0.166332	0.174129	0.181908	0.189670	0.197413	0.205136	0.212840	0.220523	0.228184
0.3	0.235823	0.243439	0.251032	0.258600	0.266144	0.273661	0.281153	0.288618	0.296055	0.303464
0.4	0.310844	0.318194	0.325515	0.332804	0.340063	0.347290	0.354484	0.361645	0.368773	0.375866
0.5	0.382925	0.389949	0.396937	0.403888	0.410803	0.417681	0.424521	0.431322	0.438086	0.444810
0.6	0.451494	0.458138	0.464742	0.471306	0.477828	0.484308	0.490746	0.497142	0.503496	0.509806
0.7	0.516073	0.522296	0.528475	0.534610	0.540700	0.546746	0.552746	0.558700	0.564609	0.570472
0.8	0.576289	0.582060	0.587784	0.593461	0.599092	0.604675	0.610211	0.615700	0.621141	0.626534
0.9	0.631880	0.637178	0.642428	0.647629	0.652783	0.657888	0.662945	0.667954	0.672914	0.677826
1.0	0.682690	0.687505	0.692272	0.696990	0.701660	0.706282	0.710856	0.715381	0.719858	0.724287
1.1	0.728668	0.733001	0.737287	0.741524	0.745714	0.749856	0.753952	0.757999	0.762000	0.765954
1.2	0.769861	0.773721	0.777535	0.781303	0.785025	0.788701	0.792331	0.795916	0.799455	0.802950
1.3	0.806399	0.809805	0.813165	0.816482	0.819755	0.822984	0.826170	0.829313	0.832414	0.835471
1.4	0.838487	0.841461	0.844393	0.847283	0.850133	0.852942	0.855710	0.858439	0.861127	0.863776
1.5	0.866386	0.868957	0.871489	0.873984	0.876440	0.878859	0.881240	0.883585	0.885894	0.888166
1.6	0.890402	0.892603	0.894768	0.896899	0.898995	0.901057	0.903086	0.905081	0.907043	0.908972
1.7	0.910869	0.912735	0.914568	0.916370	0.918141	0.919882	0.921593	0.923273	0.924924	0.926546
1.8	0.928140	0.929705	0.931241	0.932750	0.934232	0.935687	0.937115	0.938517	0.939892	0.941242
1.9	0.942567	0.943867	0.945142	0.946394	0.947621	0.948824	0.950005	0.951162	0.952297	0.953409
2.0	0.954500	0.955569	0.956617	0.957644	0.958650	0.959636	0.960602	0.961548	0.962475	0.963383
2.1	0.964272	0.965142	0.965994	0.966829	0.967646	0.968445	0.969228	0.969994	0.970743	0.971476
2.2	0.972194	0.972895	0.973582	0.974253	0.974909	0.975551	0.976179	0.976793	0.977393	0.977979
2.3	0.978552	0.979112	0.979660	0.980194	0.980717	0.981227	0.981725	0.982212	0.982688	0.983152
2.4	0.983605	0.984048	0.984480	0.984902	0.985313	0.985715	0.986107	0.986489	0.986862	0.987226

(continued)

Table A.2 (continued)

	0	1	2	3	4	5	6	7	8	9
2.5	0.987581	0.987927	0.988265	0.988594	0.988915	0.989228	0.989533	0.989831	0.990120	0.990403
2.6	0.990678	0.990946	0.991207	0.991462	0.991710	0.991951	0.992186	0.992415	0.992638	0.992855
2.7	0.993066	0.993272	0.993472	0.993667	0.993857	0.994041	0.994220	0.994395	0.994565	0.994730
2.8	0.994890	0.995046	0.995198	0.995346	0.995489	0.995628	0.995764	0.995896	0.996024	0.996148
2.9	0.996269	0.996386	0.996500	0.996611	0.996718	0.996823	0.996924	0.997022	0.997118	0.997211
3.0	0.997301	0.997388	0.997473	0.997555	0.997635	0.997712	0.997787	0.997860	0.997930	0.997999
3.1	0.998065	0.998130	0.998192	0.998252	0.998311	0.998368	0.998423	0.998476	0.998528	0.998578
3.2	0.998626	0.998673	0.998719	0.998763	0.998805	0.998846	0.998886	0.998925	0.998962	0.998999
3.3	0.999034	0.999067	0.999100	0.999132	0.999163	0.999192	0.999221	0.999249	0.999276	0.999301
3.4	0.999327	0.999351	0.999374	0.999397	0.999419	0.999440	0.999460	0.999480	0.999499	0.999517
3.5	0.999535	0.999552	0.999569	0.999585	0.999600	0.999615	0.999630	0.999643	0.999657	0.999670
3.6	0.999682	0.999694	0.999706	0.999717	0.999728	0.999738	0.999748	0.999758	0.999767	0.999776
3.7	0.999785	0.999793	0.999801	0.999809	0.999816	0.999824	0.999831	0.999837	0.999844	0.999850
3.8	0.999856	0.999861	0.999867	0.999872	0.999877	0.999882	0.999887	0.999892	0.999896	0.999900
3.9	0.999904	0.999908	0.999912	0.999915	0.999919	0.999922	0.999925	0.999929	0.999932	0.999934
4.0	0.999937	0.999940	0.999942	0.999945	0.999947	0.999949	0.999951	0.999953	0.999955	0.999957
4.1	0.999959	0.999961	0.999963	0.999964	0.999966	0.999967	0.999969	0.999970	0.999971	0.999973
4.2	0.999974	0.999975	0.999976	0.999977	0.999978	0.999979	0.999980	0.999981	0.999982	0.999983
4.3	0.999983	0.999984	0.999985	0.999986	0.999986	0.999987	0.999987	0.999988	0.999989	0.999989
4.4	0.999990	0.999990	0.999991	0.999991	0.999991	0.999992	0.999992	0.999993	0.999993	0.999993
4.5	0.999994	0.999994	0.999994	0.999995	0.999995	0.999995	0.999995	0.999996	0.999996	0.999996
4.6	0.999996	0.999996	0.999997	0.999997	0.999997	0.999997	0.999997	0.999997	0.999998	0.999998
4.7	0.999998	0.999998	0.999998	0.999998	0.999998	0.999998	0.999998	0.999999	0.999999	0.999999
4.8	0.999999	0.999999	0.999999	0.999999	0.999999	0.999999	0.999999	0.999999	0.999999	0.999999
4.9	0.999999	1.000000	1.000000	1.000000	1.000000	1.000000	1.000000	1.000000	1.000000	1.000000

and it is therefore related to the integral $A(z)$ by

$$B(z) = \frac{1}{2} + \frac{A(z)}{2}. \tag{A.7}$$

The values of $B(z)$ are tabulated in Table A.3. Each number in the table corresponds to a number z given by the number in the left column (e.g., 0.0, 0.1, etc.), and for which the second decimal digit is given by the number in the top column (e.g., the value of 0.503990 corresponds to $z = 0.01$).

Critical values of the standard Gaussian distribution functions corresponding to selected values of the integrals $A(z)$ and $B(z)$ are shown in Table A.4. They indicate the value of the variable z required to include a given probability, and are useful for either two-sided or one-sided rejection regions in hypothesis testing.

A.2   Upper and Lower Limits for a Poisson Distribution

The Gehrels approximation described in [16] can be used to calculate upper and lower limits for a Poisson distribution, when n_{obs} counts are recorded. The confidence level is described by the parameter S, corresponding to the number of standard deviations σ for a Gaussian distribution; for example, $S = 1$ corresponds to an 84.1 % confidence level, $S = 2$ to a 97.7 %, and $S = 3$ corresponds to 99.9 %; see Table 5.2 for correspondence between values of S and probability. The upper and lower limits are described, in the simplest approximation, by

$$\begin{cases} \lambda_{up} = n_{obs} + \dfrac{S^2 + 3}{4} + S\sqrt{n_{obs} + \dfrac{3}{4}} \\ \lambda_{lo} = n_{obs} \left(1 - \dfrac{1}{9 n_{obs}} - \dfrac{S}{3\sqrt{n_{obs}}} \right)^3 \end{cases} \tag{A.8}$$

and more accurate approximations are provided in [16] (Tables A.5 and A.6).

A.3   The χ^2 Distribution

The probability distribution function for a χ^2 variable is defined in (7.11) as

$$f_{\chi^2}(z) = \left(\frac{1}{2} \right)^{f/2} \frac{1}{\Gamma(f/2)} e^{-\frac{z}{2}} z^{\frac{f}{2}-1},$$

Table A.3 Values of the integral $B(z)$ as a function of z

	0	1	2	3	4	5	6	7	8	9
0.0	0.500000	0.503990	0.507979	0.511967	0.515954	0.519939	0.523922	0.527903	0.531882	0.535857
0.1	0.539828	0.543796	0.547759	0.551717	0.555670	0.559618	0.563560	0.567495	0.571424	0.575346
0.2	0.579260	0.583166	0.587065	0.590954	0.594835	0.598707	0.602568	0.606420	0.610262	0.614092
0.3	0.617912	0.621720	0.625516	0.629300	0.633072	0.636831	0.640577	0.644309	0.648028	0.651732
0.4	0.655422	0.659097	0.662758	0.666402	0.670032	0.673645	0.677242	0.680823	0.684387	0.687933
0.5	0.691463	0.694975	0.698469	0.701944	0.705402	0.708841	0.712261	0.715661	0.719043	0.722405
0.6	0.725747	0.729069	0.732371	0.735653	0.738914	0.742154	0.745373	0.748571	0.751748	0.754903
0.7	0.758037	0.761148	0.764238	0.767305	0.770350	0.773373	0.776373	0.779350	0.782305	0.785236
0.8	0.788145	0.791030	0.793892	0.796731	0.799546	0.802338	0.805106	0.807850	0.810571	0.813267
0.9	0.815940	0.818589	0.821214	0.823815	0.826392	0.828944	0.831473	0.833977	0.836457	0.838913
1.0	0.841345	0.843753	0.846136	0.848495	0.850830	0.853141	0.855428	0.857691	0.859929	0.862144
1.1	0.864334	0.866501	0.868643	0.870762	0.872857	0.874928	0.876976	0.879000	0.881000	0.882977
1.2	0.884931	0.886861	0.888768	0.890652	0.892513	0.894351	0.896166	0.897958	0.899728	0.901475
1.3	0.903200	0.904902	0.906583	0.908241	0.909878	0.911492	0.913085	0.914657	0.916207	0.917736
1.4	0.919244	0.920731	0.922197	0.923642	0.925067	0.926471	0.927855	0.929220	0.930564	0.931888
1.5	0.933193	0.934479	0.935745	0.936992	0.938220	0.939430	0.940620	0.941793	0.942947	0.944083
1.6	0.945201	0.946301	0.947384	0.948450	0.949498	0.950529	0.951543	0.952541	0.953522	0.954486
1.7	0.955435	0.956367	0.957284	0.958185	0.959071	0.959941	0.960797	0.961637	0.962462	0.963273
1.8	0.964070	0.964853	0.965621	0.966375	0.967116	0.967844	0.968558	0.969259	0.969946	0.970621
1.9	0.971284	0.971934	0.972571	0.973197	0.973811	0.974412	0.975003	0.975581	0.976149	0.976705
2.0	0.977250	0.977785	0.978309	0.978822	0.979325	0.979818	0.980301	0.980774	0.981238	0.981692
2.1	0.982136	0.982571	0.982997	0.983415	0.983823	0.984223	0.984614	0.984997	0.985372	0.985738
2.2	0.986097	0.986448	0.986791	0.987127	0.987455	0.987776	0.988090	0.988397	0.988697	0.988990
2.3	0.989276	0.989556	0.989830	0.990097	0.990359	0.990614	0.990863	0.991106	0.991344	0.991576
2.4	0.991803	0.992024	0.992240	0.992451	0.992657	0.992858	0.993054	0.993245	0.993431	0.993613

2.5	0.993791	0.993964	0.994133	0.994297	0.994458	0.994614	0.994767	0.994915	0.995060	0.995202
2.6	0.995339	0.995473	0.995604	0.995731	0.995855	0.995976	0.996093	0.996208	0.996319	0.996428
2.7	0.996533	0.996636	0.996736	0.996834	0.996928	0.997021	0.997110	0.997198	0.997282	0.997365
2.8	0.997445	0.997523	0.997599	0.997673	0.997745	0.997814	0.997882	0.997948	0.998012	0.998074
2.9	0.998135	0.998193	0.998250	0.998306	0.998359	0.998412	0.998462	0.998511	0.998559	0.998606
3.0	0.998651	0.998694	0.998737	0.998778	0.998818	0.998856	0.998894	0.998930	0.998965	0.999000
3.1	0.999033	0.999065	0.999096	0.999126	0.999156	0.999184	0.999212	0.999238	0.999264	0.999289
3.2	0.999313	0.999337	0.999359	0.999381	0.999403	0.999423	0.999443	0.999463	0.999481	0.999499
3.3	0.999517	0.999534	0.999550	0.999566	0.999582	0.999596	0.999611	0.999625	0.999638	0.999651
3.4	0.999663	0.999676	0.999687	0.999699	0.999710	0.999720	0.999730	0.999740	0.999750	0.999759
3.5	0.999768	0.999776	0.999785	0.999793	0.999800	0.999808	0.999815	0.999822	0.999829	0.999835
3.6	0.999841	0.999847	0.999853	0.999859	0.999864	0.999869	0.999874	0.999879	0.999884	0.999888
3.7	0.999893	0.999897	0.999901	0.999905	0.999908	0.999912	0.999915	0.999919	0.999922	0.999925
3.8	0.999928	0.999931	0.999934	0.999936	0.999939	0.999941	0.999944	0.999946	0.999948	0.999950
3.9	0.999952	0.999954	0.999956	0.999958	0.999960	0.999961	0.999963	0.999964	0.999966	0.999967
4.0	0.999969	0.999970	0.999971	0.999973	0.999974	0.999975	0.999976	0.999977	0.999978	0.999979
4.1	0.999980	0.999981	0.999981	0.999982	0.999983	0.999984	0.999985	0.999985	0.999986	0.999986
4.2	0.999987	0.999988	0.999988	0.999989	0.999989	0.999990	0.999990	0.999991	0.999991	0.999991
4.3	0.999992	0.999992	0.999993	0.999993	0.999993	0.999994	0.999994	0.999994	0.999994	0.999995
4.4	0.999995	0.999995	0.999995	0.999996	0.999996	0.999996	0.999996	0.999997	0.999997	0.999997
4.5	0.999997	0.999997	0.999997	0.999997	0.999998	0.999998	0.999998	0.999998	0.999998	0.999998
4.6	0.999998	0.999998	0.999999	0.999999	0.999999	0.999999	0.999999	0.999999	0.999999	0.999999
4.7	0.999999	0.999999	0.999999	0.999999	0.999999	0.999999	0.999999	0.999999	0.999999	1.000000
4.8	1.000000	1.000000	1.000000	1.000000	1.000000	1.000000	1.000000	1.000000	1.000000	1.000000
4.9	1.000000	1.000000	1.000000	1.000000	1.000000	1.000000	1.000000	1.000000	1.000000	1.000000

Table A.4 Table of critical values of the standard Gaussian distribution to include a given probability, for two-sided confidence intervals (-z, z) of the integral $A(z)$, and for one-sided intervals (-∞, z) of the integral $B(z)$

Probability	Two-sided z	One-sided z
0.01	0.013	−2.326
0.05	0.063	−1.645
0.10	0.126	−1.282
0.20	0.253	−0.842
0.30	0.385	−0.524
0.40	0.524	−0.253
0.50	0.674	−0.000
0.60	0.842	0.253
0.70	1.036	0.524
0.80	1.282	0.842
0.90	1.645	1.282
0.95	1.960	1.645
0.99	2.576	2.326
0.999	3.290	3.090
0.9999	3.890	3.718

Table A.5 Selected upper limits for a Poisson variable using the Gehrels approximation

	Upper limits		
	Poisson parameter S or confidence level		
	$S = 1$	$S = 2$	$S = 3$
n_{obs}	(1-σ, or 84.1 %)	(2-σ, or 97.7 %)	(3-σ, or 99.9 %)
0	1.87	3.48	5.60
1	3.32	5.40	7.97
2	4.66	7.07	9.97
3	5.94	8.62	11.81
4	7.18	10.11	13.54
5	8.40	11.55	15.19
6	9.60	12.95	16.79
7	10.78	14.32	18.35
8	11.96	15.67	19.87
9	13.12	16.99	21.37
10	14.28	18.31	22.84
20	25.56	30.86	36.67
30	36.55	42.84	49.64
40	47.38	54.52	62.15
50	58.12	66.00	74.37
60	68.79	77.34	86.38
70	79.41	88.57	98.23
80	89.99	99.72	109.96
90	100.53	110.80	121.58
100	111.04	121.82	133.11

Table A.6 Selected lower limits for a Poisson variable using the Gehrels approximation

	Lower limits		
	Poisson parameter S or confidence level		
	$S = 1$	$S = 2$	$S = 3$
n_{obs}	$(1\text{-}\sigma, \text{ or } 84.1\%)$	$(2\text{-}\sigma, \text{ or } 97.7\%)$	$(3\text{-}\sigma, \text{ or } 99.9\%)$
1	0.17	0.01	0.00
2	0.71	0.21	0.03
3	1.37	0.58	0.17
4	2.09	1.04	0.42
5	2.85	1.57	0.75
6	3.63	2.14	1.13
7	4.42	2.75	1.56
8	5.24	3.38	2.02
9	6.06	4.04	2.52
10	6.90	4.71	3.04
20	15.57	12.08	9.16
30	24.56	20.07	16.16
40	33.70	28.37	23.63
50	42.96	36.88	31.40
60	52.28	45.53	39.38
70	61.66	54.28	47.52
80	71.08	63.13	55.79
90	80.53	72.04	64.17
100	90.02	81.01	72.63

where f is the number of degrees of freedom. The critical value or p-quantile of the distribution is given by

$$P_{\chi^2}(z \le \chi^2_{crit}) = \int_0^{\chi^2_{crit}} f_{\chi^2}(z)dz = p \qquad (A.9)$$

or, equivalently,

$$P_{\chi^2}(z \ge \chi^2_{crit}) = \int_{\chi^2_{crit}}^{\infty} f_{\chi^2}(z)dz = 1 - p. \qquad (A.10)$$

The critical value is a function of the number of degrees of freedom f and the level of probability p. Normally p is intended as a large number, such as 0.68, 0.90, or 0.99, meaning that there is just a 32, 10, or 1 % probability to have values higher than the critical value χ^2_{crit}.

As described in Sect. 7.2, the χ^2 distribution has the following mean and variance:

$$\begin{cases} \mu = f \\ \sigma^2 = 2f. \end{cases}$$

It is convenient to tabulate the value of reduced χ^2, or χ^2_{crit}/f, that corresponds to a given probability level, as function of the number of degrees of freedom. Selected critical values of the χ^2 distribution are reported in Table A.7. When using this table, remember to multiply the tabulated reduced χ^2 by the number of defrees of freedom f to obtain the value of χ^2.

If Z is a χ^2-distributed variable with f degrees of freedoms,

$$\lim_{f\to\infty} \frac{Z-f}{\sqrt{2f}} = N(0,1). \tag{A.11}$$

In fact, a χ^2 variable is obtained as the sum of independent distributions (Sect. 7.2), to which the central theorem limit applies (Sect. 4.3). For a large number of degrees of freedom, the standard Gaussian distribution can be used to supplement Table A.7 according to (A.11). For example, for $p = 0.99$, the one-sided critical value of the standard Gaussian is approximately 2.326, according to Table A.4. Using this value into (A.11) for $f = 200$ would give a critical value for the χ^2 distribution of 1.2326 (compare to 1.247 from Table A.7). The values of $f = \infty$ in Table A.7 is obtained using the Gaussian approximation, according to (A.11).

A.4   The F Distribution

The F distribution with f_1, f_2 degrees of freedom is defined in (7.22) as

$$f_F(z) = \frac{\Gamma\left(\dfrac{f_1+f_2}{2}\right)}{\Gamma\left(\dfrac{f_1}{2}\right)\Gamma\left(\dfrac{f_2}{2}\right)} \left(\frac{f_1}{f_2}\right)^{\frac{f_1}{2}} \frac{z^{\frac{f_1}{2}-1}}{\left(1+z\dfrac{f_1}{f_2}\right)^{\frac{f_1+f_2}{2}}}.$$

The critical value F_{crit} that includes a probability p is given by

$$P(z \geq F_{crit}) = \int_{F_{crit}}^{\infty} f_F(z)dz = 1 - p, \tag{A.12}$$

and it is a function of the degrees of freedom f_1 and f_2. In Table A.8 are reported the critical values for various probability levels p, for a fixed value $f_1 = 1$, and as function of f_2. Tables A.9, A.10, A.11, A.12, A.13, A.14, and A.15 have the critical values as function of both f_1 and f_2.

Asymptotic values when f_1 and f_2 approach infinity can be found using (7.25), reported here for convenience:

$$\begin{cases} \lim_{f_2\to\infty} f_F(z,f_1,f_2) = f_{\chi^2}(x,f_1) \text{ where } x = f_1 z \\ \lim_{f_1\to\infty} f_F(z,f_1,f_2) = f_{\chi^2}(x,f_2) \text{ where } x = f_2/z. \end{cases}$$

Table A.7 Critical values of the χ^2 distribution

Probability p to have a value of reduced χ^2 below the critical value

f	0.01	0.05	0.10	0.20	0.30	0.40	0.50	0.60	0.70	0.80	0.90	0.95	0.99
1	0.00016	0.00390	0.01580	0.0642	0.1485	0.2750	0.4549	0.7083	1.0742	1.6424	2.7055	3.8415	6.6349
2	0.0101	0.0513	0.1054	0.2231	0.3567	0.5108	0.6931	0.9163	1.2040	1.6094	2.3026	2.9957	4.6052
3	0.0383	0.1173	0.1948	0.3351	0.4746	0.6231	0.7887	0.9821	1.2216	1.5472	2.0838	2.6049	3.7816
4	0.0743	0.1777	0.2659	0.4122	0.5487	0.6882	0.8392	1.0112	1.2196	1.4972	1.9449	2.3719	3.3192
5	0.1109	0.2291	0.3221	0.4685	0.6000	0.7311	0.8703	1.0264	1.2129	1.4578	1.8473	2.2141	3.0172
6	0.1454	0.2726	0.3674	0.5117	0.6379	0.7617	0.8914	1.0351	1.2052	1.4263	1.7741	2.0986	2.8020
7	0.1770	0.3096	0.4047	0.5460	0.6673	0.7847	0.9065	1.0405	1.1976	1.4005	1.7167	2.0096	2.6393
8	0.2058	0.3416	0.4362	0.5742	0.6909	0.8028	0.9180	1.0438	1.1906	1.3788	1.6702	1.9384	2.5113
9	0.2320	0.3695	0.4631	0.5978	0.7104	0.8174	0.9270	1.0460	1.1841	1.3602	1.6315	1.8799	2.4073
10	0.256	0.394	0.487	0.618	0.727	0.830	0.934	1.047	1.178	1.344	1.599	1.831	2.321
11	0.278	0.416	0.507	0.635	0.741	0.840	0.940	1.048	1.173	1.330	1.570	1.789	2.248
12	0.298	0.436	0.525	0.651	0.753	0.848	0.945	1.049	1.168	1.318	1.546	1.752	2.185
13	0.316	0.453	0.542	0.664	0.764	0.856	0.949	1.049	1.163	1.307	1.524	1.720	2.130
14	0.333	0.469	0.556	0.676	0.773	0.863	0.953	1.049	1.159	1.296	1.505	1.692	2.082
15	0.349	0.484	0.570	0.687	0.781	0.869	0.956	1.049	1.155	1.287	1.487	1.666	2.039
16	0.363	0.498	0.582	0.697	0.789	0.874	0.959	1.049	1.151	1.279	1.471	1.643	2.000
17	0.377	0.510	0.593	0.706	0.796	0.879	0.961	1.048	1.148	1.271	1.457	1.623	1.965
18	0.390	0.522	0.604	0.714	0.802	0.883	0.963	1.048	1.145	1.264	1.444	1.604	1.934
19	0.402	0.532	0.613	0.722	0.808	0.887	0.965	1.048	1.142	1.258	1.432	1.587	1.905
20	0.413	0.543	0.622	0.729	0.813	0.890	0.967	1.048	1.139	1.252	1.421	1.571	1.878
30	0.498	0.616	0.687	0.779	0.850	0.915	0.978	1.044	1.118	1.208	1.342	1.459	1.696
40	0.554	0.663	0.726	0.809	0.872	0.928	0.983	1.041	1.104	1.182	1.295	1.394	1.592
50	0.594	0.695	0.754	0.829	0.886	0.937	0.987	1.038	1.094	1.163	1.263	1.350	1.523

(continued)

Table A.7 (continued)

Probability p to have a value of reduced χ^2 below the critical value

f	0.01	0.05	0.10	0.20	0.30	0.40	0.50	0.60	0.70	0.80	0.90	0.95	0.99
60	0.625	0.720	0.774	0.844	0.897	0.944	0.989	1.036	1.087	1.150	1.240	1.318	1.473
70	0.649	0.739	0.790	0.856	0.905	0.949	0.990	1.034	1.081	1.139	1.222	1.293	1.435
80	0.669	0.755	0.803	0.865	0.911	0.952	0.992	1.032	1.076	1.130	1.207	1.274	1.404
90	0.686	0.768	0.814	0.873	0.917	0.955	0.993	1.031	1.073	1.123	1.195	1.257	1.379
100	0.700	0.780	0.823	0.880	0.921	0.958	0.993	1.030	1.069	1.117	1.185	1.243	1.358
200	0.782	0.841	0.874	0.915	0.945	0.971	0.997	1.022	1.050	1.083	1.130	1.170	1.247
300	0.820	0.870	0.897	0.931	0.956	0.977	0.998	1.019	1.041	1.068	1.106	1.138	1.200
400	0.843	0.887	0.910	0.940	0.962	0.980	0.998	1.016	1.036	1.059	1.092	1.119	1.172
500	0.86	0.90	0.92	0.95	0.97	0.98	1.00	1.01	1.03	1.05	1.08	1.10	1.15
1000	0.90	0.93	0.94	0.96	0.98	0.99	1.00	1.01	1.02	1.04	1.06	1.07	1.11
∞	0.90	0.93	0.94	0.96	0.98	0.99	1.00	1.01	1.02	1.04	1.06	1.07	1.10

Table A.8 Critical values of F statistics for $f_1 = 1$ degrees of freedom

f_2	Probability p to have a value of F below the critical value						
	0.50	0.60	0.70	0.80	0.90	0.95	0.99
1	1.000	1.894	3.852	9.472	39.863	161.448	4052.182
2	0.667	1.125	1.922	3.556	8.526	18.513	98.503
3	0.585	0.957	1.562	2.682	5.538	10.128	34.116
4	0.549	0.885	1.415	2.351	4.545	7.709	21.198
5	0.528	0.846	1.336	2.178	4.060	6.608	16.258
6	0.515	0.820	1.286	2.073	3.776	5.987	13.745
7	0.506	0.803	1.253	2.002	3.589	5.591	12.246
8	0.499	0.790	1.228	1.951	3.458	5.318	11.259
9	0.494	0.780	1.209	1.913	3.360	5.117	10.561
10	0.490	0.773	1.195	1.883	3.285	4.965	10.044
20	0.472	0.740	1.132	1.757	2.975	4.351	8.096
30	0.466	0.729	1.112	1.717	2.881	4.171	7.562
40	0.463	0.724	1.103	1.698	2.835	4.085	7.314
50	0.462	0.721	1.097	1.687	2.809	4.034	7.171
60	0.461	0.719	1.093	1.679	2.791	4.001	7.077
70	0.460	0.717	1.090	1.674	2.779	3.978	7.011
80	0.459	0.716	1.088	1.670	2.769	3.960	6.963
90	0.459	0.715	1.087	1.667	2.762	3.947	6.925
100	0.458	0.714	1.085	1.664	2.756	3.936	6.895
200	0.457	0.711	1.080	1.653	2.731	3.888	6.763
∞	0.455	0.708	1.074	1.642	2.706	3.842	6.635

For example, the critical values of the F distribution for $f_1 = 1$ and in the limit of large f_2 are obtained from the first row of Table A.7.

A.5   The Student's t Distribution

The Student t distribution is given by (7.34),

$$f_T(t) = \frac{1}{\sqrt{f\pi}} \frac{\Gamma((f+1)/2)}{\Gamma(f/2)} \times \left(1 + \frac{t^2}{f}\right)^{-\frac{1}{2}(f+1)},$$

where f is the number of degrees of freedom. The probability p that the absolute value of a t variable exceeds a critical value T_{crit} is given by

$$P(|t| \leq T_{crit}) = P(|\bar{x} - \mu| \leq T_{crit} \cdot s/\sqrt{n}) = \int_{-T_{crit}}^{T_{crit}} f_T(t)dt = 1 - p. \qquad (A.13)$$

Table A.9 Critical values of F statistic that include $p = 0.50$ probability

f_2	f_1									
	2	4	6	8	10	20	40	60	80	100
1	1.500	1.823	1.942	2.004	2.042	2.119	2.158	2.172	2.178	2.182
2	1.000	1.207	1.282	1.321	1.345	1.393	1.418	1.426	1.430	1.433
3	0.881	1.063	1.129	1.163	1.183	1.225	1.246	1.254	1.257	1.259
4	0.828	1.000	1.062	1.093	1.113	1.152	1.172	1.178	1.182	1.184
5	0.799	0.965	1.024	1.055	1.073	1.111	1.130	1.136	1.139	1.141
6	0.780	0.942	1.000	1.030	1.048	1.084	1.103	1.109	1.113	1.114
7	0.767	0.926	0.983	1.013	1.030	1.066	1.085	1.091	1.094	1.096
8	0.757	0.915	0.971	1.000	1.017	1.053	1.071	1.077	1.080	1.082
9	0.749	0.906	0.962	0.990	1.008	1.043	1.061	1.067	1.070	1.072
10	0.743	0.899	0.954	0.983	1.000	1.035	1.053	1.059	1.062	1.063
20	0.718	0.868	0.922	0.950	0.966	1.000	1.017	1.023	1.026	1.027
30	0.709	0.858	0.912	0.939	0.955	0.989	1.006	1.011	1.014	1.016
40	0.705	0.854	0.907	0.934	0.950	0.983	1.000	1.006	1.008	1.010
50	0.703	0.851	0.903	0.930	0.947	0.980	0.997	1.002	1.005	1.007
60	0.701	0.849	0.901	0.928	0.945	0.978	0.994	1.000	1.003	1.004
70	0.700	0.847	0.900	0.927	0.943	0.976	0.993	0.998	1.001	1.003
80	0.699	0.846	0.899	0.926	0.942	0.975	0.992	0.997	1.000	1.002
90	0.699	0.845	0.898	0.925	0.941	0.974	0.991	0.996	0.999	1.001
100	0.698	0.845	0.897	0.924	0.940	0.973	0.990	0.996	0.998	1.000
200	0.696	0.842	0.894	0.921	0.937	0.970	0.987	0.992	0.995	0.997
∞	0.693	0.839	0.891	0.918	0.934	0.967	0.983	0.989	0.992	0.993

These two-sided critical values are tabulated in Tables A.16, A.17, A.18, A.19, A.20, A.21, and A.22 for selected values of f, as function of the critical value T_{crit}. In these tables, the left column indicates the value of T_{crit} to the first decimal digit, and the values on the top column are the second decimal digit.

Table A.23 provides a comparison of the probability p for five critical values, $T_{crit} = 1$ through 5, as function of f. The case of $f = \infty$ corresponds to a standard Gaussian.

A.6 The Linear Correlation Coefficient r

The linear correlation coefficient is defined as

$$r^2 = \frac{\left(N \sum x_i y_i - \sum x_i \sum y_i\right)^2}{\left(N \sum x_i^2 - \left(\sum x_i\right)^2\right)\left(N \sum y_i^2 - \left(\sum y_i\right)^2\right)} \tag{A.14}$$

Table A.10 Critical values of F statistic that include $p = 0.60$ probability

f_2	f_1 2	4	6	8	10	20	40	60	80	100
1	2.625	3.093	3.266	3.355	3.410	3.522	3.579	3.598	3.608	3.613
2	1.500	1.718	1.796	1.835	1.859	1.908	1.933	1.941	1.945	1.948
3	1.263	1.432	1.489	1.518	1.535	1.570	1.588	1.593	1.596	1.598
4	1.162	1.310	1.359	1.383	1.397	1.425	1.439	1.444	1.446	1.448
5	1.107	1.243	1.287	1.308	1.320	1.345	1.356	1.360	1.362	1.363
6	1.072	1.200	1.241	1.260	1.272	1.293	1.303	1.307	1.308	1.309
7	1.047	1.171	1.209	1.227	1.238	1.257	1.266	1.269	1.270	1.271
8	1.030	1.150	1.186	1.203	1.213	1.231	1.239	1.241	1.242	1.243
9	1.016	1.133	1.168	1.185	1.194	1.210	1.217	1.219	1.220	1.221
10	1.006	1.120	1.154	1.170	1.179	1.194	1.200	1.202	1.203	1.204
20	0.960	1.064	1.093	1.106	1.112	1.122	1.124	1.124	1.124	1.124
30	0.945	1.046	1.074	1.085	1.090	1.097	1.097	1.097	1.096	1.096
40	0.938	1.037	1.064	1.075	1.080	1.085	1.084	1.083	1.082	1.081
50	0.933	1.032	1.058	1.068	1.073	1.078	1.076	1.074	1.073	1.072
60	0.930	1.029	1.054	1.064	1.069	1.073	1.070	1.068	1.066	1.065
70	0.928	1.026	1.052	1.061	1.066	1.069	1.066	1.064	1.062	1.061
80	0.927	1.024	1.049	1.059	1.064	1.067	1.063	1.060	1.059	1.057
90	0.926	1.023	1.048	1.057	1.062	1.065	1.061	1.058	1.056	1.054
100	0.925	1.021	1.047	1.056	1.060	1.063	1.059	1.056	1.054	1.052
200	0.921	1.016	1.041	1.050	1.054	1.055	1.050	1.046	1.043	1.041
∞	0.916	1.011	1.035	1.044	1.047	1.048	1.041	1.036	1.032	1.029

and it is equal to the product bb', where b is the best-fit slope of the linear regression of Y on X, and b' is the slope of the linear regression of X on Y. The probability distribution function of r, under the hypothesis that the variables X and Y are not correlated, is given by

$$f_r(r) = \frac{1}{\sqrt{\pi}} \frac{\Gamma(\frac{f+1}{2})}{\Gamma(\frac{f}{2})} \left(\frac{1}{1-r^2} \right)^{-\frac{f-2}{2}} \tag{A.15}$$

where N is the size of the sample, and $f = N- 2$ is the effective number of degrees of freedom of the dataset.

In Table A.24 we report the critical values of r calculated from the following equation,

$$1-p = \int_{-r_{crit}}^{r_{crit}} f_r(r)dr \tag{A.16}$$

Table A.11 Critical values of F statistic that include $p = 0.70$ probability

f_2	f_1									
	2	4	6	8	10	20	40	60	80	100
1	5.056	5.830	6.117	6.267	6.358	6.544	6.639	6.671	6.687	6.697
2	2.333	2.561	2.640	2.681	2.705	2.754	2.779	2.787	2.791	2.794
3	1.847	1.985	2.028	2.048	2.061	2.084	2.096	2.100	2.102	2.103
4	1.651	1.753	1.781	1.793	1.800	1.812	1.818	1.819	1.820	1.821
5	1.547	1.629	1.648	1.656	1.659	1.665	1.666	1.666	1.667	1.667
6	1.481	1.551	1.565	1.570	1.571	1.572	1.570	1.570	1.569	1.569
7	1.437	1.499	1.509	1.511	1.511	1.507	1.504	1.502	1.501	1.501
8	1.405	1.460	1.467	1.468	1.466	1.460	1.455	1.452	1.451	1.450
9	1.380	1.431	1.436	1.435	1.433	1.424	1.417	1.414	1.413	1.412
10	1.361	1.408	1.412	1.409	1.406	1.395	1.387	1.384	1.382	1.381
20	1.279	1.311	1.305	1.297	1.290	1.268	1.252	1.245	1.242	1.240
30	1.254	1.280	1.271	1.261	1.253	1.226	1.206	1.197	1.192	1.189
40	1.241	1.264	1.255	1.243	1.234	1.205	1.182	1.172	1.167	1.163
50	1.233	1.255	1.245	1.233	1.223	1.192	1.167	1.156	1.150	1.146
60	1.228	1.249	1.238	1.226	1.215	1.183	1.157	1.146	1.139	1.135
70	1.225	1.245	1.233	1.221	1.210	1.177	1.150	1.138	1.131	1.127
80	1.222	1.242	1.230	1.217	1.206	1.172	1.144	1.132	1.125	1.120
90	1.220	1.239	1.227	1.214	1.203	1.168	1.140	1.127	1.120	1.115
100	1.219	1.237	1.225	1.212	1.200	1.165	1.137	1.123	1.116	1.111
200	1.211	1.228	1.215	1.201	1.189	1.152	1.121	1.106	1.097	1.091
∞	1.204	1.220	1.205	1.191	1.178	1.139	1.104	1.087	1.076	1.069

where p is the probability for a given value of the correlation coefficient to exceed, in absolute value, the critical value r_{crit}. The critical values are function of the number of degrees of freedom, and of the probability p.

To evaluate the probability distribution function in the case of large f, a convenient approximation can be given using the asymptotic expansion for the Gamma function (see [1]):

$$\Gamma(az + b) \simeq \sqrt{2\pi} e^{-az} (az)^{az+b-1/2}. \tag{A.17}$$

For large values of f, the ratio of the Gamma functions can therefore be approximated as

$$\frac{\Gamma\left(\dfrac{f+1}{2}\right)}{\Gamma\left(\dfrac{f}{2}\right)} \simeq \sqrt{\dfrac{f}{2}}.$$

Table A.12 Critical values of F statistic that include $p = 0.80$ probability

f_2	f_1 2	4	6	8	10	20	40	60	80	100
1	12.000	13.644	14.258	14.577	14.772	15.171	15.374	15.442	15.477	15.497
2	4.000	4.236	4.317	4.358	4.382	4.432	4.456	4.465	4.469	4.471
3	2.886	2.956	2.971	2.976	2.979	2.983	2.984	2.984	2.984	2.984
4	2.472	2.483	2.473	2.465	2.460	2.445	2.436	2.433	2.431	2.430
5	2.259	2.240	2.217	2.202	2.191	2.166	2.151	2.146	2.143	2.141
6	2.130	2.092	2.062	2.042	2.028	1.995	1.976	1.969	1.965	1.963
7	2.043	1.994	1.957	1.934	1.918	1.879	1.857	1.849	1.844	1.842
8	1.981	1.923	1.883	1.856	1.838	1.796	1.770	1.761	1.756	1.753
9	1.935	1.870	1.826	1.798	1.778	1.732	1.704	1.694	1.689	1.686
10	1.899	1.829	1.782	1.752	1.732	1.682	1.653	1.642	1.636	1.633
20	1.746	1.654	1.596	1.558	1.531	1.466	1.424	1.408	1.399	1.394
30	1.699	1.600	1.538	1.497	1.468	1.395	1.347	1.328	1.318	1.312
40	1.676	1.574	1.509	1.467	1.437	1.360	1.308	1.287	1.276	1.269
50	1.662	1.558	1.492	1.449	1.418	1.338	1.284	1.262	1.249	1.241
60	1.653	1.548	1.481	1.437	1.406	1.324	1.268	1.244	1.231	1.223
70	1.647	1.540	1.473	1.429	1.397	1.314	1.256	1.231	1.218	1.209
80	1.642	1.535	1.467	1.422	1.390	1.306	1.247	1.222	1.208	1.199
90	1.639	1.531	1.463	1.418	1.385	1.300	1.240	1.214	1.200	1.191
100	1.636	1.527	1.459	1.414	1.381	1.296	1.234	1.208	1.193	1.184
200	1.622	1.512	1.443	1.396	1.363	1.274	1.209	1.180	1.163	1.152
∞	1.609	1.497	1.426	1.379	1.344	1.252	1.182	1.150	1.130	1.117

A.7 The Kolmogorov–Smirnov Test

The one-sample Kolmogorov–Smirnov statistic D_N is defined in (13.7) as

$$D_N = \max_x |F_N(x) - F(x)|,$$

where $F(x)$ is the parent distribution, and $F_N(x)$ the sample distribution.
The cumulative distribution of the test statistic can be approximated by

$$P(D_N < z/(\sqrt{N} + 0.12 + 0.11/\sqrt{N})) \simeq \Phi(z).$$

where

$$\Phi(z) = \sum_{r=-\infty}^{\infty} (-1)^r e^{-2r^2 z^2}.$$

Table A.13 Critical values of F statistic that include $p = 0.90$ probability

f_2	f_1 2	4	6	8	10	20	40	60	80	100
1	49.500	55.833	58.204	59.439	60.195	61.740	62.529	62.794	62.927	63.007
2	9.000	9.243	9.326	9.367	9.392	9.441	9.466	9.475	9.479	9.481
3	5.462	5.343	5.285	5.252	5.230	5.184	5.160	5.151	5.147	5.144
4	4.325	4.107	4.010	3.955	3.920	3.844	3.804	3.790	3.782	3.778
5	3.780	3.520	3.404	3.339	3.297	3.207	3.157	3.140	3.132	3.126
6	3.463	3.181	3.055	2.983	2.937	2.836	2.781	2.762	2.752	2.746
7	3.257	2.960	2.827	2.752	2.703	2.595	2.535	2.514	2.504	2.497
8	3.113	2.806	2.668	2.589	2.538	2.425	2.361	2.339	2.328	2.321
9	3.006	2.693	2.551	2.469	2.416	2.298	2.232	2.208	2.196	2.189
10	2.924	2.605	2.461	2.377	2.323	2.201	2.132	2.107	2.095	2.087
20	2.589	2.249	2.091	1.999	1.937	1.794	1.708	1.677	1.660	1.650
30	2.489	2.142	1.980	1.884	1.820	1.667	1.573	1.538	1.519	1.507
40	2.440	2.091	1.927	1.829	1.763	1.605	1.506	1.467	1.447	1.434
50	2.412	2.061	1.895	1.796	1.729	1.568	1.465	1.424	1.402	1.389
60	2.393	2.041	1.875	1.775	1.707	1.544	1.437	1.395	1.372	1.358
70	2.380	2.027	1.860	1.760	1.691	1.526	1.418	1.374	1.350	1.335
80	2.370	2.016	1.849	1.748	1.680	1.513	1.403	1.358	1.334	1.318
90	2.362	2.008	1.841	1.739	1.671	1.503	1.391	1.346	1.320	1.304
100	2.356	2.002	1.834	1.732	1.663	1.494	1.382	1.336	1.310	1.293
200	2.329	1.973	1.804	1.701	1.631	1.458	1.339	1.289	1.261	1.242
∞	2.303	1.945	1.774	1.670	1.599	1.421	1.295	1.240	1.207	1.185

and it is independent of the form of the parent distribution $F(x)$. For large values of N, we can use the asymptotic equation

$$P(D_N < z/\sqrt{N}) = \Phi(z).$$

In Table A.25 are listed the critical values of $\sqrt{N}D_N$ for various levels of probability. Values of the Kolmogorov–Smirnov statistic above the critical value indicate a rejection of the null hypothesis that the data are drawn from the parent model.

The two-sample Kolmogorov–Smirnov statistic is

$$D_{NM} = \max_x |F_M(x) - G_N(x)|$$

where $F_M(x)$ and $G_N(x)$ are the sample cumulative distribution of two independent sets of observations of size M and N. This statistic has the same distribution as the one-sample Kolmogorov-Smirnov statistic, with the substitution of $MN/(M+N)$ in place of N, and in the limit of large M and N, (13.12).

Table A.14 Critical values of F statistic that include $p = 0.95$ probability

f_2	f_1									
	2	4	6	8	10	20	40	60	80	100
1	199.500	224.583	233.986	238.883	241.882	248.013	251.143	252.196	252.724	253.041
2	19.000	19.247	19.330	19.371	19.396	19.446	19.471	19.479	19.483	19.486
3	9.552	9.117	8.941	8.845	8.786	8.660	8.594	8.572	8.561	8.554
4	6.944	6.388	6.163	6.041	5.964	5.803	5.717	5.688	5.673	5.664
5	5.786	5.192	4.950	4.818	4.735	4.558	4.464	4.431	4.415	4.405
6	5.143	4.534	4.284	4.147	4.060	3.874	3.774	3.740	3.722	3.712
7	4.737	4.120	3.866	3.726	3.636	3.444	3.340	3.304	3.286	3.275
8	4.459	3.838	3.581	3.438	3.347	3.150	3.043	3.005	2.986	2.975
9	4.256	3.633	3.374	3.230	3.137	2.936	2.826	2.787	2.768	2.756
10	4.103	3.478	3.217	3.072	2.978	2.774	2.661	2.621	2.601	2.588
20	3.493	2.866	2.599	2.447	2.348	2.124	1.994	1.946	1.922	1.907
30	3.316	2.690	2.421	2.266	2.165	1.932	1.792	1.740	1.712	1.695
40	3.232	2.606	2.336	2.180	2.077	1.839	1.693	1.637	1.608	1.589
50	3.183	2.557	2.286	2.130	2.026	1.784	1.634	1.576	1.544	1.525
60	3.150	2.525	2.254	2.097	1.992	1.748	1.594	1.534	1.502	1.481
70	3.128	2.503	2.231	2.074	1.969	1.722	1.566	1.504	1.471	1.450
80	3.111	2.486	2.214	2.056	1.951	1.703	1.545	1.482	1.448	1.426
90	3.098	2.473	2.201	2.043	1.938	1.688	1.528	1.464	1.429	1.407
100	3.087	2.463	2.191	2.032	1.927	1.677	1.515	1.450	1.414	1.392
200	3.041	2.417	2.144	1.985	1.878	1.623	1.455	1.385	1.346	1.321
∞	2.996	2.372	2.099	1.938	1.831	1.571	1.394	1.318	1.273	1.243

Table A.15 Critical values of F statistic that include $p = 0.99$ probability

f_2	f_1									
	2	4	6	8	10	20	40	60	80	100
1	4999.500	5624.583	5858.986	5981.070	6055.847	6208.730	6286.782	6313.030	6326.197	6334.110
2	99.000	99.249	99.333	99.374	99.399	99.449	99.474	99.482	99.487	99.489
3	30.817	28.710	27.911	27.489	27.229	26.690	26.411	26.316	26.269	26.240
4	18.000	15.977	15.207	14.799	14.546	14.020	13.745	13.652	13.605	13.577
5	13.274	11.392	10.672	10.289	10.051	9.553	9.291	9.202	9.157	9.130
6	10.925	9.148	8.466	8.102	7.874	7.396	7.143	7.057	7.013	6.987
7	9.547	7.847	7.191	6.840	6.620	6.155	5.908	5.824	5.781	5.755
8	8.649	7.006	6.371	6.029	5.814	5.359	5.116	5.032	4.989	4.963
9	8.022	6.422	5.802	5.467	5.257	4.808	4.567	4.483	4.441	4.415
10	7.559	5.994	5.386	5.057	4.849	4.405	4.165	4.082	4.039	4.014
20	5.849	4.431	3.871	3.564	3.368	2.938	2.695	2.608	2.563	2.535
30	5.390	4.018	3.473	3.173	2.979	2.549	2.299	2.208	2.160	2.131
40	5.178	3.828	3.291	2.993	2.801	2.369	2.114	2.019	1.969	1.938
50	5.057	3.720	3.186	2.890	2.698	2.265	2.006	1.909	1.857	1.825
60	4.977	3.649	3.119	2.823	2.632	2.198	1.936	1.836	1.783	1.749
70	4.922	3.600	3.071	2.777	2.585	2.150	1.886	1.784	1.730	1.696
80	4.881	3.563	3.036	2.742	2.551	2.115	1.849	1.746	1.690	1.655
90	4.849	3.535	3.009	2.715	2.524	2.088	1.820	1.716	1.659	1.623
100	4.824	3.513	2.988	2.694	2.503	2.067	1.797	1.692	1.634	1.598
200	4.713	3.414	2.893	2.601	2.411	1.971	1.695	1.584	1.521	1.481
∞	4.605	3.319	2.802	2.511	2.321	1.878	1.592	1.473	1.404	1.358

Table A.16 Integral of Student's function ($f = 1$), or probability p, as function of critical value T_{crit}

	0	1	2	3	4	5	6	7	8	9
0.0	0.000000	0.006366	0.012731	0.019093	0.025451	0.031805	0.038151	0.044491	0.050821	0.057142
0.1	0.063451	0.069748	0.076031	0.082299	0.088551	0.094786	0.101003	0.107201	0.113378	0.119533
0.2	0.125666	0.131775	0.137860	0.143920	0.149953	0.155958	0.161936	0.167884	0.173803	0.179691
0.3	0.185547	0.191372	0.197163	0.202921	0.208645	0.214334	0.219988	0.225605	0.231187	0.236731
0.4	0.242238	0.247707	0.253138	0.258530	0.263883	0.269197	0.274472	0.279706	0.284900	0.290054
0.5	0.295167	0.300240	0.305272	0.310262	0.315212	0.320120	0.324987	0.329813	0.334597	0.339340
0.6	0.344042	0.348702	0.353321	0.357899	0.362436	0.366932	0.371387	0.375801	0.380175	0.384508
0.7	0.388800	0.393053	0.397266	0.401438	0.405572	0.409666	0.413721	0.417737	0.421714	0.425653
0.8	0.429554	0.433417	0.437242	0.441030	0.444781	0.448495	0.452173	0.455814	0.459420	0.462990
0.9	0.466525	0.470025	0.473490	0.476920	0.480317	0.483680	0.487010	0.490306	0.493570	0.496801
1.0	0.500000	0.503167	0.506303	0.509408	0.512481	0.515524	0.518537	0.521520	0.524474	0.527398
1.1	0.530293	0.533159	0.535997	0.538807	0.541589	0.544344	0.547071	0.549772	0.552446	0.555094
1.2	0.557716	0.560312	0.562883	0.565429	0.567950	0.570447	0.572919	0.575367	0.577792	0.580193
1.3	0.582571	0.584927	0.587259	0.589570	0.591858	0.594124	0.596369	0.598592	0.600795	0.602976
1.4	0.605137	0.607278	0.609398	0.611499	0.613580	0.615641	0.617684	0.619707	0.621712	0.623698
1.5	0.625666	0.627616	0.629548	0.631462	0.633359	0.635239	0.637101	0.638947	0.640776	0.642589
1.6	0.644385	0.646165	0.647930	0.649678	0.651411	0.653129	0.654832	0.656519	0.658192	0.659850
1.7	0.661494	0.663124	0.664739	0.666340	0.667928	0.669502	0.671062	0.672609	0.674143	0.675663
1.8	0.677171	0.678666	0.680149	0.681619	0.683077	0.684522	0.685956	0.687377	0.688787	0.690185
1.9	0.691572	0.692947	0.694311	0.695664	0.697006	0.698337	0.699657	0.700967	0.702266	0.703555
2.0	0.704833	0.706101	0.707359	0.708607	0.709846	0.711074	0.712293	0.713502	0.714702	0.715893
2.1	0.717074	0.718246	0.719410	0.720564	0.721709	0.722846	0.723974	0.725093	0.726204	0.727307
2.2	0.728401	0.729487	0.730565	0.731635	0.732696	0.733750	0.734797	0.735835	0.736866	0.737889
2.3	0.738905	0.739914	0.740915	0.741909	0.742895	0.743875	0.744847	0.745813	0.746772	0.747723
2.4	0.748668	0.749607	0.750539	0.751464	0.752383	0.753295	0.754201	0.755101	0.755994	0.756881

(continued)

Table A.16 (continued)

	0	1	2	3	4	5	6	7	8	9
2.5	0.757762	0.758638	0.759507	0.760370	0.761227	0.762078	0.762924	0.763764	0.764598	0.765427
2.6	0.766250	0.767068	0.767880	0.768687	0.769488	0.770284	0.771075	0.771861	0.772642	0.773417
2.7	0.774188	0.774953	0.775714	0.776469	0.777220	0.777966	0.778707	0.779443	0.780175	0.780902
2.8	0.781625	0.782342	0.783056	0.783765	0.784469	0.785169	0.785865	0.786556	0.787243	0.787926
2.9	0.788605	0.789279	0.789949	0.790616	0.791278	0.791936	0.792590	0.793240	0.793887	0.794529
3.0	0.795168	0.795802	0.796433	0.797060	0.797684	0.798304	0.798920	0.799532	0.800141	0.800746
3.1	0.801348	0.801946	0.802541	0.803133	0.803720	0.804305	0.804886	0.805464	0.806039	0.806610
3.2	0.807178	0.807743	0.808304	0.808863	0.809418	0.809970	0.810519	0.811065	0.811608	0.812148
3.3	0.812685	0.813219	0.813750	0.814278	0.814803	0.815325	0.815845	0.816361	0.816875	0.817386
3.4	0.817894	0.818400	0.818903	0.819403	0.819900	0.820395	0.820887	0.821376	0.821863	0.822348
3.5	0.822829	0.823308	0.823785	0.824259	0.824731	0.825200	0.825667	0.826131	0.826593	0.827053
3.6	0.827510	0.827965	0.828418	0.828868	0.829316	0.829761	0.830205	0.830646	0.831085	0.831521
3.7	0.831956	0.832388	0.832818	0.833246	0.833672	0.834096	0.834517	0.834937	0.835354	0.835770
3.8	0.836183	0.836594	0.837004	0.837411	0.837816	0.838220	0.838621	0.839020	0.839418	0.839813
3.9	0.840207	0.840599	0.840989	0.841377	0.841763	0.842147	0.842530	0.842911	0.843290	0.843667
4.0	0.844042	0.844416	0.844788	0.845158	0.845526	0.845893	0.846258	0.846621	0.846983	0.847343
4.1	0.847701	0.848057	0.848412	0.848766	0.849118	0.849468	0.849816	0.850163	0.850509	0.850853
4.2	0.851195	0.851536	0.851875	0.852213	0.852549	0.852883	0.853217	0.853548	0.853879	0.854208
4.3	0.854535	0.854861	0.855185	0.855508	0.855830	0.856150	0.856469	0.856787	0.857103	0.857417
4.4	0.857731	0.858043	0.858353	0.858663	0.858971	0.859277	0.859583	0.859887	0.860190	0.860491
4.5	0.860791	0.861090	0.861388	0.861684	0.861980	0.862274	0.862566	0.862858	0.863148	0.863437
4.6	0.863725	0.864012	0.864297	0.864582	0.864865	0.865147	0.865428	0.865707	0.865986	0.866263
4.7	0.866539	0.866815	0.867089	0.867362	0.867633	0.867904	0.868174	0.868442	0.868710	0.868976
4.8	0.869242	0.869506	0.869769	0.870031	0.870292	0.870553	0.870812	0.871070	0.871327	0.871583
4.9	0.871838	0.872092	0.872345	0.872597	0.872848	0.873098	0.873347	0.873596	0.873843	0.874089

Table A.17 Integral of Student's function ($f = 2$), or probability p, as function of critical value T_{crit}

	0	1	2	3	4	5	6	7	8	9
0.0	0.000000	0.007071	0.014141	0.021208	0.028273	0.035333	0.042388	0.049437	0.056478	0.063511
0.1	0.070535	0.077548	0.084549	0.091538	0.098513	0.105474	0.112420	0.119349	0.126261	0.133154
0.2	0.140028	0.146882	0.153715	0.160526	0.167313	0.174078	0.180817	0.187532	0.194220	0.200881
0.3	0.207514	0.214119	0.220695	0.227241	0.233756	0.240239	0.246691	0.253110	0.259496	0.265848
0.4	0.272166	0.278448	0.284695	0.290906	0.297080	0.303218	0.309318	0.315380	0.321403	0.327388
0.5	0.333333	0.339240	0.345106	0.350932	0.356718	0.362462	0.368166	0.373829	0.379450	0.385029
0.6	0.390567	0.396062	0.401516	0.406926	0.412295	0.417620	0.422903	0.428144	0.433341	0.438496
0.7	0.443607	0.448676	0.453702	0.458684	0.463624	0.468521	0.473376	0.478187	0.482956	0.487682
0.8	0.492366	0.497008	0.501607	0.506164	0.510679	0.515152	0.519583	0.523973	0.528322	0.532629
0.9	0.536895	0.541121	0.545306	0.549450	0.553555	0.557619	0.561644	0.565629	0.569575	0.573482
1.0	0.577351	0.581180	0.584972	0.588725	0.592441	0.596120	0.599761	0.603366	0.606933	0.610465
1.1	0.613960	0.617420	0.620844	0.624233	0.627588	0.630907	0.634193	0.637444	0.640662	0.643846
1.2	0.646997	0.650115	0.653201	0.656255	0.659276	0.662266	0.665225	0.668153	0.671050	0.673917
1.3	0.676753	0.679560	0.682337	0.685085	0.687804	0.690494	0.693156	0.695790	0.698397	0.700975
1.4	0.703527	0.706051	0.708549	0.711021	0.713466	0.715886	0.718280	0.720649	0.722993	0.725312
1.5	0.727607	0.729878	0.732125	0.734348	0.736547	0.738724	0.740878	0.743009	0.745118	0.747204
1.6	0.749269	0.751312	0.753334	0.755335	0.757314	0.759273	0.761212	0.763130	0.765029	0.766908
1.7	0.768767	0.770607	0.772428	0.774230	0.776013	0.777778	0.779525	0.781254	0.782965	0.784658
1.8	0.786334	0.787993	0.789635	0.791260	0.792868	0.794460	0.796036	0.797596	0.799140	0.800668
1.9	0.802181	0.803679	0.805161	0.806628	0.808081	0.809519	0.810943	0.812352	0.813748	0.815129
2.0	0.816497	0.817851	0.819192	0.820519	0.821833	0.823134	0.824423	0.825698	0.826961	0.828212
2.1	0.829450	0.830677	0.831891	0.833094	0.834285	0.835464	0.836632	0.837788	0.838934	0.840068
2.2	0.841191	0.842304	0.843406	0.844497	0.845578	0.846649	0.847710	0.848760	0.849801	0.850831
2.3	0.851852	0.852864	0.853865	0.854858	0.855841	0.856815	0.857780	0.858735	0.859682	0.860621
2.4	0.861550	0.862471	0.863384	0.864288	0.865183	0.866071	0.866950	0.867822	0.868685	0.869541

(continued)

Table A.17 (continued)

	0	1	2	3	4	5	6	7	8	9
2.5	0.870389	0.871229	0.872061	0.872887	0.873704	0.874515	0.875318	0.876114	0.876902	0.877684
2.6	0.878459	0.879227	0.879988	0.880743	0.881490	0.882232	0.882966	0.883695	0.884417	0.885132
2.7	0.885842	0.886545	0.887242	0.887933	0.888618	0.889298	0.889971	0.890639	0.891301	0.891957
2.8	0.892608	0.893253	0.893893	0.894527	0.895156	0.895780	0.896398	0.897011	0.897619	0.898222
2.9	0.898820	0.899413	0.900001	0.900584	0.901163	0.901736	0.902305	0.902869	0.903429	0.903984
3.0	0.904534	0.905080	0.905622	0.906159	0.906692	0.907220	0.907745	0.908265	0.908780	0.909292
3.1	0.909800	0.910303	0.910803	0.911298	0.911790	0.912278	0.912762	0.913242	0.913718	0.914191
3.2	0.914660	0.915125	0.915586	0.916044	0.916499	0.916950	0.917397	0.917841	0.918282	0.918719
3.3	0.919153	0.919583	0.920010	0.920434	0.920855	0.921273	0.921687	0.922098	0.922507	0.922912
3.4	0.923314	0.923713	0.924109	0.924502	0.924892	0.925279	0.925664	0.926045	0.926424	0.926800
3.5	0.927173	0.927543	0.927911	0.928276	0.928639	0.928998	0.929355	0.929710	0.930062	0.930411
3.6	0.930758	0.931103	0.931445	0.931784	0.932121	0.932456	0.932788	0.933118	0.933445	0.933771
3.7	0.934094	0.934414	0.934733	0.935049	0.935363	0.935675	0.935984	0.936292	0.936597	0.936900
3.8	0.937201	0.937500	0.937797	0.938092	0.938385	0.938676	0.938965	0.939252	0.939537	0.939820
3.9	0.940101	0.940380	0.940657	0.940933	0.941206	0.941478	0.941748	0.942016	0.942282	0.942547
4.0	0.942809	0.943070	0.943330	0.943587	0.943843	0.944097	0.944350	0.944601	0.944850	0.945098
4.1	0.945343	0.945588	0.945831	0.946072	0.946311	0.946550	0.946786	0.947021	0.947255	0.947487
4.2	0.947717	0.947946	0.948174	0.948400	0.948625	0.948848	0.949070	0.949290	0.949509	0.949727
4.3	0.949943	0.950158	0.950372	0.950584	0.950795	0.951005	0.951213	0.951420	0.951626	0.951830
4.4	0.952034	0.952235	0.952436	0.952636	0.952834	0.953031	0.953227	0.953422	0.953615	0.953807
4.5	0.953998	0.954188	0.954377	0.954565	0.954752	0.954937	0.955121	0.955305	0.955487	0.955668
4.6	0.955848	0.956027	0.956205	0.956381	0.956557	0.956732	0.956905	0.957078	0.957250	0.957420
4.7	0.957590	0.957759	0.957926	0.958093	0.958259	0.958424	0.958587	0.958750	0.958912	0.959073
4.8	0.959233	0.959392	0.959551	0.959708	0.959865	0.960020	0.960175	0.960329	0.960481	0.960634
4.9	0.960785	0.960935	0.961085	0.961233	0.961381	0.961528	0.961674	0.961820	0.961964	0.962108

Table A.18 Integral of Student's function (f = 3), or probability p, as function of critical value T_{crit}

	0	1	2	3	4	5	6	7	8	9
0.0	0.000000	0.007351	0.014701	0.022049	0.029394	0.036735	0.044071	0.051401	0.058725	0.066041
0.1	0.073348	0.080645	0.087932	0.095207	0.102469	0.109718	0.116953	0.124172	0.131375	0.138562
0.2	0.145730	0.152879	0.160009	0.167118	0.174205	0.181271	0.188313	0.195332	0.202326	0.209294
0.3	0.216237	0.223152	0.230040	0.236900	0.243730	0.250531	0.257301	0.264040	0.270748	0.277423
0.4	0.284065	0.290674	0.297248	0.303788	0.310293	0.316761	0.323194	0.329590	0.335948	0.342269
0.5	0.348552	0.354796	0.361002	0.367168	0.373294	0.379380	0.385426	0.391431	0.397395	0.403318
0.6	0.409199	0.415038	0.420835	0.426590	0.432302	0.437972	0.443599	0.449182	0.454723	0.460220
0.7	0.465673	0.471083	0.476449	0.481772	0.487051	0.492286	0.497477	0.502624	0.507727	0.512786
0.8	0.517801	0.522773	0.527700	0.532584	0.537424	0.542220	0.546973	0.551682	0.556348	0.560970
0.9	0.565549	0.570085	0.574579	0.579029	0.583437	0.587802	0.592125	0.596406	0.600645	0.604842
1.0	0.608998	0.613112	0.617186	0.621218	0.625209	0.629160	0.633071	0.636942	0.640773	0.644565
1.1	0.648317	0.652030	0.655705	0.659341	0.662939	0.666499	0.670021	0.673506	0.676953	0.680364
1.2	0.683738	0.687076	0.690378	0.693644	0.696875	0.700071	0.703232	0.706358	0.709450	0.712508
1.3	0.715533	0.718524	0.721482	0.724407	0.727300	0.730161	0.732990	0.735787	0.738553	0.741289
1.4	0.743993	0.746667	0.749311	0.751925	0.754510	0.757066	0.759593	0.762091	0.764561	0.767002
1.5	0.769416	0.771803	0.774163	0.776495	0.778801	0.781081	0.783335	0.785563	0.787766	0.789943
1.6	0.792096	0.794223	0.796327	0.798406	0.800462	0.802494	0.804503	0.806488	0.808451	0.810392
1.7	0.812310	0.814206	0.816080	0.817933	0.819764	0.821575	0.823365	0.825134	0.826883	0.828611
1.8	0.830320	0.832010	0.833680	0.835331	0.836962	0.838576	0.840170	0.841747	0.843305	0.844846
1.9	0.846369	0.847874	0.849363	0.850834	0.852289	0.853727	0.855148	0.856554	0.857943	0.859316
2.0	0.860674	0.862017	0.863344	0.864656	0.865954	0.867236	0.868504	0.869758	0.870998	0.872223
2.1	0.873435	0.874633	0.875818	0.876989	0.878147	0.879292	0.880425	0.881544	0.882651	0.883746
2.2	0.884828	0.885899	0.886957	0.888004	0.889039	0.890062	0.891074	0.892075	0.893065	0.894044
2.3	0.895012	0.895969	0.896916	0.897853	0.898779	0.899695	0.900600	0.901496	0.902383	0.903259
2.4	0.904126	0.904983	0.905831	0.906670	0.907500	0.908321	0.909133	0.909936	0.910730	0.911516

(continued)

Table A.18 (continued)

	0	1	2	3	4	5	6	7	8	9
2.5	0.912294	0.913063	0.913824	0.914576	0.915321	0.916057	0.916786	0.917507	0.918221	0.918926
2.6	0.919625	0.920315	0.920999	0.921675	0.922344	0.923007	0.923662	0.924310	0.924951	0.925586
2.7	0.926214	0.926836	0.927451	0.928060	0.928662	0.929258	0.929848	0.930432	0.931010	0.931582
2.8	0.932147	0.932708	0.933262	0.933811	0.934354	0.934891	0.935423	0.935950	0.936471	0.936987
2.9	0.937498	0.938004	0.938504	0.939000	0.939490	0.939976	0.940456	0.940932	0.941403	0.941870
3.0	0.942332	0.942789	0.943241	0.943689	0.944133	0.944572	0.945007	0.945438	0.945864	0.946287
3.1	0.946705	0.947119	0.947529	0.947935	0.948337	0.948735	0.949129	0.949520	0.949906	0.950289
3.2	0.950669	0.951044	0.951416	0.951785	0.952149	0.952511	0.952869	0.953223	0.953575	0.953922
3.3	0.954267	0.954608	0.954946	0.955281	0.955613	0.955942	0.956267	0.956590	0.956909	0.957226
3.4	0.957539	0.957850	0.958157	0.958462	0.958764	0.959064	0.959360	0.959654	0.959945	0.960234
3.5	0.960519	0.960803	0.961083	0.961361	0.961637	0.961910	0.962180	0.962448	0.962714	0.962977
3.6	0.963238	0.963497	0.963753	0.964007	0.964259	0.964508	0.964755	0.965000	0.965243	0.965484
3.7	0.965722	0.965959	0.966193	0.966426	0.966656	0.966884	0.967110	0.967335	0.967557	0.967777
3.8	0.967996	0.968212	0.968427	0.968640	0.968851	0.969060	0.969268	0.969473	0.969677	0.969879
3.9	0.970079	0.970278	0.970475	0.970670	0.970864	0.971056	0.971246	0.971435	0.971622	0.971808
4.0	0.971992	0.972174	0.972355	0.972535	0.972713	0.972889	0.973064	0.973238	0.973410	0.973581
4.1	0.973750	0.973918	0.974084	0.974250	0.974413	0.974576	0.974737	0.974897	0.975055	0.975212
4.2	0.975368	0.975523	0.975676	0.975829	0.975980	0.976129	0.976278	0.976425	0.976571	0.976716
4.3	0.976860	0.977003	0.977144	0.977285	0.977424	0.977562	0.977699	0.977835	0.977970	0.978104
4.4	0.978237	0.978369	0.978500	0.978630	0.978758	0.978886	0.979013	0.979139	0.979263	0.979387
4.5	0.979510	0.979632	0.979753	0.979873	0.979992	0.980110	0.980228	0.980344	0.980460	0.980574
4.6	0.980688	0.980801	0.980913	0.981024	0.981135	0.981244	0.981353	0.981461	0.981568	0.981674
4.7	0.981780	0.981884	0.981988	0.982091	0.982194	0.982295	0.982396	0.982496	0.982596	0.982694
4.8	0.982792	0.982890	0.982986	0.983082	0.983177	0.983271	0.983365	0.983458	0.983550	0.983642
4.9	0.983733	0.983823	0.983913	0.984002	0.984091	0.984178	0.984266	0.984352	0.984438	0.984523

Table A.19 Integral of Student's function (f = 5), or probability p, as function of critical value T_{crit}

	0	1	2	3	4	5	6	7	8	9
0.0	0.000000	0.007592	0.015183	0.022772	0.030359	0.037942	0.045520	0.053093	0.060659	0.068219
0.1	0.075770	0.083312	0.090844	0.098366	0.105875	0.113372	0.120856	0.128325	0.135780	0.143218
0.2	0.150640	0.158043	0.165429	0.172795	0.180141	0.187466	0.194769	0.202050	0.209308	0.216542
0.3	0.223751	0.230935	0.238092	0.245223	0.252326	0.259401	0.266446	0.273463	0.280448	0.287403
0.4	0.294327	0.301218	0.308077	0.314902	0.321694	0.328451	0.335173	0.341860	0.348510	0.355124
0.5	0.361701	0.368241	0.374742	0.381206	0.387630	0.394015	0.400361	0.406667	0.412932	0.419156
0.6	0.425340	0.431482	0.437582	0.443641	0.449657	0.455630	0.461561	0.467449	0.473293	0.479094
0.7	0.484851	0.490565	0.496234	0.501859	0.507440	0.512976	0.518468	0.523915	0.529317	0.534674
0.8	0.539986	0.545253	0.550476	0.555653	0.560785	0.565871	0.570913	0.575910	0.580861	0.585768
0.9	0.590629	0.595445	0.600217	0.604944	0.609626	0.614263	0.618855	0.623404	0.627908	0.632367
1.0	0.636783	0.641154	0.645482	0.649767	0.654007	0.658205	0.662359	0.666471	0.670539	0.674566
1.1	0.678549	0.682491	0.686391	0.690249	0.694066	0.697841	0.701576	0.705270	0.708923	0.712536
1.2	0.716109	0.719643	0.723136	0.726591	0.730007	0.733384	0.736723	0.740023	0.743286	0.746512
1.3	0.749700	0.752851	0.755965	0.759043	0.762085	0.765092	0.768062	0.770998	0.773899	0.776765
1.4	0.779596	0.782394	0.785158	0.787889	0.790587	0.793252	0.795885	0.798485	0.801054	0.803591
1.5	0.806097	0.808572	0.811016	0.813430	0.815814	0.818168	0.820493	0.822789	0.825056	0.827295
1.6	0.829505	0.831688	0.833843	0.835970	0.838071	0.840145	0.842192	0.844213	0.846209	0.848179
1.7	0.850124	0.852043	0.853938	0.855809	0.857655	0.859478	0.861277	0.863053	0.864805	0.866535
1.8	0.868243	0.869928	0.871591	0.873233	0.874853	0.876452	0.878030	0.879587	0.881124	0.882640
1.9	0.884137	0.885614	0.887072	0.888510	0.889930	0.891330	0.892712	0.894076	0.895422	0.896750
2.0	0.898061	0.899354	0.900630	0.901889	0.903132	0.904358	0.905567	0.906761	0.907938	0.909101
2.1	0.910247	0.911378	0.912495	0.913596	0.914683	0.915755	0.916813	0.917857	0.918887	0.919904
2.2	0.920906	0.921896	0.922872	0.923835	0.924786	0.925723	0.926649	0.927562	0.928462	0.929351
2.3	0.930228	0.931093	0.931947	0.932789	0.933620	0.934440	0.935249	0.936048	0.936835	0.937613
2.4	0.938380	0.939136	0.939883	0.940620	0.941347	0.942064	0.942772	0.943470	0.944159	0.944839

(continued)

Table A.19 (continued)

	0	1	2	3	4	5	6	7	8	9
2.5	0.945510	0.946172	0.946826	0.947470	0.948107	0.948734	0.949354	0.949965	0.950568	0.951164
2.6	0.951751	0.952331	0.952903	0.953467	0.954024	0.954574	0.955116	0.955652	0.956180	0.956702
2.7	0.957216	0.957724	0.958225	0.958720	0.959208	0.959690	0.960166	0.960635	0.961098	0.961556
2.8	0.962007	0.962452	0.962892	0.963326	0.963754	0.964177	0.964594	0.965006	0.965412	0.965814
2.9	0.966210	0.966601	0.966987	0.967368	0.967744	0.968115	0.968482	0.968843	0.969200	0.969553
3.0	0.969901	0.970245	0.970584	0.970919	0.971250	0.971576	0.971898	0.972217	0.972531	0.972841
3.1	0.973147	0.973450	0.973748	0.974043	0.974334	0.974621	0.974905	0.975185	0.975462	0.975735
3.2	0.976005	0.976272	0.976535	0.976795	0.977051	0.977305	0.977555	0.977802	0.978046	0.978287
3.3	0.978525	0.978760	0.978992	0.979221	0.979448	0.979672	0.979893	0.980111	0.980326	0.980539
3.4	0.980749	0.980957	0.981162	0.981365	0.981565	0.981763	0.981958	0.982151	0.982342	0.982530
3.5	0.982716	0.982900	0.983081	0.983261	0.983438	0.983613	0.983786	0.983957	0.984126	0.984292
3.6	0.984457	0.984620	0.984781	0.984940	0.985097	0.985252	0.985406	0.985557	0.985707	0.985855
3.7	0.986001	0.986146	0.986288	0.986429	0.986569	0.986707	0.986843	0.986977	0.987111	0.987242
3.8	0.987372	0.987500	0.987627	0.987753	0.987877	0.987999	0.988120	0.988240	0.988359	0.988475
3.9	0.988591	0.988705	0.988818	0.988930	0.989041	0.989150	0.989258	0.989364	0.989470	0.989574
4.0	0.989677	0.989779	0.989880	0.989979	0.990078	0.990175	0.990271	0.990366	0.990461	0.990554
4.1	0.990646	0.990737	0.990826	0.990915	0.991003	0.991090	0.991176	0.991261	0.991345	0.991429
4.2	0.991511	0.991592	0.991673	0.991752	0.991831	0.991909	0.991986	0.992062	0.992137	0.992211
4.3	0.992285	0.992358	0.992430	0.992501	0.992572	0.992641	0.992710	0.992778	0.992846	0.992913
4.4	0.992979	0.993044	0.993108	0.993172	0.993236	0.993298	0.993360	0.993421	0.993482	0.993542
4.5	0.993601	0.993660	0.993718	0.993775	0.993832	0.993888	0.993944	0.993999	0.994053	0.994107
4.6	0.994160	0.994213	0.994265	0.994317	0.994368	0.994418	0.994468	0.994518	0.994567	0.994615
4.7	0.994663	0.994711	0.994758	0.994804	0.994850	0.994896	0.994941	0.994986	0.995030	0.995073
4.8	0.995117	0.995160	0.995202	0.995244	0.995285	0.995327	0.995367	0.995408	0.995447	0.995487
4.9	0.995526	0.995565	0.995603	0.995641	0.995678	0.995715	0.995752	0.995789	0.995825	0.995860

Table A.20 Integral of Student's function (f = 10), or probability p, as function of critical value T_{crit}

	0	1	2	3	4	5	6	7	8	9
0.0	0.000000	0.007782	0.015563	0.023343	0.031120	0.038893	0.046662	0.054426	0.062184	0.069936
0.1	0.077679	0.085414	0.093140	0.100856	0.108560	0.116253	0.123933	0.131600	0.139252	0.146889
0.2	0.154511	0.162116	0.169703	0.177272	0.184822	0.192352	0.199861	0.207350	0.214816	0.222260
0.3	0.229679	0.237075	0.244446	0.251791	0.259110	0.266402	0.273666	0.280902	0.288109	0.295286
0.4	0.302433	0.309549	0.316633	0.323686	0.330706	0.337692	0.344645	0.351563	0.358447	0.365295
0.5	0.372107	0.378882	0.385621	0.392322	0.398985	0.405610	0.412197	0.418744	0.425251	0.431718
0.6	0.438145	0.444531	0.450876	0.457179	0.463441	0.469660	0.475837	0.481971	0.488061	0.494109
0.7	0.500113	0.506072	0.511988	0.517860	0.523686	0.529468	0.535205	0.540897	0.546543	0.552144
0.8	0.557700	0.563209	0.568673	0.574091	0.579463	0.584788	0.590067	0.595300	0.600487	0.605627
0.9	0.610721	0.615768	0.620769	0.625724	0.630632	0.635494	0.640309	0.645078	0.649800	0.654477
1.0	0.659107	0.663691	0.668230	0.672722	0.677169	0.681569	0.685925	0.690235	0.694499	0.698719
1.1	0.702893	0.707023	0.711108	0.715149	0.719145	0.723097	0.727006	0.730870	0.734691	0.738469
1.2	0.742204	0.745896	0.749545	0.753152	0.756717	0.760240	0.763721	0.767161	0.770559	0.773917
1.3	0.777235	0.780511	0.783748	0.786945	0.790103	0.793221	0.796301	0.799341	0.802344	0.805308
1.4	0.808235	0.811124	0.813976	0.816792	0.819570	0.822313	0.825019	0.827690	0.830326	0.832927
1.5	0.835493	0.838025	0.840523	0.842987	0.845417	0.847815	0.850180	0.852512	0.854813	0.857081
1.6	0.859319	0.861525	0.863700	0.865845	0.867959	0.870044	0.872099	0.874125	0.876122	0.878091
1.7	0.880031	0.881943	0.883828	0.885685	0.887516	0.889319	0.891097	0.892848	0.894573	0.896273
1.8	0.897948	0.899598	0.901223	0.902825	0.904402	0.905955	0.907485	0.908992	0.910477	0.911938
1.9	0.913378	0.914796	0.916191	0.917566	0.918919	0.920252	0.921564	0.922856	0.924128	0.925380
2.0	0.926612	0.927826	0.929020	0.930196	0.931353	0.932492	0.933613	0.934717	0.935803	0.936871
2.1	0.937923	0.938958	0.939977	0.940979	0.941965	0.942936	0.943891	0.944830	0.945755	0.946664
2.2	0.947559	0.948440	0.949306	0.950158	0.950996	0.951821	0.952632	0.953430	0.954215	0.954987
2.3	0.955746	0.956493	0.957228	0.957950	0.958661	0.959360	0.960047	0.960723	0.961388	0.962042
2.4	0.962685	0.963317	0.963939	0.964550	0.965151	0.965743	0.966324	0.966896	0.967458	0.968010

(continued)

Table A.20 (continued)

	0	1	2	3	4	5	6	7	8	9
2.5	0.968554	0.969088	0.969613	0.970130	0.970637	0.971136	0.971627	0.972110	0.972584	0.973050
2.6	0.973509	0.973960	0.974403	0.974838	0.975266	0.975687	0.976101	0.976508	0.976908	0.977301
2.7	0.977687	0.978067	0.978440	0.978807	0.979168	0.979522	0.979871	0.980213	0.980550	0.980881
2.8	0.981206	0.981526	0.981840	0.982149	0.982452	0.982750	0.983044	0.983332	0.983615	0.983893
2.9	0.984167	0.984436	0.984700	0.984960	0.985215	0.985466	0.985712	0.985955	0.986193	0.986427
3.0	0.986657	0.986883	0.987105	0.987323	0.987538	0.987749	0.987956	0.988160	0.988360	0.988556
3.1	0.988750	0.988940	0.989126	0.989310	0.989490	0.989667	0.989842	0.990013	0.990181	0.990346
3.2	0.990509	0.990668	0.990825	0.990979	0.991131	0.991280	0.991426	0.991570	0.991711	0.991850
3.3	0.991987	0.992121	0.992253	0.992383	0.992510	0.992635	0.992758	0.992879	0.992998	0.993115
3.4	0.993229	0.993342	0.993453	0.993562	0.993669	0.993774	0.993878	0.993979	0.994079	0.994177
3.5	0.994274	0.994369	0.994462	0.994554	0.994644	0.994732	0.994819	0.994905	0.994989	0.995071
3.6	0.995153	0.995232	0.995311	0.995388	0.995464	0.995538	0.995611	0.995683	0.995754	0.995824
3.7	0.995892	0.995959	0.996025	0.996090	0.996154	0.996217	0.996278	0.996339	0.996398	0.996457
3.8	0.996515	0.996571	0.996627	0.996682	0.996735	0.996788	0.996840	0.996891	0.996941	0.996991
3.9	0.997039	0.997087	0.997134	0.997180	0.997226	0.997270	0.997314	0.997357	0.997399	0.997441
4.0	0.997482	0.997522	0.997562	0.997601	0.997639	0.997677	0.997714	0.997750	0.997786	0.997821
4.1	0.997856	0.997890	0.997923	0.997956	0.997989	0.998020	0.998052	0.998082	0.998113	0.998142
4.2	0.998172	0.998201	0.998229	0.998257	0.998284	0.998311	0.998337	0.998363	0.998389	0.998414
4.3	0.998439	0.998463	0.998487	0.998511	0.998534	0.998557	0.998579	0.998601	0.998623	0.998644
4.4	0.998665	0.998686	0.998706	0.998726	0.998746	0.998765	0.998784	0.998803	0.998821	0.998840
4.5	0.998857	0.998875	0.998892	0.998909	0.998926	0.998942	0.998958	0.998974	0.998990	0.999005
4.6	0.999020	0.999035	0.999050	0.999064	0.999078	0.999092	0.999106	0.999120	0.999133	0.999146
4.7	0.999159	0.999172	0.999184	0.999196	0.999208	0.999220	0.999232	0.999243	0.999255	0.999266
4.8	0.999277	0.999288	0.999298	0.999309	0.999319	0.999329	0.999339	0.999349	0.999358	0.999368
4.9	0.999377	0.999387	0.999396	0.999404	0.999413	0.999422	0.999430	0.999439	0.999447	0.999455

Table A.21 Integral of Student's function (f = 20), or probability p, as function of critical value T_{crit}

	0	1	2	3	4	5	6	7	8	9
0.0	0.000000	0.007880	0.015758	0.023636	0.031510	0.039382	0.047249	0.055111	0.062968	0.070818
0.1	0.078660	0.086494	0.094320	0.102135	0.109940	0.117733	0.125514	0.133282	0.141036	0.148776
0.2	0.156500	0.164208	0.171899	0.179572	0.187227	0.194863	0.202478	0.210073	0.217647	0.225199
0.3	0.232727	0.240232	0.247713	0.255169	0.262599	0.270003	0.277379	0.284728	0.292049	0.299341
0.4	0.306604	0.313836	0.321037	0.328207	0.335345	0.342450	0.349522	0.356561	0.363565	0.370534
0.5	0.377468	0.384366	0.391228	0.398053	0.404841	0.411591	0.418302	0.424975	0.431608	0.438202
0.6	0.444756	0.451269	0.457742	0.464174	0.470563	0.476911	0.483217	0.489480	0.495700	0.501877
0.7	0.508010	0.514099	0.520145	0.526146	0.532102	0.538013	0.543879	0.549700	0.555476	0.561206
0.8	0.566889	0.572527	0.578119	0.583664	0.589162	0.594614	0.600019	0.605378	0.610689	0.615953
0.9	0.621171	0.626341	0.631463	0.636539	0.641567	0.646548	0.651482	0.656368	0.661207	0.665999
1.0	0.670744	0.675441	0.680091	0.684694	0.689250	0.693759	0.698222	0.702637	0.707006	0.711328
1.1	0.715604	0.719833	0.724016	0.728154	0.732245	0.736291	0.740291	0.744245	0.748155	0.752019
1.2	0.755839	0.759614	0.763344	0.767031	0.770673	0.774271	0.777826	0.781338	0.784807	0.788233
1.3	0.791616	0.794957	0.798256	0.801513	0.804728	0.807903	0.811036	0.814129	0.817181	0.820193
1.4	0.823165	0.826098	0.828992	0.831846	0.834662	0.837440	0.840180	0.842882	0.845546	0.848174
1.5	0.850765	0.853319	0.855837	0.858320	0.860767	0.863179	0.865556	0.867899	0.870207	0.872482
1.6	0.874723	0.876932	0.879107	0.881250	0.883361	0.885440	0.887487	0.889503	0.891489	0.893444
1.7	0.895369	0.897264	0.899130	0.900967	0.902775	0.904554	0.906305	0.908029	0.909725	0.911394
1.8	0.913036	0.914651	0.916241	0.917804	0.919342	0.920855	0.922343	0.923806	0.925245	0.926660
1.9	0.928052	0.929420	0.930765	0.932088	0.933388	0.934666	0.935922	0.937157	0.938370	0.939563
2.0	0.940735	0.941886	0.943018	0.944130	0.945222	0.946295	0.947350	0.948385	0.949402	0.950402
2.1	0.951383	0.952347	0.953293	0.954222	0.955135	0.956031	0.956911	0.957774	0.958622	0.959455
2.2	0.960272	0.961074	0.961861	0.962634	0.963392	0.964136	0.964866	0.965583	0.966286	0.966976
2.3	0.967653	0.968317	0.968969	0.969608	0.970235	0.970850	0.971453	0.972044	0.972624	0.973194
2.4	0.973752	0.974299	0.974835	0.975361	0.975877	0.976383	0.976879	0.977365	0.977842	0.978309

(continued)

Table A.21 (continued)

	0	1	2	3	4	5	6	7	8	9
2.5	0.978767	0.979216	0.979656	0.980087	0.980510	0.980924	0.981330	0.981727	0.982117	0.982499
2.6	0.982873	0.983240	0.983599	0.983951	0.984296	0.984634	0.984965	0.985289	0.985607	0.985918
2.7	0.986222	0.986521	0.986813	0.987099	0.987380	0.987654	0.987923	0.988186	0.988444	0.988696
2.8	0.988943	0.989185	0.989422	0.989654	0.989881	0.990103	0.990321	0.990534	0.990743	0.990947
2.9	0.991146	0.991342	0.991533	0.991721	0.991904	0.992083	0.992259	0.992431	0.992599	0.992764
3.0	0.992925	0.993082	0.993236	0.993387	0.993535	0.993679	0.993820	0.993959	0.994094	0.994226
3.1	0.994356	0.994482	0.994606	0.994727	0.994846	0.994962	0.995075	0.995186	0.995294	0.995400
3.2	0.995504	0.995606	0.995705	0.995802	0.995897	0.995990	0.996081	0.996169	0.996256	0.996341
3.3	0.996424	0.996505	0.996585	0.996662	0.996738	0.996812	0.996885	0.996956	0.997025	0.997093
3.4	0.997159	0.997224	0.997287	0.997349	0.997410	0.997469	0.997527	0.997583	0.997638	0.997692
3.5	0.997745	0.997797	0.997847	0.997897	0.997945	0.997992	0.998038	0.998083	0.998127	0.998170
3.6	0.998212	0.998253	0.998293	0.998333	0.998371	0.998408	0.998445	0.998481	0.998516	0.998550
3.7	0.998583	0.998616	0.998648	0.998679	0.998709	0.998739	0.998768	0.998797	0.998824	0.998851
3.8	0.998878	0.998904	0.998929	0.998954	0.998978	0.999002	0.999025	0.999047	0.999069	0.999091
3.9	0.999112	0.999132	0.999152	0.999172	0.999191	0.999210	0.999228	0.999246	0.999263	0.999280
4.0	0.999297	0.999313	0.999329	0.999345	0.999360	0.999375	0.999389	0.999403	0.999417	0.999430
4.1	0.999444	0.999456	0.999469	0.999481	0.999493	0.999505	0.999516	0.999528	0.999539	0.999549
4.2	0.999560	0.999570	0.999580	0.999590	0.999599	0.999608	0.999617	0.999626	0.999635	0.999643
4.3	0.999652	0.999660	0.999667	0.999675	0.999683	0.999690	0.999697	0.999704	0.999711	0.999718
4.4	0.999724	0.999731	0.999737	0.999743	0.999749	0.999755	0.999760	0.999766	0.999771	0.999776
4.5	0.999782	0.999787	0.999792	0.999796	0.999801	0.999806	0.999810	0.999815	0.999819	0.999823
4.6	0.999827	0.999831	0.999835	0.999839	0.999842	0.999846	0.999850	0.999853	0.999856	0.999860
4.7	0.999863	0.999866	0.999869	0.999872	0.999875	0.999878	0.999881	0.999884	0.999886	0.999889
4.8	0.999891	0.999894	0.999896	0.999899	0.999901	0.999903	0.999906	0.999908	0.999910	0.999912
4.9	0.999914	0.999916	0.999918	0.999920	0.999922	0.999923	0.999925	0.999927	0.999928	0.999930

Table A.22 Integral of Student's function (f = 50), or probability p, as function of critical value T_{crit}

	0	1	2	3	4	5	6	7	8	9
0.0	0.000000	0.007939	0.015877	0.023814	0.031748	0.039678	0.047605	0.055527	0.063443	0.071353
0.1	0.079256	0.087150	0.095036	0.102912	0.110778	0.118632	0.126474	0.134304	0.142120	0.149922
0.2	0.157708	0.165479	0.173233	0.180970	0.188689	0.196388	0.204069	0.211729	0.219367	0.226985
0.3	0.234579	0.242151	0.249698	0.257221	0.264719	0.272191	0.279637	0.287055	0.294445	0.301807
0.4	0.309140	0.316443	0.323715	0.330957	0.338167	0.345345	0.352490	0.359602	0.366680	0.373723
0.5	0.380732	0.387705	0.394642	0.401542	0.408406	0.415232	0.422020	0.428770	0.435481	0.442152
0.6	0.448784	0.455376	0.461927	0.468437	0.474906	0.481333	0.487717	0.494060	0.500359	0.506616
0.7	0.512829	0.518998	0.525123	0.531204	0.537240	0.543231	0.549177	0.555077	0.560932	0.566742
0.8	0.572505	0.578222	0.583892	0.589516	0.595093	0.600623	0.606106	0.611542	0.616931	0.622272
0.9	0.627566	0.632812	0.638010	0.643161	0.648263	0.653318	0.658326	0.663285	0.668196	0.673059
1.0	0.677875	0.682642	0.687362	0.692033	0.696657	0.701233	0.705762	0.710243	0.714676	0.719062
1.1	0.723400	0.727691	0.731935	0.736132	0.740282	0.744386	0.748442	0.752452	0.756416	0.760334
1.2	0.764206	0.768032	0.771812	0.775547	0.779237	0.782882	0.786482	0.790037	0.793548	0.797015
1.3	0.800438	0.803818	0.807154	0.810447	0.813697	0.816904	0.820070	0.823193	0.826274	0.829314
1.4	0.832312	0.835270	0.838187	0.841064	0.843901	0.846698	0.849455	0.852174	0.854853	0.857495
1.5	0.860098	0.862663	0.865190	0.867681	0.870135	0.872552	0.874933	0.877278	0.879587	0.881862
1.6	0.884101	0.886306	0.888477	0.890614	0.892718	0.894788	0.896826	0.898831	0.900805	0.902746
1.7	0.904656	0.906535	0.908383	0.910201	0.911989	0.913747	0.915476	0.917176	0.918847	0.920490
1.8	0.922105	0.923693	0.925253	0.926786	0.928293	0.929773	0.931227	0.932656	0.934060	0.935439
1.9	0.936793	0.938123	0.939429	0.940711	0.941970	0.943206	0.944420	0.945611	0.946780	0.947927
2.0	0.949053	0.950158	0.951243	0.952306	0.953350	0.954374	0.955378	0.956363	0.957329	0.958276
2.1	0.959205	0.960116	0.961009	0.961884	0.962742	0.963584	0.964408	0.965216	0.966008	0.966784
2.2	0.967544	0.968289	0.969019	0.969734	0.970434	0.971120	0.971791	0.972449	0.973093	0.973724
2.3	0.974341	0.974946	0.975538	0.976117	0.976684	0.977239	0.977782	0.978313	0.978833	0.979342
2.4	0.979840	0.980327	0.980803	0.981269	0.981725	0.982171	0.982607	0.983033	0.983450	0.983857

(continued)

Table A.22 (continued)

	0	1	2	3	4	5	6	7	8	9
2.5	0.984255	0.984645	0.985026	0.985398	0.985761	0.986117	0.986464	0.986804	0.987135	0.987459
2.6	0.987776	0.988085	0.988387	0.988683	0.988971	0.989252	0.989527	0.989796	0.990058	0.990314
2.7	0.990564	0.990807	0.991046	0.991278	0.991505	0.991726	0.991942	0.992153	0.992359	0.992560
2.8	0.992756	0.992947	0.993133	0.993315	0.993493	0.993666	0.993835	0.993999	0.994160	0.994316
2.9	0.994469	0.994618	0.994763	0.994904	0.995042	0.995176	0.995307	0.995435	0.995559	0.995681
3.0	0.995799	0.995914	0.996026	0.996135	0.996242	0.996345	0.996447	0.996545	0.996641	0.996734
3.1	0.996825	0.996914	0.997000	0.997084	0.997165	0.997245	0.997323	0.997398	0.997471	0.997543
3.2	0.997612	0.997680	0.997746	0.997810	0.997872	0.997933	0.997992	0.998050	0.998106	0.998160
3.3	0.998213	0.998264	0.998314	0.998363	0.998411	0.998457	0.998501	0.998545	0.998587	0.998628
3.4	0.998669	0.998707	0.998745	0.998782	0.998818	0.998853	0.998886	0.998919	0.998951	0.998982
3.5	0.999012	0.999042	0.999070	0.999098	0.999125	0.999151	0.999176	0.999201	0.999225	0.999248
3.6	0.999270	0.999292	0.999314	0.999334	0.999354	0.999374	0.999393	0.999411	0.999429	0.999447
3.7	0.999463	0.999480	0.999496	0.999511	0.999526	0.999540	0.999554	0.999568	0.999581	0.999594
3.8	0.999607	0.999619	0.999631	0.999642	0.999653	0.999664	0.999674	0.999684	0.999694	0.999704
3.9	0.999713	0.999722	0.999731	0.999739	0.999747	0.999755	0.999763	0.999770	0.999777	0.999784
4.0	0.999791	0.999798	0.999804	0.999810	0.999816	0.999822	0.999828	0.999833	0.999839	0.999844
4.1	0.999849	0.999854	0.999858	0.999863	0.999867	0.999871	0.999875	0.999879	0.999883	0.999887
4.2	0.999891	0.999894	0.999898	0.999901	0.999904	0.999907	0.999910	0.999913	0.999916	0.999919
4.3	0.999921	0.999924	0.999926	0.999929	0.999931	0.999933	0.999936	0.999938	0.999940	0.999942
4.4	0.999944	0.999945	0.999947	0.999949	0.999951	0.999952	0.999954	0.999955	0.999957	0.999958
4.5	0.999960	0.999961	0.999962	0.999964	0.999965	0.999966	0.999967	0.999968	0.999969	0.999970
4.6	0.999971	0.999972	0.999973	0.999974	0.999975	0.999976	0.999977	0.999977	0.999978	0.999979
4.7	0.999980	0.999980	0.999981	0.999982	0.999982	0.999983	0.999983	0.999984	0.999985	0.999985
4.8	0.999986	0.999986	0.999987	0.999987	0.999988	0.999988	0.999988	0.999989	0.999989	0.999990
4.9	0.999990	0.999990	0.999991	0.999991	0.999991	0.999992	0.999992	0.999992	0.999992	0.999993

Table A.23 Comparison of integrals of Student's function at different critical values.

f	Critical value				
	$T = 1$	$T = 2$	$T = 3$	$T = 4$	$T = 5$
1	0.500000	0.704833	0.795168	0.844042	0.874334
2	0.577351	0.816497	0.904534	0.942809	0.962251
3	0.608998	0.860674	0.942332	0.971992	0.984608
4	0.626099	0.883884	0.960058	0.983870	0.992510
5	0.636783	0.898061	0.969901	0.989677	0.995896
6	0.644083	0.907574	0.975992	0.992881	0.997548
7	0.649384	0.914381	0.980058	0.994811	0.998435
8	0.653407	0.919484	0.982929	0.996051	0.998948
9	0.656564	0.923448	0.985044	0.996890	0.999261
10	0.659107	0.926612	0.986657	0.997482	0.999463
11	0.661200	0.929196	0.987921	0.997914	0.999598
12	0.662951	0.931345	0.988934	0.998239	0.999691
13	0.664439	0.933160	0.989762	0.998488	0.999757
14	0.665718	0.934712	.990449	0.998684	0.999806
15	0.666830	0.936055	0.991028	0.998841	0.999842
16	0.667805	0.937228	0.991521	0.998968	0.999870
17	0.668668	0.938262	0.991946	0.999073	0.999891
18	0.669435	0.939179	0.992315	0.999161	0.999908
19	0.670123	0.939998	0.992639	0.999234	0.999921
20	0.670744	0.940735	0.992925	0.999297	0.999932
21	0.671306	0.941400	0.993179	0.999351	0.999940
22	0.671817	0.942005	0.993406	0.999397	0.999948
23	0.672284	0.942556	0.993610	0.999438	0.999954
24	0.672713	0.943061	0.993795	0.999474	0.999959
25	0.673108	0.943524	0.993962	0.999505	0.999963
26	0.673473	0.943952	0.994115	0.999533	0.999967
27	0.673811	0.944348	0.994255	0.999558	0.999970
28	0.674126	0.944715	0.994383	0.999580	0.999973
29	0.674418	0.945057	0.994501	0.999600	0.999975
30	0.674692	0.945375	0.994610	0.999619	0.999977
31	0.674948	0.945673	0.994712	0.999635	0.999979
32	0.675188	0.945952	0.994806	0.999650	0.999981
33	0.675413	0.946214	0.994893	0.999664	0.999982
34	0.675626	0.946461	0.994975	0.999677	0.999983
35	0.675826	0.946693	0.995052	0.999688	0.999984
36	0.676015	0.946912	0.995123	0.999699	0.999985
37	0.676194	0.947119	0.995191	0.999709	0.999986
38	0.676364	0.947315	0.995254	0.999718	0.999987
39	0.676525	0.947501	0.995314	0.999727	0.999988
40	0.676678	0.947678	0.995370	0.999735	0.999989

(continued)

Table A.23 (continued)

f	Critical value				
	$T = 1$	$T = 2$	$T = 3$	$T = 4$	$T = 5$
41	0.676824	0.947846	0.995424	0.999742	0.999989
42	0.676963	0.948006	0.995474	0.999749	0.999990
43	0.677095	0.948158	0.995522	0.999755	0.999990
44	0.677222	0.948304	0.995568	0.999761	0.999991
45	0.677343	0.948443	0.995611	0.999767	0.999991
46	0.677458	0.948576	0.995652	0.999773	0.999992
47	0.677569	0.948703	0.995691	0.999778	0.999992
48	0.677675	0.948824	0.995729	0.999782	0.999992
49	0.677777	0.948941	0.995765	0.999787	0.999993
50	0.677875	0.949053	0.995799	0.999791	0.999993
∞	0.682690	0.954500	0.997301	0.999937	1.000000

Table A.24 Critical values of the linear correlation coefficient

f	Probability p to have an absolute value of r below the critical value						
	0.50	0.60	0.70	0.80	0.90	0.95	0.99
2	0.500	0.600	0.700	0.800	0.900	0.950	0.990
3	0.404	0.492	0.585	0.687	0.805	0.878	0.959
4	0.347	0.426	0.511	0.608	0.729	0.811	0.917
5	0.309	0.380	0.459	0.551	0.669	0.754	0.875
6	0.281	0.347	0.420	0.507	0.621	0.707	0.834
7	0.260	0.321	0.390	0.472	0.582	0.666	0.798
8	0.242	0.300	0.365	0.443	0.549	0.632	0.765
9	0.228	0.282	0.344	0.419	0.521	0.602	0.735
10	0.216	0.268	0.327	0.398	0.497	0.576	0.708
20	0.152	0.189	0.231	0.284	0.360	0.423	0.537
30	0.124	0.154	0.189	0.233	0.296	0.349	0.449
40	0.107	0.133	0.164	0.202	0.257	0.304	0.393
50	0.096	0.119	0.147	0.181	0.231	0.273	0.354
60	0.087	0.109	0.134	0.165	0.211	0.250	0.325
70	0.081	0.101	0.124	0.153	0.195	0.232	0.302
80	0.076	0.094	0.116	0.143	0.183	0.217	0.283
90	0.071	0.089	0.109	0.135	0.173	0.205	0.267
100	0.068	0.084	0.104	0.128	0.164	0.195	0.254
200	0.048	0.060	0.073	0.091	0.116	0.138	0.181
300	0.039	0.049	0.060	0.074	0.095	0.113	0.148
500	0.030	0.038	0.046	0.057	0.073	0.087	0.114
1000	0.021	0.027	0.033	0.041	0.052	0.062	0.081

Table A.25 Critical values of the Kolmogorov–Smirnov statistic D_N

N	Probability p to have $D_N \times \sqrt{N}$ below the critical value						
	0.50	0.60	0.70	0.80	0.90	0.95	0.99
1	0.750	0.800	0.850	0.900	0.950	0.975	0.995
2	0.707	0.782	0.866	0.967	1.098	1.191	1.314
3	0.753	0.819	0.891	0.978	1.102	1.226	1.436
4	0.762	0.824	0.894	0.985	1.130	1.248	1.468
5	0.765	0.827	0.902	0.999	1.139	1.260	1.495
6	0.767	0.833	0.910	1.005	1.146	1.272	1.510
7	0.772	0.838	0.914	1.009	1.154	1.279	1.523
8	0.776	0.842	0.917	1.013	1.159	1.285	1.532
9	0.779	0.844	0.920	1.017	1.162	1.290	1.540
10	0.781	0.846	0.923	1.020	1.166	1.294	1.546
15	0.788	0.855	0.932	1.030	1.177	1.308	1.565
20	0.793	0.860	0.937	1.035	1.184	1.315	1.576
25	0.796	0.863	0.941	1.039	1.188	1.320	1.583
30	0.799	0.866	0.943	1.042	1.192	1.324	1.588
35	0.801	0.868	0.946	1.045	1.194	1.327	1.591
40	0.803	0.869	0.947	1.046	1.196	1.329	1.594
45	0.804	0.871	0.949	1.048	1.198	1.331	1.596
50	0.805	0.872	0.950	1.049	1.199	1.332	1.598
60	0.807	0.874	0.952	1.051	1.201	1.335	1.601
70	0.808	0.875	0.953	1.053	1.203	1.337	1.604
80	0.810	0.877	0.955	1.054	1.205	1.338	1.605
90	0.811	0.878	0.956	1.055	1.206	1.339	1.607
100	0.811	0.879	0.957	1.056	1.207	1.340	1.608
200	0.816	0.883	0.961	1.061	1.212	1.346	1.614
300	0.818	0.885	0.964	1.063	1.214	1.348	1.617
500	0.820	0.887	0.966	1.065	1.216	1.350	1.620
1000	0.822	0.890	0.968	1.067	1.218	1.353	1.622
∞	0.828	0.895	0.973	1.073	1.224	1.358	1.628

References

1. Abramowitz, M., Stegun, I.A.: Handbook of Mathematical Functions: With Formulas, Graphs, and Mathematical Tables. Dover, New York (1970)
2. Akritas, M.G., Bershady, M.A.: Linear regression for astronomical data with measurement errors and intrinsic scatter. Astrophys. J. **470**, 706 (1996). doi:10.1086/177901
3. Bayes, T., Price, R.: An essay towards solving a problem in the doctrine of chances. Philos. Trans. R. Soc. Lond. **53** (1763)
4. Bonamente, M., Swartz, D.A., Weisskopf, M.C., Murray, S.S.: Swift XRT observations of the possible dark galaxy VIRGOHI 21. Astrophys. J. Lett. **686**, L71–L74 (2008). doi:10.1086/592819
5. Bonamente, M., Hasler, N., Bulbul, E., Carlstrom, J.E., Culverhouse, T.L., Gralla, M., Greer, C., Hawkins, D., Hennessy, R., Joy, M., Kolodziejczak, J., Lamb, J.W., Landry, D., Leitch, E.M., Marrone, D.P., Miller, A., Mroczkowski, T., Muchovej, S., Plagge, T., Pryke, C., Sharp, M., Woody, D.: Comparison of pressure profiles of massive relaxed galaxy clusters using the Sunyaev–Zel'dovich and X-ray data. New. J. Phys. **14**(2), 025010 (2012). doi:10.1088/1367-2630/14/2/025010
6. Brooks, S.P., Gelman, A.: General methods for monitoring convergence of iterative simulations. J. Comput. Graph. Stat. **7**, 434–455 (1998)
7. Bulmer, M.G.: Principles of Statistics. Dover, New York (1967)
8. Carlin, B., Gelfand, A., Smith, A.: Hierarchical Bayesian analysis for changepoint problems. Appl. Stat. **41**, 389–405 (1992)
9. Cash, W.: Parameter estimation in astronomy through application of the likelihood ratio. Astrophys. J. **228**, 939–947 (1979)
10. Cowan, G.: Statistical Data Analysis. Oxford University Press, Oxford (1998)
11. Cramer, H.: Mathematical Methods of Statistics. Princeton University Press, Princeton (1946)
12. Emslie, A.G., Massone, A.M.: Bayesian confidence limits of electron spectra obtained through regularized inversion of solar hard X-ray spectra. Astrophys. J. **759**, 122 (2012)
13. Fisher, R.A.: On a distribution yielding the error functions of several well known statistics. Proc. Int. Congr. Math. **2**, 805–813 (1924)
14. Fisher, R.A.: The use of multiple measurements in taxonomic problems. Ann. Eugenics **7**, 179–188 (1936)
15. Gamerman, D.: Markov Chain Monte Carlo. Chapman and Hall/CRC, London/New York (1997)
16. Gehrels, N.: Confidence limits for small numbers of events in astrophysical data. Astrophys. J. **303**, 336–346 (1986). doi:10.1086/164079

© Springer Science+Busines Media New York 2017

M. Bonamente, *Statistics and Analysis of Scientific Data*, Graduate Texts in Physics, DOI 10.1007/978-1-4939-6572-4

17. Gelman, A., Rubin, D.: Inference from iterative simulation using multiple sequences. Stat. Sci. **7**, 457–511 (1992)
18. Gosset, W.S.: The probable error of a mean. Biometrika **6**, 1–25 (1908)
19. Hastings, W.K.: Monte Carlo sampling methods using Markov chains and their applications. Biometrika **57**(1), 97–109 (1970). doi:10.1093/biomet/57.1.97. http://biomet.oxfordjournals.org/content/57/1/97.abstract
20. Helmert, F.R.: Die genauigkeit der formel von peters zur berechnung des wahrscheinlichen fehlers director beobachtungen gleicher genauigkeit. Astron. Nachr. **88**, 192–218 (1876)
21. Hubble, E., Humason, M.: The velocity-distance relation among extra-galactic nebulae. Astrophys. J. **74**, 43 (1931)
22. Isobe, T., Feigelson, Isobe, T., Feigelson, E.D., Akritas, M.G., Babu, G.J.: Linear regression in astronomy. Astrophys. J. **364**, 104 (1990)
23. Jeffreys, H.: Theory of Probability. Oxford University Press, London (1939)
24. Kelly, B.C.: Some aspects of measurement error in linear regression of astronomical data. Astrophys. J. **665**, 1489–1506 (2007). doi: 10.1086/519947
25. Kolmogorov, A.: Sulla determinazione empirica di una legge di distribuzione. Giornale dell' Istituto Italiano degli Attuari **4**, 1–11 (1933)
26. Kolmogorov, A.N.: Foundations of the Theory of Probability. Chelsea, New York (1950)
27. Lampton, M., Margon, B., Bowyer, S.: Parameter estimation in X-ray astronomy. Astrophys. J. **208**, 177–190 (1976). doi:10.1086/154592
28. Lewis, S.: gibbsit. http://lib.stat.cmu.edu/S/gibbsit
29. Madsen, R.W., Moeschberger, M.L.: Introductory Statistics for Business and Economics. Prentice-Hall, Englewood Cliffs (1983)
30. Marsaglia, G., Tsang, W., Wang, J.: Evaluating Kolmogorov's distribution. J. Stat. Softw. **8**, 1–4 (2003)
31. Mendel, G.: Versuche über plflanzenhybriden (experiments in plant hybridization). Verhandlungen des naturforschenden Vereines in Brünn, pp. 3–47 (1865)
32. Metropolis, N., Rosenbluth, A.W., Rosenbluth, M.N., Teller, A.H., Teller, E.: Equation of state calculations by fast computing machines. J. Chem. Phys. **21**, 1087–1092 (1953). doi: 10.1063/1.1699114
33. Pearson, K., Lee, A.: On the laws on inheritance in men. Biometrika **2**, 357–462 (1903)
34. Plummer, M., Best, N., Cowles, K., Vines, K.: CODA: convergence diagnosis and output analysis for MCMC. R News **6**(1), 7–11 (2006). http://CRAN.R-project.org/doc/Rnews/
35. Press, W., Teukolski, S., Vetterling, W., Flannery, B.: Numerical Recipes 3rd Edition: The Art of Scientific Computing. Cambridge University Press, Cambridge (2007)
36. Protassov, R., van Dyk, D.A., Connors, A., Kashyap, V.L., Siemiginowska, A.: Statistics, handle with care: detecting multiple model components with the likelihood ratio test. Astrophys. J. **571**, 545–559 (2002). doi:1086/339856
37. Raftery, A., Lewis, S.: How many iterations in the gibbs sampler? Bayesian Stat. **4**, 763–773 (1992)
38. Ross, S.M.: Introduction to Probability Models. Academic, San Diego (2003)
39. Thomson, J.J.: Cathode rays. Philos. Mag. **44**, 293 (1897)
40. Tremaine, S., Gebhardt, K., Bender, R., Bower, G., Dressler, A., Faber, S.M., Filippenko, A.V., Green, R., Grillmair, C., Ho, L.C., Kormendy, J., Lauer, T.R., Magorrian, J., Pinkney, J., Richstone, D.: The slope of the black hole mass versus velocity dispersion correlation. Astrophys. J. **574**, 740–753 (2002). doi:10.1086/341002
41. Von Eye, A., Schuster, C.: Regression Analysis for Social Sciences. Academic, New York (1998)
42. Wilks, S.S.: Mathematical Statistics. Princeton University Press, Princeton (1943)

Index

© Springer Science+Busines Media New York 2017
M. Bonamente, *Statistics and Analysis of Scientific Data*, Graduate Texts
in Physics, DOI 10.1007/978-1-4939-6572-4

Printed in the United States
By Bookmasters